Smart Cities

Wireless Communications and Networking Technologies: Classifications, Advancement, and Applications

Series Editor: D.K. Lobiyal, R.S. Rao and Vishal Jain

The series addresses different algorithms, architecture, standards and protocols, tools and methodologies, which could be beneficial in implementing next generation mobile network for the communication. Aimed at senior undergraduate students, graduate students, academic researchers, and professionals, the proposed series will focus on the fundamentals and advances of wireless communication and networking, and their such as mobile ad-hoc network (MANET), wireless sensor network (WSN), wireless mess network (WMN), vehicular ad-hoc networks (VANET), vehicular cloud network (VCN), vehicular sensor network (VSN) reliable cooperative network (RCN), mobile opportunistic network (MON), delay tolerant networks (DTN), flying ad-hoc network (FANET) and wireless body sensor network (WBSN).

Cloud Computing Enabled Big-Data Analytics in Wireless Ad-hoc Networks

Sanjoy Das, Ram Shringar Rao, Indrani Das, Vishal Jain and Nanhay Singh

Smart Cities

Concepts, Practices, and Applications

Krishna Kumar, Gaurav Saini, Duc Manh Nguyen, Narendra Kumar and Rachna Shah

For more information about this series, please visit: https://www.routledge.com/Wireless%20Communications%20and%20Networking%20Technologies/book-series/WCANT

Smart Cities

Concepts, Practices, and Applications

Edited by

Krishna Kumar
Gaurav Saini
Duc Manh Nguyen
Narendra Kumar
Rachna Shah

CRC Press
Taylor & Francis Group
Boca Raton New York London

CRC Press is an imprint of the
Taylor & Francis Group, an **informa** business

First edition published 2022
by CRC Press
6000 Broken Sound Parkway NW, Suite 300, Boca Raton, FL 33487-2742

and by CRC Press
4 Park Square, Milton Park, Abingdon, Oxon, OX14 4RN

CRC Press is an imprint of Taylor & Francis Group, LLC

Library of Congress Cataloging-in-Publication Data
Names: Kumar, Krishna, Engineer editor.
Title: Smart cities : concepts, practices, and applications / edited by
Krishna Kumar, Gaurav Saini, Duc Manh Nguyen, Narendra Kumar, Rachna Shah.
Other titles: Smart cities (CRC Press)
Description: First edition. | Boca Raton : CRC Press, [2022] | Series:
Wireless communications and networking technologies | Includes
bibliographical references and index.
Identifiers: LCCN 2021056099 (print) | LCCN 2021056100 (ebook) | ISBN
9781032190327 (hbk) | ISBN 9781032262284 (pbk) | ISBN 9781003287186 (ebk)
Subjects: LCSH: Smart cities.
Classification: LCC TD159.4 .S59 2022 (print) | LCC TD159.4 (ebook) |
DDC 307.76--dc23/eng/20220125
LC record available at https://lccn.loc.gov/2021056099
LC ebook record available at https://lccn.loc.gov/2021056100

ISBN: 978-1-032-19032-7 (hbk)
ISBN: 978-1-032-26228-4 (pbk)
ISBN: 978-1-003-28718-6 (ebk)

DOI: 10.1201/9781003287186

Typeset in Palatino
by SPi Technologies India Pvt Ltd (Straive)

Contents

Preface

A "smart city" is one that has developed technological infrastructure that enables it to collect, aggregate, and analyze real-time data to improve the lives of its residents. Most of the people across India and other developing countries are migrating into the cities to experience urban life. The people moving toward the urban areas are more ambitious to experience the developments in cities where the facilities are more user-friendly and comfortable. Smart cities have the better Information and Communication Technology (ICT) platform used to strengthen city management and promote economic development. ICT is a paradigm in which objects are interconnected on a network that can collect, store and distribute data based on the current situation and events. Smart cities mainly have three key features, i.e., physical & technological infrastructure, environmental monitoring & sensitivity, and smart citizen services. The key challenges in implementing a Smart City framework are retrofitting existing city infrastructure, financing issues, technical constraints, and the reliability of utility services. Moreover, a smart city can be applied as structures and processes, as multidisciplinary collaborative government environments.

The idea of monitoring or controlling our environment is a significant fear for many people. Trust is key to the success of smart cities. Building a smart city requires high standards of transparency and oversight to ensure that the data is being used legally, responsibly, and in the interest of the public. General Data Protection Regulation (GDPR) measure can be used to ensure the privacy which is maintained without sacrificing the benefits of efficiency or public safety. A higher degree of service digitalization, privacy & data protection, uses of technologies to reduce the environmental problems, traffic management, and waste management for sustainable utilization of natural resources.

The conceptualization of smart cities came into existence, and many developing countries are working toward these development concepts. Keeping this in view, the present book aimed to explore the various aspect of smart cities and their architecture along with the application of the latest technologies viz. IoT, AI, and Clouding Computing. The concept of smart cities, their development, technological advancements, and issue related to them have been discussed in detail.

Readers

This book is helpful for researchers, academicians, and developers working in the area of sustainable smart city development, intelligent transport systems, sewage & waste management, and sustainable renewable energy development.

The main features of the book are:

- It has covered all the latest developments and future aspects of smart cities.
- This book is very useful for the new researchers and developers working in the field to quickly know-how the best performing methods.
- The book is concisely written, lucid, comprehensive, application-based, graphical, schematics, and covers all the aspects of smart city development.

Chapter Organization

Chapter 1: It gives an overview of smart cities, integrated approaches toward smart cities, and global practices for smart cities. This chapter will help to make a generalized understanding of this subject.

Chapter 2: Describes a smart city analytical framework of Vietnam. It provides a case study to make understanding about the latest development and challenges faced by Vietnam.

Chapter 3: Explains the case study on Brazilian slums to analyze the contribution of games in the design of Smart Cities.

Chapter 4: Gives an overview of IoT-based framework toward a feasible, safe, and reliable Smart City using Drone Surveillance. This chapter also discusses the application of IoT-enabled smart cities and has proposed one effective smart city model.

Chapter 5: Developed an End-to-End framework for dynamic crime profiling of places. It helps to minimize unwanted activities and improve safety and security.

Chapter 6: Gives an overview of intelligent transportation and traffic management systems. This chapter also explains the existing American, Canadian, and European ITS architecture.

Chapter 7: This chapter explains the role of IoT and AI in the development of smart cities.

Chapter 8: Provides a case study on Southern India to understand the flood management policies in megacities. It also provides a comparative study on water policies and suggested various recommendations based on water policies.

Chapter 9: Provides a case study on maintenance methodologies embraced by O&M at Kochi Metro Rail Limited, India.

Chapter 10: Gives an architecture of smart lights to minimize energy consumption using IoT.

Chapter 11: Reviewed the flexible communication technologies utilized for developing Smart Cities. The explanation about the concept of industry 4.0 (i.e., IoT 4.0) has also been covered.

Chapter 12: Gives an overview of sewage management, their sources, effects, and treatment technologies.

Chapter 13: Describe the steps for the fabrication of mullite ceramic using industrial waste.

Chapter 14: Gives an overview of renewable energy technologies which are being used for the sustainable development of societies.

About the Book

This book presents the principles, techniques, design, and implementation of smart cities, with a balance in the presentation of theoretical and practical aspects. The book gives a clear analysis of smart cities models and explains all the relevant concepts on smart city frameworks like the role of games in the design of the smart city, drone surveillance, criminal profiling, intelligent transport systems & traffic management, flood management policies, maintenance methodologies, sewage management, application of AI and IoT, and sustainable energy generation.

The book is primarily intended for undergraduates, postgraduates, researchers, developers, and policymakers working in the area of smart city development.

Editors

Krishna Kumar is presently working as a Research and Development Engineer at UJVN Ltd., before joining UJVNL, he has worked as Assistant Professor at BTKIT, Dwarahat (India). He received his B.E. (Electronics and Communication Engineering) from Govind Ballabh Pant Engineering College, Pauri Garhwal, and M.Tech (Digital Systems) from Motilal Nehru NIT Allahabad. He is also pursuing his Ph.D. from Indian Institute of Technology, Roorkee. He has more than 11 years of experience and has published numerous research papers in international journals. His research area includes Renewable Energy, Artificial Intelligence, and IoT.

Gaurav Saini is presently working as Assistant Professor at the Indian Institute of Engineering Science and Technology, Shibpur, India. He has obtained his Doctorate and Masters in renewable energy from the Indian Institute of Technology Roorkee and graduated in Mechanical Engineering from Gautam Buddha Technical University Lucknow, India. He has more than five years of teaching and research experience in renewable energy and related technology. He has published several research publications in refereed journals and shared his research at national and international platforms. His research field includes Renewable energy technology, Hydraulic turbines, Computational fluid dynamics, and Optimization.

Duc Manh Nguyen is presently working as a Researcher Engineer at G2touch, South Korea. He has received his bachelor's degree from Ha-Noi University of Science and Technology (HUST) and a Ph.D. from the University of Ulsan, South Korea. He has also worked for Samsung Electronic Viet-Nam and Viet-Nam Posts and Telecommunications Group. He has published numerous papers in international journals and conferences. His present research area includes networking and communication technologies.

Narendra Kumar is presently working as Assistant Professor in the School of Computing at DIT University, Dehradun, India. He is an M.Tech (Computer Science) from B.I.T. Mesra Ranchi, Jharkhand, and Ph.D. (Computer Science) from Deen Dayal Upadhyaya Gorakhpur University, Gorakhpur. He has more than 12 years of teaching experience. He has published numerous research papers in international journals and conferences. His present research area includes Data science and IoT.

Rachna Shah is presently working as a Scientist at the National Informatics Centre, Govt. of India. She did her B.E. (Computer Science and Engineering) from Kumaon Engineering College, Dwarahat, India (A Govt. of Uttarakhand Institution). She has more than 10 years of experience in software development, architecture planning, and e-Governance. She has received various Excellence awards from Govt. of India and Govt. of Uttarakhand. She has published numerous research papers in national and international journals. Her research area includes Data Science, IoT, and Open-Source Software Development.

Contributors

Tasneem Ahmed
Advanced Computing and Research
    Laboratory
Department of Computer
    Application
Integral University
Lucknow, India

S. Anandakumar
Department of Civil Engineering
Kongu Engineering College
Perundurai, India

Harish Chandra Arora
Academy of Scientific and
    Innovative Research
Council of Scientific and Industrial
    Research – Central Building
    Research Institute (CSIR–CBRI)
Roorkee, India

Carolin Arul
Centre for Water Resources
Anna University
Chennai, India

Daniel Oliveira Cruz
Anhembi Morumbi University
São Paulo, Brazil

Dipankar Das
Department of Material Science and
    Engineering
Tripura University (A Central
    University)
Agartala, India

Mohammad Faisal
Advanced Computing and Research
    Laboratory
Department of Computer
    Application
Integral University
Lucknow, India

Neeraj Goel
Department of Computer Science
    and Engineering
Indian Institute of Technology,
    Ropar
Punjab, India

R. D. Gomathi
Department of Electronics and
    Communication Engineering
Kongu Engineering College
Perundurai, India

Shailendra Kumar Gupta
Department of Computer Science
    and Engineering
Indian Institute of Technology,
    Ropar
Punjab, India

Pooja N. Kakani
L&T Smart World
Chennai, India

Nishant Raj Kapoor
Academy of Scientific and
    Innovative Research
Council of Scientific and Industrial
    Research – Central Building
    Research Institute (CSIR–CBRI)
Roorkee, India

Javed Ahmad Khan
Department of Information
    Technology
Government Girls Polytechnic
    College
Ballia, India

Bui Huy Khoi
Industrial University of Ho Chi
    Minh City
Ho Chi Minh City, Vietnam

Aman Kumar
Academy of Scientific and
    Innovative Research
Council of Scientific and Industrial
    Research – Central Building
    Research Institute (CSIR–CBRI)
Roorkee, India

Ashok Kumar
Academy of Scientific and
    Innovative Research
Council of Scientific and Industrial
    Research – Central Building
    Research Institute (CSIR–CBRI)
Roorkee, India

Krishna Kumar
Department of Hydro and
    Renewable Energy
Indian Institute of Technology
Roorkee, India

Nguyen Viet Lam
Industrial University of Ho Chi
    Minh City
Ho Chi Minh City, Vietnam

C. Maheswari
Department of Mechatronics
    Engineering
Kongu Engineering College
Perundurai, India

Monika
Department of Computer Science
Shaheed Rajguru College of Applied
    Sciences for Women
University of Delhi
Delhi, India

Indhumathi Natarajan
Department of Electronics and
    Communication Engineering
Kongu Engineering College
Perundurai, India

Sergio Nesteriuk
Anhembi Morumbi University
São Paulo, Brazil

Priyanka Prabhakaran
Department of Civil Engineering
Kongu Engineering College
Perundurai, India

R. Shyam Shankaran
L&T Smart World
Chennai, India

Rahul
Department of Software
    Engineering
Delhi Technological University
Delhi, India

Logesh Rajendran
L&T Smart World
Chennai, India

A. S. Ramya
Kongu Engineering College
Perundurai, India

Kritesh Rauniyar
Department of Computer Science
  and Engineering
Delhi Technological University
Delhi, India

Y. Rekha
KCG College of Technology
Chennai, India

Prasanta Kumar Rout
Department of Material Science and
  Engineering
Tripura University (A Central
  University)
Agartala, India

Romit Roy
Department of Material Science and
  Engineering
Tripura University (A Central
  University)
Suryamaninagar, Agartala, Tripura,
  India

Gaurav Saini
School of Advanced Materials
Green Energy and Sensor Systems
Indian Institute of Engineering
Science and Technology
Shibpur Howrah, India

Karuna Saini
Department of Chemistry
Gurukula Kangri Vishwavidyalaya
Haridwar, India

Mukesh Saini
Department of Computer Science
  and Engineering
Indian Institute of Technology,
  Ropar
Punjab, India

Maheswaran Shanmugam
Department of Electronics and
  Communication Engineering
Kongu Engineering College
Perundurai, India

Sathesh Shanmugam
Department of Electronics and
  Communication Engineering
Kongu Engineering College
Perundurai, India

Shivom Sharma
Department of Mechanical
  Engineering
IMS Engineering College
Ghaziabad, India

Shreyanshu Shekhar
Department of Computer Science
  and Engineering
Indian Institute of Technology,
  Ropar
Punjab, India

S. Suriya
Jerusalem College of Engineering
Chennai, India

Nayyar Ali Usmani
Hanu Software Solutions
Noida, India

Balasubramaniam Vivek
Department of Electronics and
  Communication Engineering
Kongu Engineering College
Perundurai, India

1

Smart Cities: A Step toward Sustainable Development

Aman Kumar and Nishant Raj Kapoor

AcSIR, CSIR-Central Building Research Institute, Roorkee, India

Harish Chandra Arora and Ashok Kumar

CSIR-Central Building Research Institute, Roorkee, India

CONTENTS

A great city is that which has the greatest men and women

Walt Whitman

1.1 Introduction

Smart cities have become a very popular trend worldwide in urban policies [1]. In 1974, Los Angeles (LA), a city of California State started the first urban big data project called "A Cluster Analysis of Los Angeles" in which the main focusing area was community analysis using cluster analysis, infrared aerial photography data and computer databases, altogether, to generate reports on neighborhood demography, quality of housing and help direct resources to reduce health threats and poverty [2, 3]. Amsterdam was conceptualized as the first virtual "digital city"—De Digital Stad (DDS) over the globe in 1994, in which the city developed a digital network for citizens to deliver their protests and notations to politicians [4]. In 1611, the merchants of this city used Hendrick de Keyser Centre to exchange trade information. Due to this reason, Amsterdam was the first smart city well before the concept was used and the city is also known as "A world Leader in Smart City Development" [5].

In 1997, the smart cities started appearing in literature and got attention from various industries and scientific communities. The main basis of smart cities is to collect, analyze, and provide data from people to people of a city by using intelligent technologies. From the perspective of Information and Communication Technology (ICT), there are several opinions regarding the meaning of "smart" [6]. In this era, the term "smart" is quite common and broadly used synonyms are "intelligent" and "modern." Moreover, smart is synonymous to efficient, when it is linked to devices. A smart city is a digital city that is connected with intelligent systems through the Internet and moving toward sustainability goals. These cities contain more accessibility, livability and are eco-friendly [7]. Numerous smart city projects in developed and developing countries are still under the construction stage. With the advancements in technology, experts and research scholars are not able to decide mutually the basic concepts and elements of a smart city. This infers that the researchers are unable to come to the implementation stage from the conceptual stage [1].

In the early 1900s, only 13% of the world's population was living in urban areas [8]. But in the year 2014, as reported by the United Nations (UN), this number increased to 41% and predicted the expected increase to 66% by 2050 [9]. UN reported that by 2030, the urban population is expected to increase by 60% globally and one in every three people will live in cities with at least half a million inhabitants [10]. Urbanization puts pressure on the cities to provide various essential parameters such as better life quality, enhanced level of productivity and environment [11]. This pressure can only be mitigated by the use of intelligent systems with uninterrupted Information Exchange and Communication Techniques (IECT), which enable sustainability and growth [12]. The problem associated with rapid migration led to various issues like increase in crimes, security, privacy, traffic congestion and different types of pollutions (air, water, noise, land etc.). One of the most effective solutions for these problems is smart cities.

To solve the problems of residents, the cities are enhancing their digital technologies. The top-most problems of the residents are medical infrastructures, traffic congestion, care of senior citizens, environmental monitoring, and crime prevention. These developments aim to attain a quality life with more productivity and competitiveness. To maintain the huge smart city data, the concept of 'big data' was evolved with Artificial Intelligence (AI) and the Internet of Things (IoT) [13, 14]. The process of modern smart cities incorporates data exchange by IoT, managing and transforming data using big data techniques and optimization data by AI [15]. But, in the future, smart city projects may be completely managed by robotic systems. The correlation between the above parameters in a smart city can be presented using equation [16], called "the smart city equation":

$$\text{Smart City} = \text{Digital City} + \text{AI} + \text{IoT} + \text{Cloud Computing}$$

1.2 Background of Smart Cities

History shows that technical revolutions are triggered globally after energy or finance crises. This enabled many developed and developing countries to further enhance the quality and efficiency of their cities with more sustainable growth. The path of industrial cities to smart cities [7] is shown in Figure 1.1.

1.2.1 Evolution of Smart Cities

The evolution of smart cities can be seen in Amsterdam way back at the beginning of the 17th century. They used it only for trading and information exchange. The economic crisis of 1857 was the reason behind the industrial revolution which gave birth to industrial cities. "Virtual city" was the originating term used by Graham and Aurigi [17]. Later the term "digital or information" city was in the trend. A literature on smart city concept is found during 1980s–1990s [17–20].

After Amsterdam, till 2010, USA, UAE, Singapore, South Korea, European Union, China, Japan, and Spain were the leading countries in developing smart cities. In 2011, 50 countries around the world participated in the first "Smart City Expo World Congress," which was held in Barcelona in Spain,

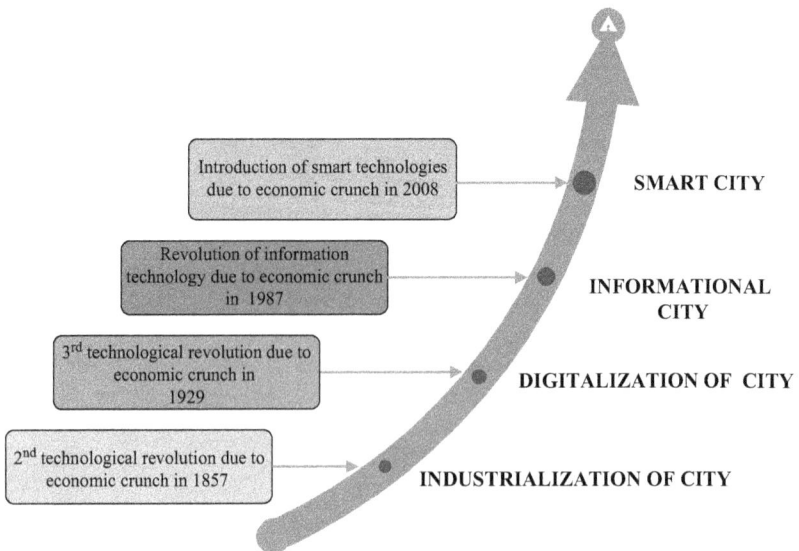

FIGURE 1.1
Smart city background.

and after that, the concept of smart cities was adopted by various developed and developing countries like China, India, United Kingdom, Canada, France, Norway, and the Netherlands. The evolution of smart cities from 1974 to 2050 is presented in Figure 1.2 [21].

1.2.2 Definitions

The term "smart city" is an omnidirectional concept [22], but its definition is quite blurred. The definition varies significantly with technology, people, governance, industry, and research experts. However, some common ideology is reported in the literature. The World Smart City Organization (WSCO) defines smart city "when investments in human, social capital, traditional (transport) as well as modern ICT infrastructure promote sustainable economic development and high quality of life, with a wise management of natural resources, through participatory actions and commitment" [23]. Some more definitions elaborating the different concepts of "Smart City" are presented in Table 1.1.

Kumar et al. 2020 [7] defines different types of cities. The types of cities are shown in Figure 1.3.

1.2.3 Basic Requirements for Sustainable Smart Cities

The basic requirements to develop any city are land, people, infrastructure, and governance.

1.2.3.1 Land

Land is the primary and most important aspect of any city. Every day the land cover is reducing due to population increase. According to the World Bank report, in the year 2016, the remaining agricultural land was just 48.6 million sq. km [30]. Smart city Singapore has an area of only 724.2 sq. km, and they have very few natural resource bays, even then that city used its land very well and it reached first place at "World Smart City Rankings" in 2020 [31]. Hence the utilization of land with proper management is important. The main basic requirements to establish a city are shown in Figure 1.4.

1.2.3.2 People

The people of a city play an important role to make cities smart cities. The smart cities mission is contributing to the improvement of the city [28]. The smart city concept for citizens includes people's values, capabilities, and talents, and creates new opportunities for the young generations and people with disabilities [32, 33].

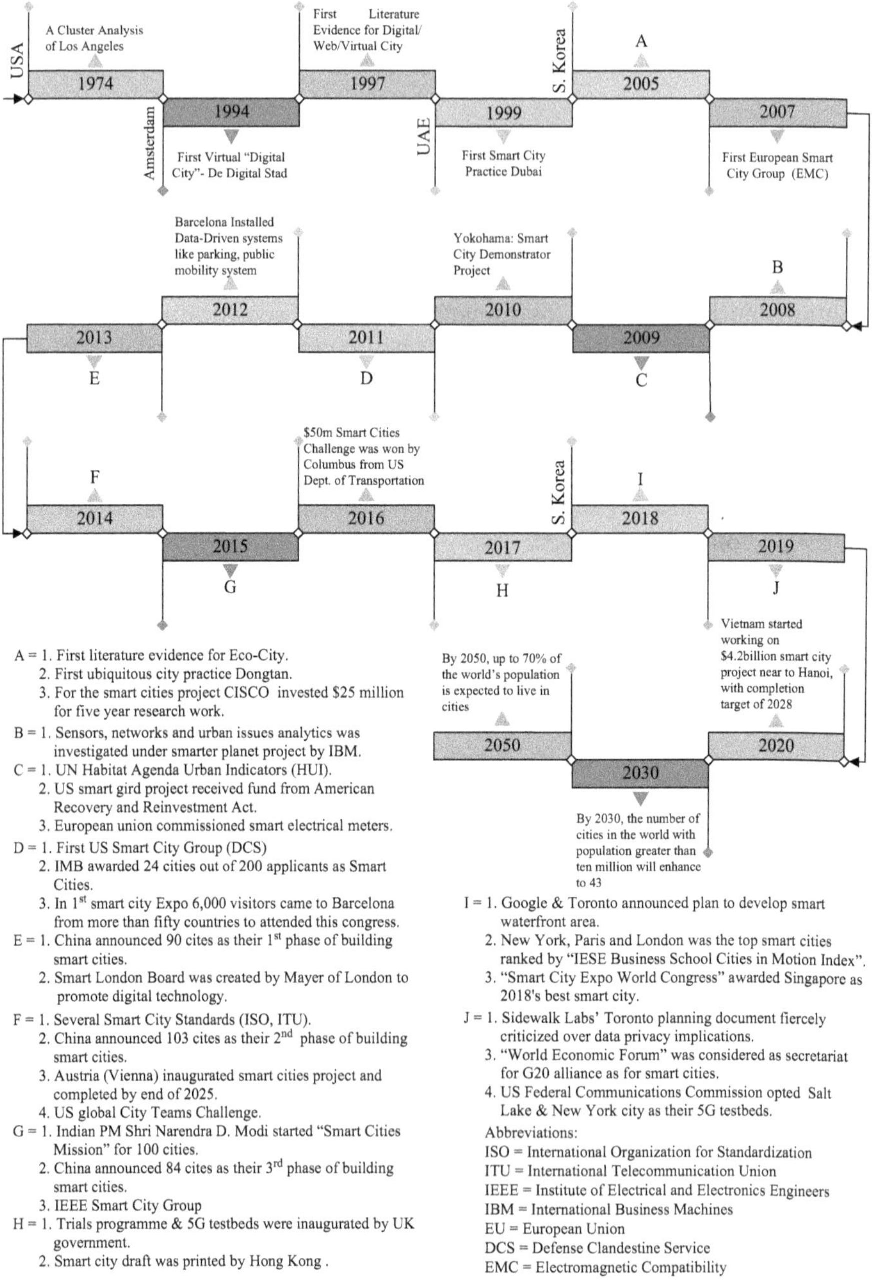
USA
1974
A Cluster Analysis of Los Angeles
Amsterdam
1994
First Virtual "Digital City"- De Digital Stad
1997
First Literature Evidence for Digital/ Web/Virtual City
UAE
1999
First Smart City Practice Dubai
S. Korea
2005
A
2007
First European Smart City Group (EMC)
2008
B
2009
C
2010
Yokohama: Smart City Demonstrator Project
2011
D
2012
Barcelona Installed Data-Driven systems like parking, public mobility system
2013
E
2014
F
2015
G
2016
$50m Smart Cities Challenge was won by Columbus from US Dept. of Transportation
2017
H
S. Korea
2018
I
2019
J
2020
Vietnam started working on $4.2billion smart city project near to Hanoi, with completion target of 2028
2030
By 2030, the number of cities in the world with population greater than ten million will enhance to 43
2050
By 2050, up to 70% of the world's population is expected to live in cities
A = 1. First literature evidence for Eco-City.
2. First ubiquitous city practice Dongtan.
3. For the smart cities project CISCO invested $25 million for five year research work.
B = 1. Sensors, networks and urban issues analytics was investigated under smarter planet project by IBM.
C = 1. UN Habitat Agenda Urban Indicators (HUI).
2. US smart gird project received fund from American Recovery and Reinvestment Act.
3. European union commissioned smart electrical meters.
D = 1. First US Smart City Group (DCS)
2. IMB awarded 24 cities out of 200 applicants as Smart Cities.
3. In 1st smart city Expo 6,000 visitors came to Barcelona from more than fifty countries to attended this congress.
E = 1. China announced 90 cites as their 1st phase of building smart cities.
2. Smart London Board was created by Mayer of London to promote digital technology.
F = 1. Several Smart City Standards (ISO, ITU).
2. China announced 103 cites as their 2nd phase of building smart cities.
3. Austria (Vienna) inaugurated smart cities project and completed by end of 2025.
4. US global City Teams Challenge.
G = 1. Indian PM Shri Narendra D. Modi started "Smart Cities Mission" for 100 cities.
2. China announced 84 cites as their 3rd phase of building smart cities.
3. IEEE Smart City Group
H = 1. Trials programme & 5G testbeds were inaugurated by UK government.
2. Smart city draft was printed by Hong Kong .
I = 1. Google & Toronto announced plan to develop smart waterfront area.
2. New York, Paris and London was the top smart cities ranked by "IESE Business School Cities in Motion Index".
3. "Smart City Expo World Congress" awarded Singapore as 2018's best smart city.
J = 1. Sidewalk Labs' Toronto planning document fiercely criticized over data privacy implications.
3. "World Economic Forum" was considered as secretariat for G20 alliance as for smart cities.
4. US Federal Communications Commission opted Salt Lake & New York city as their 5G testbeds.
Abbreviations:
ISO = International Organization for Standardization
ITU = International Telecommunication Union
IEEE = Institute of Electrical and Electronics Engineers
IBM = International Business Machines
EU = European Union
DCS = Defense Clandestine Service
EMC = Electromagnetic Compatibility

FIGURE 1.2
Evolution of cities.

TABLE 1.1

Smart City Definitions

Source	Year	Definition
Hall [24]	2000	The vision of "smart cities is the urban center of the future, made safe, secure environmentally green, and efficient because all structures—whether for power, water, transportation, etc. are designed, constructed and maintained making use of advanced, integrated materials, sensors, electronics and networks which are interfaced with computerized systems comprised of databases, tracking and decision-making algorithms."
Giffinger et al. [25]	2007	"A city well performing in a forward-looking way in economy, people, governance, mobility, environment, and living built on the smart combination of endowments and activities of self-decisive, independent and aware citizens. Smart city generally refers to the search and identification of intelligent solutions which allow modern cities to enhance the quality of the services provided to citizens."
IEEE [26]	—	"Smart city brings together technology, government and society to enable the following characteristics: smart cities, a smart economy, smart mobility, a smart environment, smart people, smart living, smart governance."
Government of India [27]	2014	"Smart city offers sustainability in terms of economic activities and employment opportunities to a wide section of its residents, regardless of their level of education, skills or income levels."
ISO [28]	2014	"Smartness of a city describes its ability to bring together all its resources effectively and seamlessly achieve the goal and fulfill the purposes it has set itself."
Smart Cities Council [29]	2016	"Smart city as one that uses ICT to enhance its livability, workability, and sustainability."
Author	2021	A smart city means, "providing comfort to the city people with security and use of renewable energy resources for living purposes."

1.2.3.3 Infrastructure

The architecture and framework of a smart city rely on the infrastructure [34]. The meaning of infrastructure changes with its context. In the context of civil engineering, infrastructure is related to buildings, roads, schools, colleges, hospitals, airports, train/metros, banks, and research facilities, etc. [35].

1.2.3.4 Governance

The smart city and governance are quite inter-linked. Smart governance ensures smart people engagement and their active participation along with dedicated service deliveries. The term "governance" is defined as "the use

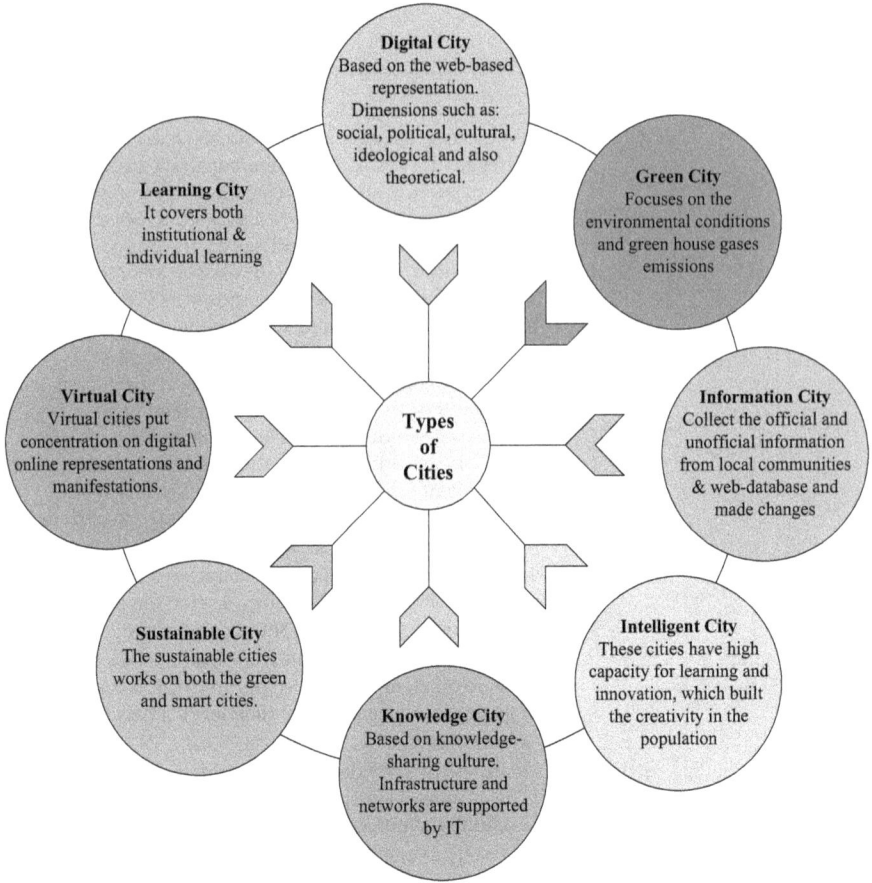

FIGURE 1.3
Types of cities.

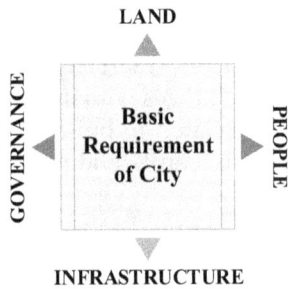

FIGURE 1.4
Basic requirements of a city.

of institutions, structures of authority and even collaboration to allocate resources and coordinate or control activity in society or the economy" [36]. The governance can be of many types such as project governance, corporate governance, technology governance, global governance, and municipal governance etc. Further, municipal governance is classified as urban and rural governance. All the city controls related to the areas such as information and technology, laws, public services, and traffic management are in the hands of governing authorities.

1.3 Integrated Approach toward Smart Cities

At present, several smart city projects have been undergoing in different countries across the world [37]. Although the practice of building smart cities is quite popular, however, there is no common ideology and principles that can define the exact meaning of what a smart city really is, various dimensions to be included in it and what technologies can be adopted for fulfilling the need. The characteristics of a smart city and key indicators are shown in Figures 1.5 and 1.6 respectively.

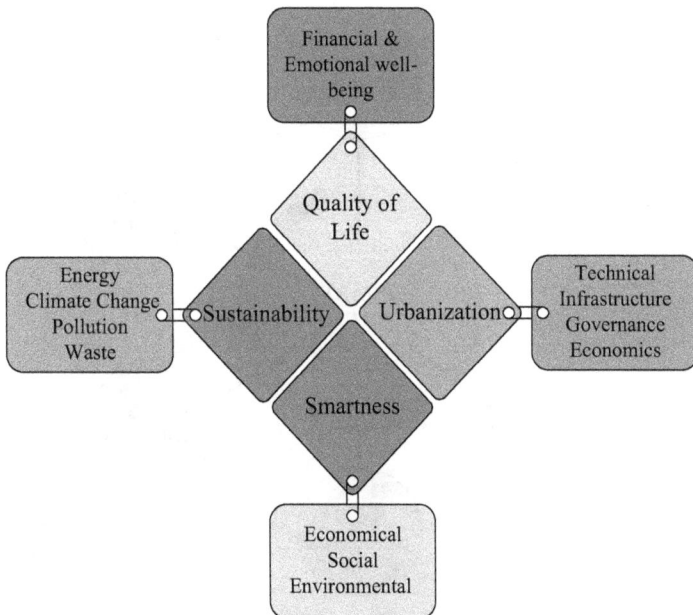

FIGURE 1.5
Smart city characteristics.

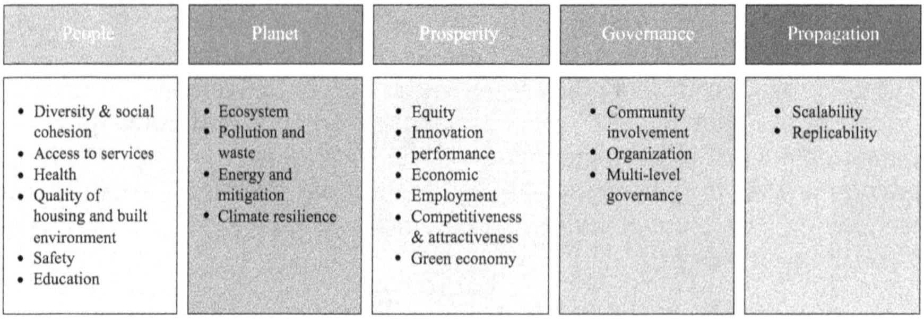

People
Planet
Prosperity
Governance
Propagation
• Diversity & social cohesion
• Access to services
• Health
• Quality of housing and built environment
• Safety
• Education
• Ecosystem
• Pollution and waste
• Energy and mitigation
• Climate resilience
• Equity
• Innovation
• performance
• Economic
• Employment
• Competitiveness & attractiveness
• Green economy
• Community involvement
• Organization
• Multi-level governance
• Scalability
• Replicability

FIGURE 1.6
Key indicators of smart city.

1.3.1 Dimensions for Smart City

The literature supports six basic dimensions, that in turn support the smart city as a substratum. The basic dimensions of a smart city include smart people, smart living, smart environment, smart governance, smart economy and smart mobility [7, 25, 37–45] as shown in Figure 1.7.

Smart People
Smart Mobility
Smart Living
SMART CITY
Smart Environment
Smart Economy
Smart Governance

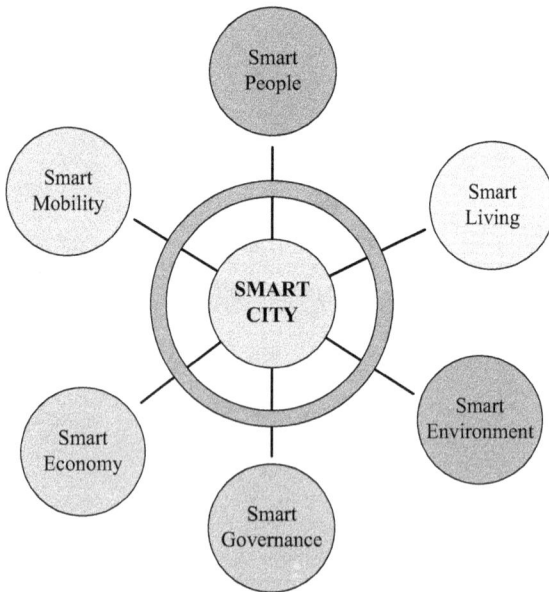

FIGURE 1.7
Steering wheel of smart city dimensions.

1.3.1.1 Smart People

The first dimension of the smart city is its "people." A city can be smart only when it is surrounded by smart citizens. The occupation defines the rate of human development and human capital of city [7]. Human and social capital (people) measures the "qualifications, affinity for life-long learning, flexibility and creativity of the people. It concerns multiple participants, such as citizens and communities (government, enterprises and universities)" [46]. Education improves the living standard of the city people. The focus of smart cities is getting the literacy rate up to graduation level for people having a special area of expertise such as in the sector of transportation, electricity, and waste management. The modes of learning in the smart cities are based on digital technologies. The people of the smart city are always ready to face the bad circumstances (like Covid-19 and bird flu) and find a unique and better solution to these issues. Smart people are multicultural and open-minded. The lifestyle and environment of smart people are healthy. Smart people are always involved in the activities like pollution-free environment, waste management, sustainability, and a better lifestyle for the inhabitants. In these activities, the city people put their views and opinions to solve the above-mentioned problems. They also have the capabilities to adopt new technologies over conventional technologies [32]. Smart people come together in urban areas, resulting in cities becoming smarter [47]. These people are those who can optimize the use of natural material by the 6R concept: Rethink, Refuse, Reduce, Reuse, Repurpose, and Recycle resources.

1.3.1.2 Smart Living

The second dimension of the smart city is its "smart living" for the people. All the infrastructure facilities are available 24/7 in smart cities for better living, security, and safety to all children, women, and elders. Smart city provides the infrastructure to the citizens enabled with early warning systems for earthquakes like natural calamities. For good quality of life and sustainability, smart cities have been focused on utilizing renewable resources [48]. In smart cities, people produce organic food for a healthy lifestyle using vertical farming techniques. In smart cities, sensors and smart equipment/utilities gather data from users and exchange data using IoT after analyzing the data using AI techniques. All the data will be managed by Big Data (BD) and Cloud Computing Techniques (CCT).

1.3.1.3 Smart Environment

The third dimension of the smart city is its "smart environment" for the people. It may be in the form of installation of sensors and other intelligent techniques on trees (intelligent trees), plants, and water, hence if someone harms the natural elements and resources, the technology will inform the concerned authorities and immediately suitable action can be taken. It protects its natural, historical, and biodiversity environmental aspects as well as maintain clean and green surroundings. It should provide enough green space and create job opportunities for all age groups. The best examples for better outer work environments are the headquarters of Facebook and Microsoft. The carbon footprint emitted by smart cities should be within the permissible limits with a concentration on energy efficiency resources. It has more electrical vehicles as compared to gas, diesel, or petrol. The smart grid should be based on renewable sources of energy such as solar and wind energy etc. The city has proper water management systems and treatment of contaminated water should be done before disposing. Rainwater harvesting should be installed in each smart home for water independence. The city should have an efficient solid and hazardous waste management system.

1.3.1.4 Smart Governance

The fourth dimension of the smart city is its "smart governance" for the people. A smart city involves e-government for the benefit of city people. The definition of e-government by the World Bank is "government agencies of information technologies (such as Wide Area Networks, the Internet and mobile computing) that have the ability to transform relations with citizens, businesses and other arms of government" [49]. It continuously enhances its capabilities to provide public services diligently, effectively, and in a very efficient manner. It practices innovation in day-to-day life for the benefit of citizens as well as exercises Responsiveness, Accountability, and Transparency (RAT) for governing the official processes.

A smart city exercises following the process smartly shown in the flow diagram (Figure 1.8a) as below.

Smart governance allows citizens to complain about their issues using digital platforms and resolve those issues and problems rapidly. A city focuses on social, economic, cultural, and environmental dimensions along with strategies for sustainable development and citizen-friendly city management systems. It should implement models like the "Triple Helix Model" for better co-operation among universities, regulatory bodies (government) and industries as shown in Figure 1.8b [50].

1.3.1.5 Smart Economy

The fifth dimension of the smart city is its "smart economy" for the people. It includes aspects like innovation and cutting-edge research in industry, science, and business. Diverse employment opportunities are available in smart cities with enhanced entrepreneurship opportunities and small businesses.

(a)

Institutions

(b)

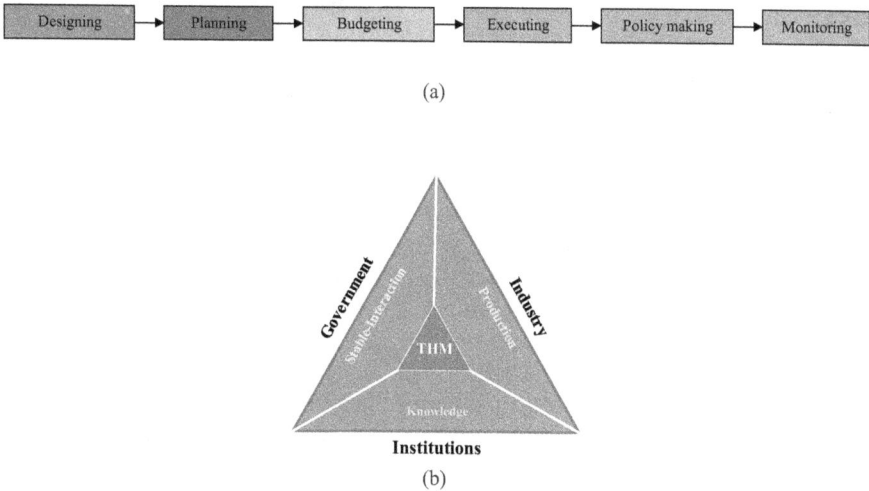

FIGURE 1.8
(a) Process of smart governance. (b) Triple Helix Model (THM).

Highly creative values and breakthrough ideas are always appreciated in smart cities. It concentrates on building strategic assets with strategic investments. The city must support its fascinating brands to compete globally. All economic activities at the local level should be smartly managed in smart cities, so all the challenges can be converted into opportunities. For sustainable development and economic boost, a smart city should focus on enhanced tourism. All the resources should be enhanced sustainably without affecting nature in a smart city for wealth creation.

1.3.1.6 Smart Mobility

The sixth dimension of the smart city is its "smart mobility" for the people. For a sustainable environment, a smart city should adopt walkable and cycle pathways. Both people and vehicles should be mobile in smart cities. It should implement an Intelligent Transportation System (ITS) to resolve traffic congestion and pedestrian traffic for effective vehicular management [51]. The city should have metro-rail, sky-train, Bus Rapid Transit System (BRTS) and other high-speed transit systems for enhanced mobility. A proper balance among transportation systems should be there in smart cities. All the routes in smart cities should be pleasurable and secured. Differently-abled people should get proper mobility options to travel in and around the smart city with ease and safety. All the mobility options should be evenly distributed in the smart city for meeting the demand of high-density and low-density living areas.

1.4 Technologies for Smart Cities

Technologies for a smart city are the crucial parameter to monitor and solve the city problems. Technology is the sole tool that can boost the journey of cities toward future smart cities. In daily life, different types of technologies are being used by the residents. Technologies like intelligent transport systems, smart health systems, smart security systems, rapid information exchange, and smart waste management systems are operated using a combination of AI, sensors, and IoT, etc. For a smart citizen of a smart city, the smartphone is the basic need after food, clothes, and home. "Smartness" is not about deploying sensors and modern communications, but it is about using the data and technology purposefully to enhance quality of life and making better decisions.

1.4.1 Internet of Things (IoT)

The term "IoT" refers to "the networked interconnection of everyday objects, many of which are equipped with ubiquitous intelligence" [52]. Kevin Ashton coined the term 'IoT' known as "internet of things" in a presentation at Massachusetts Institute of Technology (MIT) based on Auto-ID Labs work in 1999 [53]. The most compatible definition of 'IoT' is given by International Telecommunication Union (ITU), which describes IoT as "a global infrastructure for the Information Society, enabling advanced services by interconnecting (physical and virtual) things based on existing and evolving interoperable information and communication technologies" [54]. In many studies, it is investigated that IoT-induced network and cloud computing technology enables conventional communication techniques to become smart. Cloud techniques strengthen the process of information exchange and analysis, enhance storage capacity, and reduce the cost of fabrication. It was estimated that more than 100 billion devices will be connected to the Internet by the end of 2020 [55].

IoT is a part of sustainable smart cities especially for residential and commercial sectors called Residential Internet of Things (RIoT), Commercial Internet of Things (CIoT), and Industrial Internet of Things (IIoT). The various application of IoT services is presented in Figure 1.9 below.

In 2011, Andrew Attwood investigated a term for critical infrastructure called infrastructure for the smart city [56]. These infrastructures aim to protect the city from abrupt failures (mobility, energy, and services etc.) with continuous functionality. For fast, efficient, secure, and wireless communication various technologies are used under IoT such as Bluetooth, Wireless Fidelity (WiFi), ZigBee, wireless cellular network, Wireless Local Area Network (WLAN) and Radio-Frequency Identification (RFID) etc. All these technologies are companion in systems like intelligent waste management systems, street lighting, utilities, and traffic congestion.

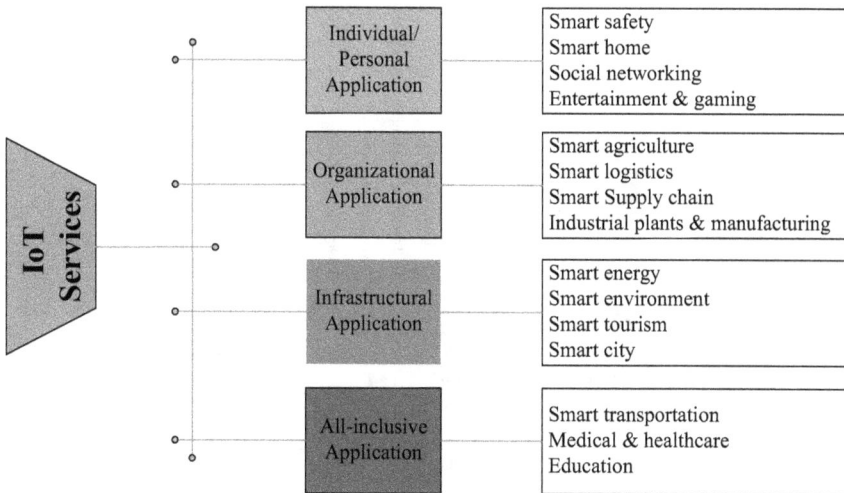

FIGURE 1.9
IoT services and application.

The implementation of the smart home was credited to the United Kingdom-based mobile operator 'Orange.' Orange announced the project "Orange at Home" in 2001 and implemented this project 32 kilometers away from London city. In this project, Orange used the latest technology and sensors to make the house working smartly. Later the Department of Digital World Research Center (DWRC) of the University of Surrey started research on the living habitants of "Orange Smart Homes." Long back in 1984, the American Association of House Builders earned the credit for using the term "smart house." This was based on the work done by hobbyists during the 1960s with the name "Wired Homes" [57]. The smart home technologies [58, 59] are shown in Figure 1.10.

Smart homes are always connected to the Internet to exchange information with inhabitants, so they can easily operate their indoor environment controlling devices such as Air Conditioner (AC), fans, lightening systems, exhaust, security alarm, and mechanical windows. from distant locations.

In smart homes, structural health monitoring is done by using smart sensors like SmartRock, SmartHub, BlueRock and SmartBox [60]. These smart sensors provide the necessary information about the structure like compressive strength of concrete, level of corrosion, the pH level of concrete, cracks, and settlements signs, seepage warnings and carbon emission at the time of construction. The data which is obtained from these sensors are used for periodic maintenance and to increase the lifespan of the structure. The

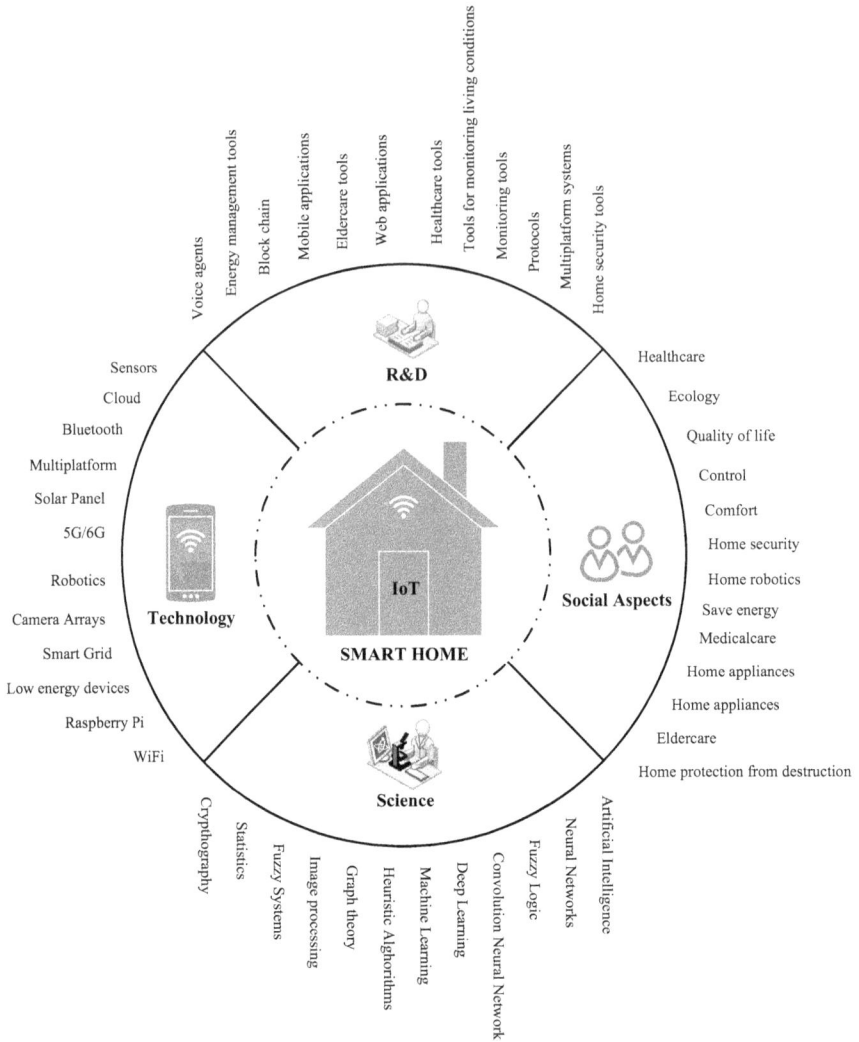

FIGURE 1.10
Smart home technologies.

compressive strength of concrete can be predicted using AI [61]. The structural damage alarming system is presented in Figure 1.11.

1.4.1.1 Challenges for Smart Homes

Various challenges encountered during the smart city up-gradation process are mentioned in literature by various authors. There are different factors as shown in Table 1.2.

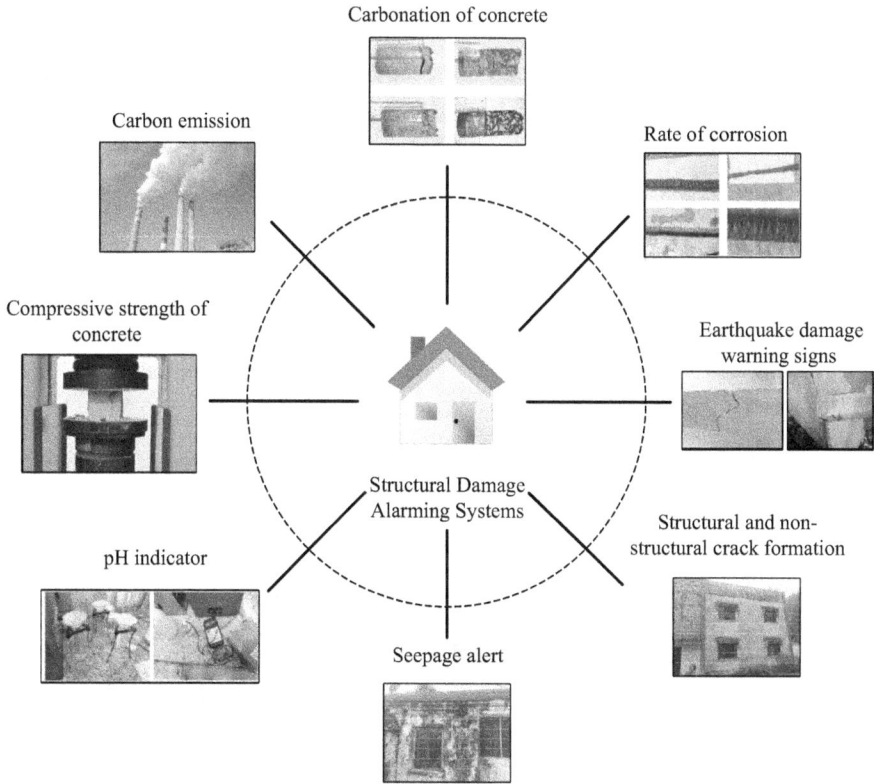

FIGURE 1.11
Smart sensors for structural health monitoring.

1.4.2 Artificial Intelligence (AI)

In the 1950s, researchers made early efforts to replicate human intelligence in machines [74]. The term AI was coined in 1955 by John McCarthy at Dartmouth College [75]. AI plays an important role in smart city development projects. In smart cities, automation is done by using robotics [14]. Along with improving the lives of the residents, maintaining the operation of the energy system is also very important. AI plays an important role in the concern of energy conservation. Researches worldwide have been focusing on the development of AI-driven smart homes and using different sustainable building materials to reduce energy consumption [76]. Some of the examples are: "ScaledHome; a scalable and reliable architecture of smart home aiming energy saving while maintaining the comfort level of the occupants," by Matteo Mendula et al. [77]; "Smart Home Prototype, EzHomeA, to remotely control different appliances" by Mohd Ashril et al. [78]. Other

TABLE 1.2

Smart Home Challenges

Author	Year	Factors
Meyers et al. [62], Venkatesh et al. [63]	2010, 2003	Less knowledge
Venkatesh [64], Balta-Ozkan et al. [65], GhaffarianHoseini et al. [66], Psychoula et al. [67], Meyers et al. [68], Ehrenhard et al. [69]	2003, 2013, 2018, 2020, 2010, 2014	High expenditure
Meyers et al. [62], Edwards et al. [70], Li et al. [71]	2010, 2001, 2012	Unavailability in the market
Jadhav et al. [72], GhaffarianHoseini et al. [66], Psychoula et al. [67], Venkatesh [64], Hosek et al. [73]	2017, 2018, 2020, 2003, 2017	Security and safety issues
Edwards et al. [70], Li et al. [71]	2001, 2012	Unreliable
GhaffarianHoseini et al. [66], Meyers et al. [68]	2018, 2010	Difficult to handle or understand
GhaffarianHoseini et al. [66], Venkatesh [64], Ehrenhard et al. [69], Li et al. [71]	2018, 2003, 2014, 2012	More complexity
Balta-Ozkan et al. [65], GhaffarianHoseini et al. [66], Psychoula et al. [67], Meyers et al. [62], Li et al. [71]	2013, 2018, 2020, 2010, 2012	Does not fit lifestyle

models by authors Christian Reinisch et al. [79], Diane J. Cook et al. [80], Sumi Helal et al. [81] Cory D. Kidd et al [82] are "ThinkHome," "MavHome," "Gator Tech Smart Home" and "Aware Home" focusing on the comfort level, sustainability, and maximizing the productivity by reducing the cost of operations respectively.

From the review of literature, it is found that different authors considered different domains of smart cities for applications based on AI. Five domains from the literature are extracted and presented in Table 1.3, where AI applications can be implemented. These domains comprise environment, government, urban settlement, economy, and social service.

Further, AI is divided into two categories [14] as shown in Figure 1.12.

1.5 Global Practices for Smart Cities

Throughout the world, the works on smart city projects have arrived at the peak. The main smart cities indexing bodies in the world are Institute for

TABLE 1.3

Domain and Their Sectors Using AI Applications

Domain	Sector of AI Applications	Raaijen and Daneva [83]	Oktaria and Kurniawan [84]	Arroub et al. [85]	Alamsyah et al. [86]
Environment	Environment	✓		✓	
	Water management				✓
	Housing				✓
	Pollution control	✓		✓	✓
	Resources	✓		✓	
	Community		✓		✓
	Public space				
	Building				✓
	Renewable energy			✓	✓
	Smart electric grid	✓			✓
Government	Transparent government	✓			✓
	Public and city administration		✓	✓	
	E-government and citizen participation		✓	✓	✓
	Emergency response				✓
	Public safety				✓
	Disaster management		✓		
	City management				
	City monitoring	✓			✓
	Crime and disaster prevention		✓		
	Environmental control		✓	✓	
	Public service				✓
	Job creation				
	Facility and infrastructure provisions controlling service	✓			
Economy	Transaction and market		✓		✓
	Logistics		✓		✓
	Advertisement				✓
	Entrepreneurship				✓

(Continued)

TABLE 1.3 (Continued)

Domain and Their Sectors Using AI Applications

Domain	Sector of AI Applications	Raaijen and Daneva [83]	Oktaria and Kurniawan [84]	Arroub et al. [85]	Alamsyah et al. [86]
	Enterprise management	✓	✓		✓
	Supply chain and commerce		✓		✓
	Warehousing		✓		
	Agriculture		✓		✓
	Industry		✓		
	Center and payment service		✓		
	Research and policy innovation	✓	✓		
Urban settlement	Water management				✓
	Energy	✓			
	Real estate		✓		
	Utilities				
	Drainage				
	Environmental road infrastructure	✓			
	Wastewater management				✓
	Information and communication				
	Open green spaces			✓	
	Financial service				
	Space for enterprises				
	Park		✓		
	Tourist		✓		✓
	Travel guidance		✓		
	Regional Information Center (RIC)				
	Lodging		✓		
	Information services for smart cities	✓			
	Traffic control services	✓	✓		
	Culture		✓		
	Parking				
	Transportation		✓		
	Automobile information service		✓		

(Continued)

TABLE 1.3 (Continued)

Domain and Their Sectors Using AI Applications

Domain	Sector of AI Applications	Raaijen and Daneva [83]	Oktaria and Kurniawan [84]	Arroub et al. [85]	Alamsyah et al. [86]
Social service	Social cohesion				✓
	Social care and welfare	✓		✓	
	Sports and Entertainment	✓			
	Learning and education	✓			
	Social service center	✓			
	E-services delivery	✓			
	Worship				
	Public transport	✓		✓	✓
	Healthcare	✓		✓	✓
	Burial	✓			

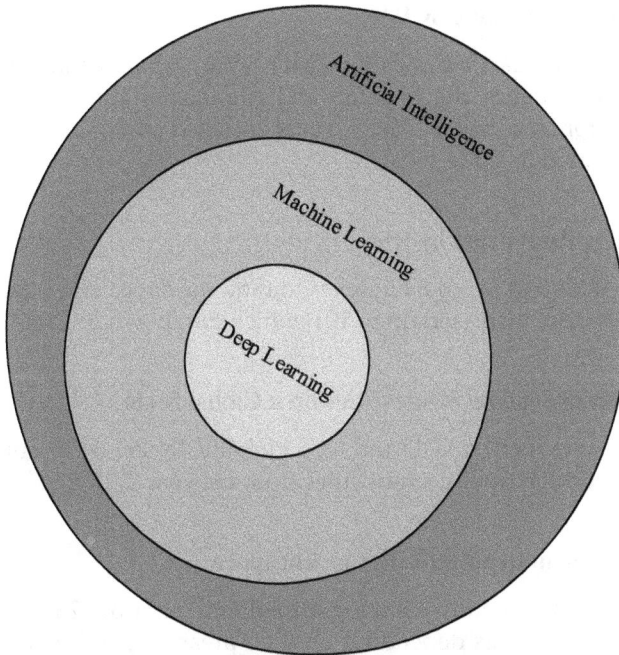

FIGURE 1.12
Types of AI.

Management Development (IMD) and Instituto de Estudios Superiores de la Empresa (IESE). According to IMD indexing parameters, the top 10 smart cities in the world are shown in Figure 1.13.

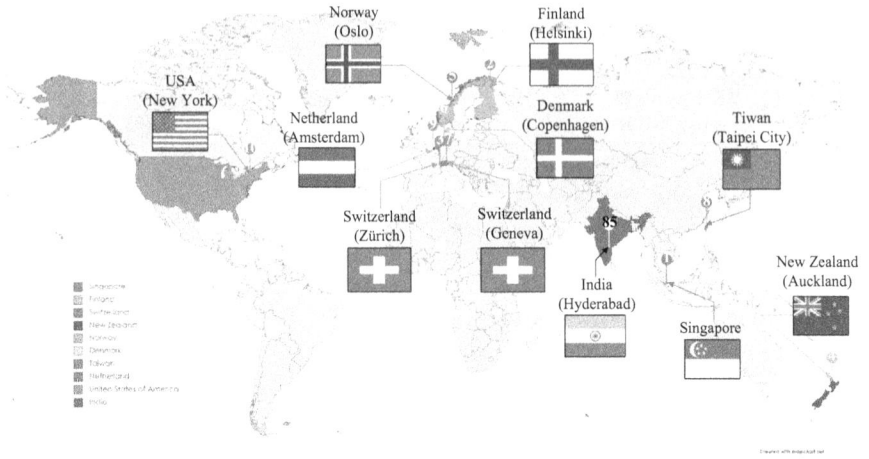

FIGURE 1.13
Top ten smart cities in the world [31].

1.5.1 Indexing Parameters by IMD

The definition of smart cities as per IMD is "an urban setting that applies technology to enhance the benefits and diminishes the shortcomings of urbanization for its citizens." The various enclosed parameters according to IMD are tabulated in Table 1.4.

1.5.2 Indexing Parameters by IESE

In 2018, the IESE had given 83 indexes to rank the smart cities. But by 2020, these indexes have increased up to 101, which are shown in Table 1.5.

1.5.3 Position of Leading Smart Cities on a Global Scale

Smart cities are ranked by IMD and IESE globally. Table 1.6 presents the positions of the world's leading smart cities from the year 2017 to 2020.

1.5.4 Development of Smart Cities in Singapore

Singapore is a small country having a total land area of 728 sq. km and it ranks 175th of countries depending on their areas [87]. The temperature of Singapore lies between 25°C to 31°C due to maritime exposure and geographical conditions [88]. America is often called 'God's Country.' Australia calls itself 'Lucky Country.' Singapore wants to be called an 'Intelligent Island' [89]. Singapore is in the scarcity of natural resources. The major resource the country has is its people. In 2015, this country was at 9th rank as per ranking

TABLE 1.4

IMD Parameters for Indexing Smart Cities

Structures	Technologies
Health and Safety	
Medical services provision is satisfactory	An application allows citizens to monitor the air quality/air pollution
Public safety	Free WiFi facilities help to access the more services
In the poorest region, the availability of basic sanitation facilities	City maintenance issues resolved online with rapid processing
Air pollution	CCTV cameras provide better security and have made residents feel safer
Rental houses availability	Online medical services and appointments faster the access
Recycling services should be adequate	Management of unwanted items using a website or smartphone application
Mobility	
Public transport is satisfactory	Mobile phone-based application provides information/ alert of traffic congestion
	Bicycle hiring has reduced congestion
	Car-sharing Apps have reduced traffic jamming
Traffic jamming is not a problem	Digital ticketing and scheduling make public transport more convenient
	Parking applications reduced journey time and provide information on parking slots
Activities	
Cultural activities (shows, bars and museums) are satisfactory	Tourism guiding application/website Digitalization of museums and show tickets
Green spaces are satisfactory	
Opportunities (Work and School)	
Minorities feel welcome	Information and Technological (IT) curriculum for city schools
Accessibility to good schools	Online studies meet the Internet connectivity speed
Businesses are creating new jobs	Web portals to finding jobs
Employment finding services are readily available	New startups and business opportunities through digital platform
Local institute provide the information for long-life learning openings	
Governance	
Residents provide feedback on local government projects	Identification of documents using online services reduces the time
Corruption free city	Voting increased using a digital platform
Citizens take part in the decision making of local government	Citizens idea portal to improve the city life
Information on local government decisions is easily accessible	The corruption of the city reduced by providing online citizen access to city finance

TABLE 1.5

IESE Indicators for Ranking Smart Cities

Social Cohesion Indicators

(i) Female-friendly	(vi) Happiness index	(xi) Price of property
(ii) Hospitals	(vii) Gini index	(xii) Homicide rate
(iii) Crime rate	(viii) Peace index	(xiii) Death rate
(iv) Slavery index	(ix) Health index	(xiv) Unemployment rate
(v) Female employment ratio	(x) Suicide rate	(xv) Terrorism

Human Capital Indicators

(i) Secondary or higher education	(v) Business schools	(viii) Per person expenditure on entertainment
(ii) Schools	(vi) Expenditure on education	(ix) Expenditure on leisure and recreation
(iii) Movement of students	(vii) Museums and art galleries	(x) Number of universities
(iv) Theaters		

Economic indicators

(i) Collaborative economy	(v) Purchasing power	(ix) GDP per capita
(ii) Ease of starting a business	(vi) Productivity	(x) Estimated GDP
(iii) Mortgage	(vii) Hourly wage in US dollars	(xi) Number of headquarters
(iv) Motivation that people must undertake early-stage entrepreneurial activity	(viii) Time required to start a business	(xii) GDP

Governance Indicators

(i) Government buildings	(v) Corruption perceptions index	(ix) Reserves
(ii) E-Government Development Index (EGDI)	(vi) ISO 37120 certification	(x) Reserves per capita
(iii) Embassies	(vii) Research centers	(xi) Employment in the public administration
(iv) Strength of legal rights index	(viii) Democracy ranking	(xii) Open data platform

Environmental Indicators

(i) Solid waste	(v) PM_{10}	(ix) Environmental performance index
(ii) Future climate	(vi) $PM_{2.5}$	(x) CO_2 emission index
(iii) CO_2 emissions	(vii) Percentage of the population with access to the water supply	(xi) Pollution index
(iv) Methane emissions	(viii) Renewable water resources	

Mobility and Transportation Indicators

(i) Bicycle rental	(vi) Scooter rental	(xi) Bike sharing
(ii) Moped rental	(vii) Bicycles per household	(xii) High-speed train

(Continued)

TABLE 1.5 (Continued)

IESE Indicators for Ranking Smart Cities

Mobility and Transportation Indicators		
(iii) Length of the metro system	(viii) Metro stations	(xiii) Traffic index
(iv) Commercial vehicles in the city	(ix) Exponential traffic index	
(v) Traffic inefficiency index	(x) Flights	
Urban Planning Indicators		
(i) Bicycles for rent	(iii) Percentage of the urban	(v) Population with adequate sanitation services
(ii) Buildings	(iv) Buildings over 35 meters high	(vi) Number of people per household
International Projection Indicators		
(i) Number of passengers per airport	(iii) Hotels	(v) Restaurant index
(ii) McDonald's	(iv) Number of conferences and meetings	(vi) Number of photos of the city uploaded online
Technology Indicators		
(i) 3G coverage	(vii) Online banking	(xiii) Social networks
(ii) Innovation index	(viii) Online video calls	(xiv) Landline subscriptions
(iii) Internet	(ix) LTE/ WiMAX	(xv) Broadband subscriptions
(iv) Mobile phone penetration ratio	(x) Personal computers	(xvi) Internet usage away from home and/or office
(v) Telephony	(xi) Mobile telephony	(xvii) Internet speed
(vi) Web Index	(xii) Wifi hotspots	

based on IESE smart cities indexing parameters [90]. Later in 2019, the country earned 1st rank with its efficient and intelligent transformation, which is motivating for other countries. In 2020, Singapore maintained its position as 1st with innovative technologies. They broadly categorized their smart city development plans into six parts: urban living, transport, strategic national projects, health, startups and businesses and digital government services.

1.5.4.1 Urban Living

In 2016, Singapore installed a trial Automated Meter Reading (AMR) initially in five hundred homes in Punggol. This trial completed its goal, which is installing a Smart Water System (SWS) in homes. This SWS provides information regarding water consumption, water leakage consumption patterns in their smartphones. Additionally, users get notifications of high usage, accidental leakage, and tips to reduce water consumption as per individual behavior. Due to the success of the trial phase, the National Water Agency of Singapore, PUB will install 3 lakh Solid Waste Management (SWM) till 2023.

TABLE 1.6

Ranking of Smart Cities

Global Rank	IMD		IESE			
	2020	2019	2020	2019	2018	2017
1.	Singapore-Singapore	Singapore-Singapore	London-UK	London-UK	New York-USA	New York-USA
2.	Helsinki-Finland	Zurich-Switzerland	New York-USA	New York-USA	London-UK	London-UK
3.	Zurich-Switzerland	Oslo-Norway	Paris-France	Amsterdam-Netherland	Paris-France	Paris-France
4.	Auckland-New Zealand	Geneva-Switzerland	Tokyo-Japan	Paris-France	Tokyo-Japan	Boston-USA
5.	Oslo-Norway	Copenhagen-Denmark	Reykjavik-Iceland	Reykjavik-Iceland	Reykjavik-Iceland	San Francisco-USA
6.	Copenhagen-Denmark	Auckland-New Zealand	Copenhagen-Denmark	Tokyo-Japan	Singapore-Singapore	Washington, D.C.-USA
7.	Geneva-Switzerland	Taipei-Taiwan	Berlin-Germany	Singapore-Singapore	Seoul-South Korea	Seoul-South Korea
8.	Taipei-Taiwan	Helsinki-Finland	Amsterdam-Netherland	Copenhagen-Denmark	Toronto-Canada	Tokyo-Japan
9.	Amsterdam-Netherland	Bilbao-Spain	Singapore-Singapore	Berlin-Germany	Hong Kong-China	Berlin-Germany
10.	New York-USA	Dusseldorf-Germany	Hong Kong-China	Vienna-Austria	Amsterdam-Netherland	Amsterdam-Netherland

Living in smart towns requires smart environment, smart planning, smart living, smart estate, and smart community. To reduce the electricity demand, citizens can look forward to the sustainability initiatives such as smart lighting. Smart lighting systems are equipped with smart sensors and work according to real-time human traffic which helps to reduce the usage of energy. SolarNova Programme introduced by the Singapore government is to deploy solar Photovoltaic (PV) systems to generate 350 MW of solar power by 2020.

Automatic pneumatic type waste collection systems, collecting waste from households through vacuum pipes and transporting it by trucks in sealed bags to the recycling units. The benefits of these pneumatic systems are that it reduces manual labor and odor. Additionally, it provides hygiene and a cleaner environment.

Singapore uses drones to inspect the roof gutters, drains, and other open sources to identify where the mosquitos can breed. Stagnant sources of water give rise to dengue, malaria, and typhoid etc. Drones in Singapore are equipped with Bti larvicide dispensing system. "Bacillus thuringiensis subspecies israelensis (Bti) bacteria are found in soil" [91], which annihilate larvae in their initial stage and thus control mosquito breeding.

Singapore developed various applications under its smart city initiatives, which enable citizens to get daily updates and forecasts of air quality, weather, lighting and Ultra-Violet Index (UVI) etc. This application named "myENV" also alerts citizens based on their specific locations about heavy rainfall, dengue-prone area, and sudden natural calamities. This application enables citizens to timely vacate their current locations according to the app alerts. In the year 2015, the Municipal Corporation (MC) of Singapore launched an application named "OneService" to resolve the municipal issues. By 2016 more than 55,000 registrations were done and approximately 32,000 cases were reported through this application. The country also provides smart systems designed for elderly people to enable them for independent living, which as a result give assurance to their children at work. 'M1 and AstraLink' are implementing real-time solutions for Smart Elderly Alert (SEA) system.

The smart city Singapore is also dedicated to its 3D digital platform, called Virtual Singapore. The capabilities of the virtual program include the 3D model of a city, virtual test-bedding, making smart applications like applications for analyzing traffic flows, emergency evacuation exit plans and more R&D activities. Along with developing a smart nation, Singapore also focuses on creating employment and reducing the travel time between the workplace and home. Every corner of Singapore is accessible, and all the amenities are allocated strategically for better living.

1.5.4.2 Health

Productivity and wellbeing in hospitals are the most considerable factors in a smart city. To target these factors Singapore implemented robotics and assistive technologies in their healthcare sector. Industrial robotic arms to carry and pack different objects, Unmanned Aerial Vehicle (UAV) to deliver medical equipment's and medicines, Automated Guided Vehicles (AGV) to transport furniture, clothes, documents, oxygen cylinders, and food are some of the advanced technologies implemented by the Ministry of Health of Singapore.

In 2017, Singapore introduced a telehealth application to provide health-related information via videoconferencing. Telehealth is defined as "the delivery and facilitation of health and health-related services including medical care, provider and patient education, health information services and self-care via telecommunications and digital communication technologies." Live video conferencing, mobile health apps, "store and forward" electronic transmission and Remote Patient Monitoring (RPM) are examples of technologies used in telehealth [92].

1.5.4.3 Transport

Smart transport systems are adopting self-driving technologies easily and efficiently. This self-driving technology not only improves the living environment but also reduces issues like manpower requirement (driver) and provides extra space. From 2017 to 2019, heavy automobile companies like Scania and Toyota worked continuously to develop driverless heavy truck technologies. The Land Transport Authorities (LTA) of Singapore developed a contactless fare payment system for public transport. Long-range RFID systems are being used with sensors to detect the passengers and deduct the fare from the passenger's travel card automatically. The country also provides "On-Demand Shuttles" (ODS) to their citizens, where they can avail autonomous shuttle facilities with a single tap on their smartphone. Passengers can modify the routes and add their favorable intermediate stoppages.

LTA gathers data from various sensors and cameras deployed in the smart transport system. Later, they analyzed the real-time traffic data obtained to find out average waiting times, issues regarding sudden crowds, traffic congestion, and peak hours. After understanding and compiling the historical data and traffic patterns, National Environment Agency (NEA) weather service and Land Transport DataMall prepare current and future traffic conditions. The main pipeline project of the Singapore smart transport authority is Fusion Analytics for Public Transport Emergency Response (FASTER) to improve transport planning.

1.5.4.4 Digital Government Services

Singapore made its governance transparent and efficient with various digital services. The 'LicenseOne' portal provides aid in granting the license and making transactions simple. This portal also provides services like renewing, amending, and terminating licenses easily from different bodies and departments to save time and money. For supporting different government systems, Centre of Excellence (CentEx) for ICT and smart systems provide data science, ICT infrastructure, AI, IoT, application development, and cybersecurity support. CentEx is also enhancing its technical capabilities in robotics, digital twins and AR/VR technologies.

The Housing and Development Board (HDB) of Singapore developed a resale portal to buy and sell flats conveniently without much legal hassle. 'LifeSG' previously known as Moments of Life, connect citizens to government services through a single platform. LifeSG complies with Public Sector Governance Act (PSGA) to safeguard the citizen's personal data. Considering the diversity of citizens, Singapore's governance decided to upgrade their services in different languages. Multilingual upgrades had already been done by HDB grant name EASE (Enhancement for Active Seniors).

Blockchain-based platform 'OpenCerts' is a unique initiative by the Ministry of Education of Singapore. Digital versions of academic documents can be issued and published using OpenCerts. This results in saving resources and money along with securing documents, certificates, and transcripts in digital formats. Parent Gateway application is an innovative approach to track the children's progress in academics. Parents can easily communicate with teachers and each can share the information and views related to children's progress. It enhances parent satisfaction and reduces the workload of teachers.

1.5.4.5 Startups and Businesses

Startups and businesses are the essential assets of any country to maintain its competence and vibrancy in the international market. Digital economy gives wings to the business and startups. 'CorpPass,' digital innovation program office, 'FinTech Sandbox,' Networked Trade Platform (NTP) and Punggol Digital District (PDD) Project are some of the current advancements done by Singapore governance. The new campus of the Singapore Institute of Technology (SIT) started a business building to increase the research collaboration between academia and industry.

1.5.4.6 Strategic National Projects

The Smart Nation Drive (SND) of Singapore includes CODeX, E-Payments, LifeSG Initiatives, National Digital Identity (NDI), SNSP, and SUM. CODeX is a Core Operations Development Environment and eXchange. It is a digital platform that provides seamless data exchange between the agencies. The CODeX is helping to reduce the bugs, increase the reliability, security, and quality of the services. These facilities provide the workspace for both private and public sectors. E-payments deliver secure and digital transactions that are

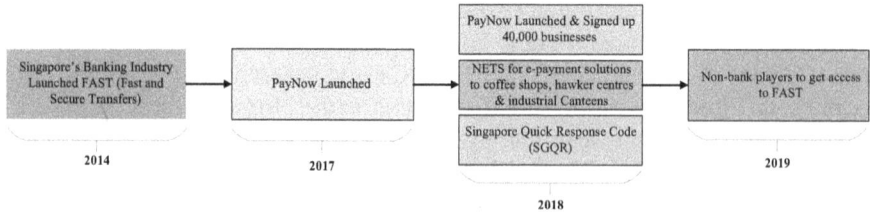

FIGURE 1.14
Strategies of Singapore National Projects.

convenient for both citizens and companies. The e-payment key milestones of Singapore Strategic National Projects (SSNP) are shown in Figure 1.14.

In the year 2020, NDI has been in operation for businesses and citizens to do their digital transactions with advanced security. SingPass (Singapore Personal Access) application is providing the access to Central Provident Fund (CPF) statement and helps residents to fill their taxes online.

Smart Lamp Post-trial work was started in the last quarter of 2019. These smart lampposts are equipped with sensors and provide updates regarding, rainfall, air quality, wind speed, pressure, humidity, and water-level. Further, the data from these sensors will be analyzed using advanced computational techniques like AI to improve the policies and process. Digital technologies increase different types of comfort namely thermal, visual, acoustical etc. [93], provide convenient transportation and manage waste in a better way. Numerous trials are going on to strengthen the intra-town connectivity and enhance citizen mobility, especially disabled persons.

1.5.5 Development of Smart Cities in India

India is called as "Land of Gods." According to Statista, India holds 7th position as per the total land area in the world [94]. According to US Census Bureau Current Population, the population of India is more than 1.3 billion and comprises 17.12% of the global population [95]. In June 2015, Prime Minister of India Shri Narendra Damodardas Modi announced a smart city project for 100 cities under the flagship of the Ministry of Housing and Urban Affairs (MoHUA) [96]. These 100 cities are presented in Figure 1.15.

The city of Hyderabad in India ranks 85th according to IMD smart city index 2020 [31]. The major core infrastructure of smart cities according to Smart City Mission (SCM) by the Ministry of Urban Development (MoUD), India includes: (i) adequate water supply; (ii) SWM, including sanitation; (iii) assured electricity supply; (iv) affordable housing, particularly for poor people; (v) management in public transport and mobility; (vi) digitalization in every sector with vigorous IT connectivity; (vii) sustainable environment; (viii) e-governance; (ix) safety and security of citizens; and (x) education and

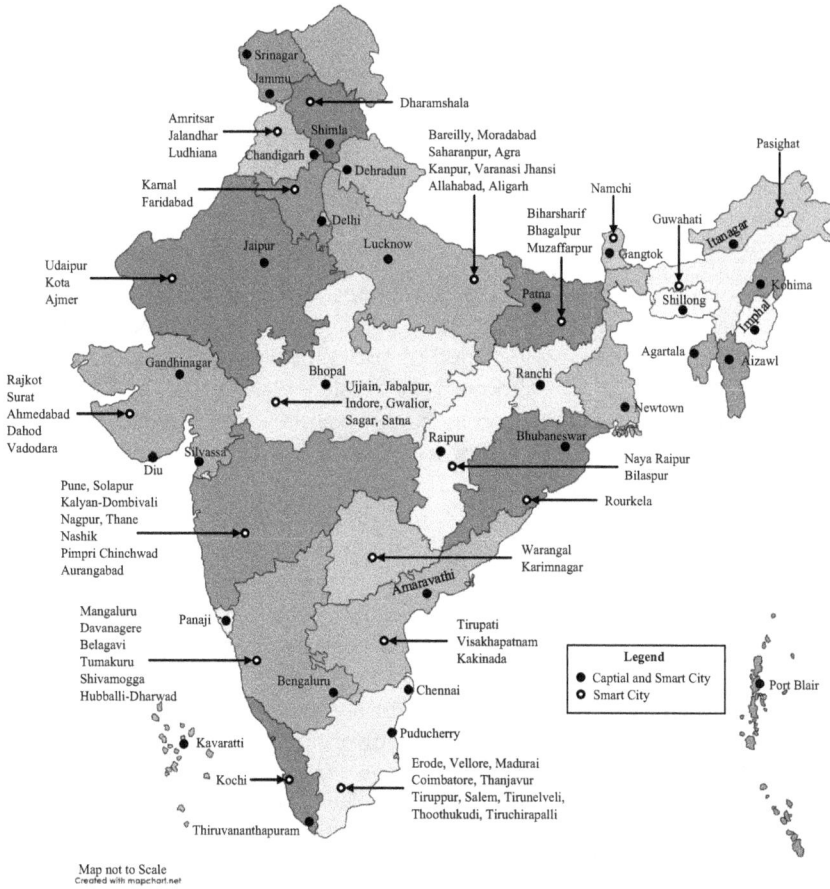

FIGURE 1.15
Selected 100 cities under smart city mission.

health. In the first round of the SCM in India, 20 cities have been selected. The selection parameters of these cities are in accordance with the "MoHUA Selection Procedure" [97]. The ease of living index is situated on three columns, containing fourteen categories and fifty indicators. The three columns are Quality of Life (QoL), Economic Ability (EcA), and Sustainability Management (SuM).

1.5.5.1 Quality of Life (QoL)

The first and most important column of the smart city is QoL. QoL is the ability of individual citizens to live with joy and comfort. Good sanitation, clean water, comfortable homes, schools [98], uninterrupted basic services, good education, and health facilities all boost the morale of citizens to prosper and work efficiently. Further, QoL was decided on the basis of health, education, mobility, water supply and SWM, housing and shelter, recreation and safety and security. The complete flow-work program for ease of living for a better quality of life is drawn in Figure 1.16.

1.5.5.2 Economic Ability (EcA)

The second column of the smart city is EcA. The economic ability focused mainly on the economic building blocks of the whole city and as an individual. The change in economic wellbeing is related to; provide better employment opportunities, growth in the wages and need for clusters etc. Further, EcA depends on the economic development, economic opportunities, and Gini coefficient.

The complete flow-work program for ease of living for better economic abilities is drawn in Figure 1.17.

1.5.5.3 Sustainability Management (SuM)

The last column of the smart city is SuM. In 1987, the United Nations Brundtland Commission (UNBC) defined sustainability as "meeting the needs of the present without compromising the ability of future generations to meet their own needs" [99]. The aim of sustainability in smart cities to provide greenery and reduce energy consumption. The main focusing areas of sustainability are the environment, green buildings, city resilience, and energy consumption. The complete flow-work program for ease of living for better sustainability management is drawn in Figure 1.18.

1.6 Conclusions

Innovative services offered by smart city helps to improve the citizen's life quality over the conventional cities. Significant work has already been done globally to upgrade cities into "smart cities." This chapter delivers a lot of information related to the origin of smart cities, their evolution, the meaning, and basic requirements of a sustainable smart city. The extent of smart technology is the need of the hour to provide a better life in urban areas, where the surge in population is quite common due to migration.

AI and IoT are the key technological advances to achieve sustainable intelligent technology goals. The IoT provides uninterrupted information exchange with security and safety to the households. AI helps in taking smart human-like decisions using image processing (deep learning) and machine

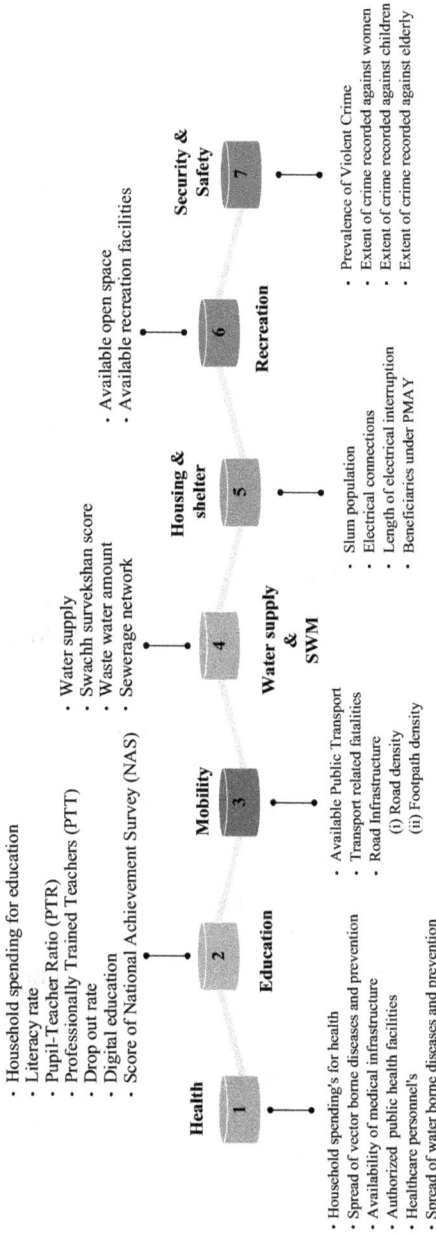

FIGURE 1.16
Flow-work program for quality of life.

- Cluster strength
- Credit availability
- Skill Development Centres
  (SDC)

Economic development   **Gini coefficient**

2

1   3

Economic opportunities

- Traded clusters   - Inequality index based
                                                                          on consumption expenditure

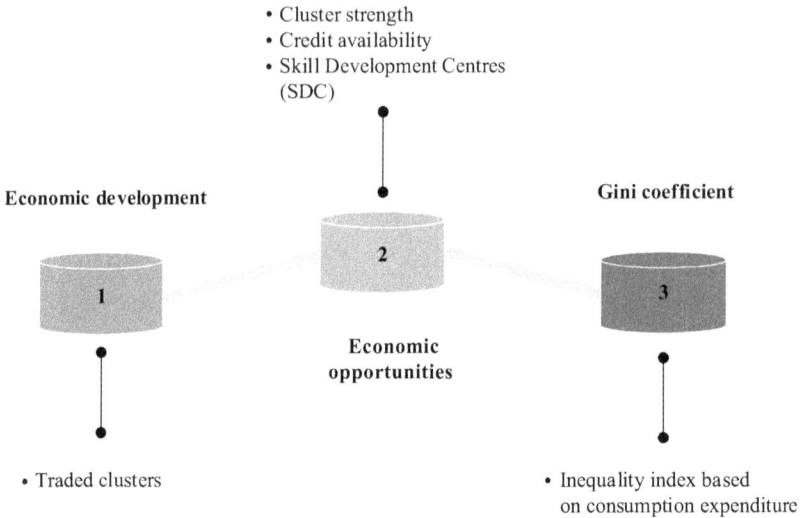

FIGURE 1.17
Flow-work program for economic ability.

learning that is further helpful in transportation engineering (traffic jamming and accident predications) and structural engineering (strength and damage predictions) etc. The combination of IoT and AI in the field of technology can be called "Brahmastra."

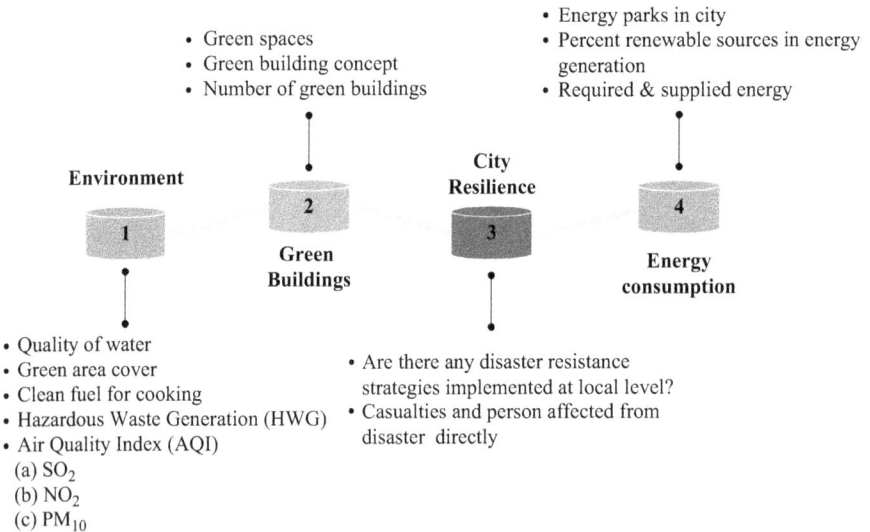

- Green spaces   - Energy parks in city
- Green building concept   - Percent renewable sources in energy
- Number of green buildings   generation
                                        - Required & supplied energy

Environment   **City Resilience**

2   3   4

1

Green Buildings   **Energy consumption**

- Quality of water
- Green area cover   - Are there any disaster resistance
- Clean fuel for cooking   strategies implemented at local level?
- Hazardous Waste Generation (HWG)   - Casualties and person affected from
- Air Quality Index (AQI)   disaster directly
  (a) SO_2
  (b) NO_2
  (c) PM_{10}

FIGURE 1.18
Flow-work program for sustainability management.

IMD and IESE are two ranking indexes that gave rankings to smart cities based on different indicators. IMD issues ranking on the basis of five indicators: health and safety, mobility, activities, opportunities, and governance. IESE ranking is based on nine indicators: social cohesion, human capital, economic, governance, environmental, urban planning, international projection, technology and mobility, and transportation. The IESE ranking is the first globally accepted ranking system to index smart cities from the year 2017 till now. However, IMD put a step in the smart city indexing system in the year 2019 and following-up till now. In both the rankings systems, Singapore and New York (USA) are the most indexed (five times) smart cities in ten leading positions.

Singapore is constantly improving its cities targeting sustainability goals with economic efficiency, social equality, and environmentally friendly policies. The country initiates the installation of AMR, SWM, Solar PVs, SEA system, UAV, AGV, RMP, ODS, and many more technologies. The country also launched various digital applications and platforms like myENV, OneService, LifeSG, OpenCerts, CorpPass, and CODeX, etc.

Currently, India is upgrading its cities with trending technologies and a total of 100 cities had been selected for this up-gradation mission and getting the "smart" tag. Sustainable smart developments of Indian cities are standings on QoL, EcA, and SuM. A complex system comprises various sensors, actuators, cameras, communications technologies, AI techniques, IoT, and big data is working as the backbone for the SCM. Mobile applications and web portals of various services have already been made and utilized by millions of people in the world's second most populated country.

Abbreviations

AC	Air Conditioner
AI	Artificial Intelligence
AGV	Automated Guided Vehicles
AMR	Automated Meter Reading
Bti	Bacillus thuringiensis subspecies israelensis
BD	Big Data
CPF	Central Provident Fund
CCT	Cloud Computing Techniques
CIoT	Commercial Internet of Things
DDS	De Digital Stad
DWRC	Department of Digital World Research Center
EASE	Enhancement for Active Seniors
FASTER	Fusion Analytics for Public Transport Emergency Response
HDB	Housing and Development Board

IIoT	Industrial Internet of Things
ICT	Information and Communication Technology
IECT	Information Exchange and Communication Techniques
IT	Information Technology
IMD	Institute for Management Development
ITS	Intelligent Transportation System
IoT	Internet of Things
LTA	Land Transport Authorities
MoHUA	Ministry of Housing and Urban Affairs
NDI	National Digital Identity
NEA	National Environment Agency
NTP	Networked Trade Platform
ODS	On-Demand Shuttles
PV	Photovoltaic
PSGA	Public Sector Governance Act
PDD	Punggol Digital District
RFID	Radio-Frequency Identification
RPM	Remote Patient Monitoring
RIoT	Residential Internet of Things
RAT	Responsiveness, Accountability, and Transparency
SIT	Singapore Institute of Technology
SSNP	Singapore Strategic National Projects
SingPass	Singapore Personal Access
SCM	Smart City Mission
SEA	Smart Elderly Alert
SND	Smart Nation Drive
SWS	Smart Water System
UVI	Ultra-Violet Index
UN	United Nations
UAV	Unmanned Aerial Vehicle
WiFi	Wireless Fidelity
WLAN	Wireless Local Area Network
WSCO	World Smart City Organization

References

1. Lee, J. Y., and Ji-In, C. 2019. The evolution of smart city policy of Korea. In *Smart City Emergence*, ed. Leonidas Anthopoulos, 173–193. Elsevier. https://doi.org/10.1016/B978-0-12-816169-2.00008-0
2. Kendig, H. 1976. Cluster analysis to classify residential areas: A Los Angeles application. *Journal of the American Institute of Planners* 42 (3):286–294. https://doi.org/10.1080/01944367608977731

3. Online Etymological Dictionary. 2019. https://www.etymonline.com/word/smart (accessed January 30, 2021)
4. Anthopoulos, L. G. 2017. *Understanding Smart Cities: A Tool for Smart Government or an Industrial Trick?*. Springer International. https://doi.org/10.1007/978-3-319-57015-0
5. Baron, G. 2012. *Amsterdam Smart City*. Amsterdam: Amsterdam Municipality. https://www.lafabriquedelacite.com/ (accessed January 24, 2021)
6. Cellary, W. 2013. Smart governance for smart industries. *Proceedings of the 7th International Conference on Theory and Practice of Electronic Governance*, Seoul, Republic of Korea. https://doi.org/10.1145/2591888.2591903
7. Kumar, A., and Rattan, J. S. 2020. A journey from conventional cities to smart cities. In *Smart Cities and Construction Technologies*, eds. S. Shirowzhan and K. Zhang, 1–14. Intechopen. https://doi.org/10.5772/intechopen.91675
8. Vasseur, J.-P., and Adam, D. 2010. Chapter 22—Smart cities and urban networks. In *Interconnecting Smart Objects with IP*, eds. J.-P. Vasseur and A. Dunkels, 335–351. Boston, MA: Morgan Kaufmann. https://doi.org/10.1016/B978-0-12-375165-2.00022-3
9. United Nations. 2014. World urbanization prospects: The 2014 revision. https://population.un.org/wup/Publications/Files/WUP2014-PressRelease.pdf (accessed January 01, 2021)
10. United Nations. 2018. World urbanization prospects: The 2018 revision. https://www.un.org/en/events/citiesday/assets/pdf/the_worlds_cities_in_2018_data_booklet.pdf (accessed January 01, 2021)
11. Voda, A. I., and Laura, D. R. 2019. Chapter 12—How can artificial intelligence respond to smart cities challenges? In *Smart Cities: Issues and Challenges*, eds. A. Visvizi and M. D. Lytras, 199–216. Elsevier. https://doi.org/10.1016/B978-0-12-816639-0.00012-0
12. Lytras, M. D., and Anna, V. 2018. Who uses smart city services and what to make of it: Toward interdisciplinary smart cities research. *Sustainability* 10 (6):1998. https://doi.org/10.3390/su10061998
13. Tolentino, J. 2017. Predictions for big data, IoT, and AI. Artificial intelligence. TNW in artificial intelligence. https://thenextweb.com/artificial-intelligence/2017/05/03/2017-predictions-for-big-data-iot-and-ai/ (accessed January 27, 2021)
14. Kumar, A., and Navdeep, M. 2021. An approach-driven: Use of artificial intelligence and its applications in civil engineering. In *Artificial Intelligence and IoT: Smart Convergence for Eco-friendly Topography*, eds. K. G. Manoharan, J. A. Nehru and S. Balasubramanian, 201–221. Singapore: Springer Singapore. https://doi.org/10.1007/978-981-33-6400-4_10
15. IEK. 2017. *Artificial Intelligence Industry Accelerated This Year*. Taiwan: IEK. https://ieknet.iek.org.tw/ieknews/news_more.aspx?nsl_id=14f5326242074c95a3f07a6c514a0081 (accessed January 27, 2021)
16. Li, D., JianJun, C., and Yuan, Y. 2015. Big data in smart cities. *Science China Information Sciences* 58 (10):1–12. https://doi.org/10.1007/s11432-015-5396-5
17. Graham, S., and Alessandro, A. 1997. Virtual cities, social polarization, and the crisis in urban public space. *Journal of Urban Technology* 4 (1):19–52. http://doi.org/10.1080/10630739708724546.

18. Lipman, A. D., Sugarman, A. D., and Cushman, R. F. 1986. *Teleports and the Intelligent City*. DOW JONES-IRWIN. http://www.sugarlaw.com/publications/teleport/teleports-all-dow-jones.pdf (accessed January 28, 2021).
19. Raynal, M. 1988. *Distributed Algorithms and Protocols*. New York City: John Wiley & Sons.
20. Hall, P. 1988. *Cities of Tomorrow: An Intellectual History of Urban Planning and Design*. Oxford: Blackwell Publishing.
21. Global Data Thematic Research. History of smart cities: Timeline. https://www.verdict.co.uk/smart-cities-timeline/#:~:text=The%20first%20smart%20city%20was,virtual%20digital%20city%20in%201994 (accessed January 27, 2021).
22. Leonidas, G. A., and Fitsilis, P. 2014. Smart cities and their roles in city competition: A classification. *International Journal of Electronic Government Research (IJEGR)* 10 (1):63–77. http://doi.org/10.4018/ijegr.2014010105
23. Caragliu, A. A., and Nijkamp, P. 2009. *Endogenous Regional Growth: The Role of Human and Cognitive Capital*. Poland: Polish Academy of Sciences.
24. Hall, R. E. 2000. The vision of a smart city. *2nd International Life Extension Technology Workshop*, Paris, France.
25. Giffinger, R., Fertner, C., Kramar, H., et al. 2007. *Smart Cities: Ranking of European Medium-Sized Cities*. Vienna: Centre of Regional Science.
26. About IEEE Smart City. IEEE smart cities. https://smartcities.ieee.org/about (accessed January 29, 2021).
27. Draft Concept Note on Smart City Scheme. Government of India—Ministry of urban development. https://web.archive.org/web/20150203073844/http://indiansmartcities.in/downloads/CONCEPT_NOTE_-3.12.2014__REVISED_AND_LATEST_.pdf (accessed January 29, 2021).
28. International Standards Organization (ISO). 2016. Sustainable development in communities. https://www.iso.org/files/live/sites/isoorg/files/archive/pdf/en/iso_37101_sustainable_development_in_communities.pdf (accessed January 30, 2021).
29. Smart Cities Council. 2016. About us. https://smartcitiescouncil.com/article/about-us-global (accessed January 29, 2021).
30. Agricultural Land Area. World data atlas. https://knoema.com/atlas/ranks/Agricultural-land-area#:~:text=The%20world's%20total%20agricultural%20land,km%20in%202016 (accessed January 30, 2021).
31. Smart City Index. 2020. IMD business school. https://www.imd.org/contentassets/4817f3697a834980b20dfdf8141c15ac/2columns-ranking-2020-big.jpg (accessed January 30, 2021).
32. Doheim, R. M., Alshimaa A. F., and Samaa, B. 2019. Chapter 17—Smart city vision and practices across the Kingdom of Saudi Arabia—A review. In *Smart Cities: Issues and Challenges*, eds. A. Visvizi and M. D. Lytras, 309–332. Elsevier. https://doi.org/10.1016/B978-0-12-816639-0.00017-X
33. Saudi Vision. 2017. Saudi vision 2030. http://vision2030.gov.sa/en/foreword (accessed 30, 2021).
34. Al-Hader, M., and Rodzi, A. 2009. The smart city infrastructure development & monitoring. *Theoretical and Empirical Researches in Urban Management* 2 (11):87–94.
35. Riva, S. E., Raffaella, R. S., Valentina, V., et al. 2017. Smart cities: Case studies. In *Smart Cities Atlas: Western and Eastern Intelligent Communities*, eds. E. R. Sanseverino, R. R. Sanseverino and V. Vaccaro, 47–140. Cham: Springer International Publishing.

36. Bell, S. 2002. *Economic Governance & Institutional Dynamics*. England, UK: Oxford University Press.
37. Yigitcanlar, T., Kamruzzaman, K., Laurie, B., et al. 2018. Understanding 'smart cities': Intertwining development drivers with desired outcomes in a multidimensional framework. *Cities* 81:145–160. https://doi.org/10.1016/j.cities.2018.04.003
38. Gupta, S., Mustafa, S. Z., and Kumar, H. 2017. Chapter 2—Smart people for smart cities: A behavioral framework for personality and roles. In *Advances in Smart Cities*, eds. A. K. Kar, M. P. Gupta and P. V. Ilavarasan, et al., 23–31. https://doi.org/10.1201/9781315156040-3
39. Pérez-delHoyo, R., and Higinio, M. 2019. Chapter 11—Knowledge society technologies for smart cities development. In *Smart Cities: Issues and Challenges*, eds. A. Visvizi and M. D. Lytras, 185–198. Elsevier. https://doi.org/10.1016/B978-0-12-816639-0.00011-9
40. Simonofski, A., Estefanía, S. A., and Yves, W. 2019. Chapter 4—Citizen participation in the design of smart cities: Methods and management framework. In *Smart Cities: Issues and Challenges*, eds. A. Visvizi and M. D. Lytras, 47–62. Elsevier. https://doi.org/10.1016/B978-0-12-816639-0.00004-1
41. Kumar, T. M. V. 2015. E-Governance for smart cities. *Advances in 21st Century Human Settlements*. https://doi.org/10.1007/978-981-287-287-6
42. Albino, V., Umberto, B., and Rosa, M. D. 2015. Smart cities: Definitions, dimensions, performance and initiatives. *Journal of Urban Technology* 22 (1):3–21. https://doi.org/10.1080/10630732.2014.942092
43. Duan, W., Rouhollah, N., and Sasan, K. 2019. Smart city concepts and dimensions. *Proceedings of the 2019 7th International Conference on Information Technology: IoT and Smart City*, Shanghai, China. https://doi.org/10.1145/3377170.3377189
44. Nam, T., and Pardo, T. A. 2011. Conceptualizing smart city with dimensions of technology, people and institutions. *The Proceedings of the 12th Annual International Conference on Digital Government Research*, USA.
45. European Investment Bank- Institute, Assessing Smart City Initiatives for the Mediterranean Region. 2017. Smart city development. https://institute.eib.org/wp-content/uploads/2017/03/4-Ascimer.pdf (accessed January 30, 2021).
46. Wang, M., Tao, Z., and Di, W. 2020. Tracking the evolution processes of smart cities in China by assessing performance and efficiency. *Technology in Society* 63:101353. https://doi.org/10.1016/j.techsoc.2020.101353
47. Kummitha, R. K. R., and Nathalie, C. 2017. How do we understand smart cities? An evolutionary perspective. *Cities* 67:43–52. https://doi.org/10.1016/j.cities.2017.04.010
48. Academic Impact. Sustainability, academic impact. https://academicimpact.un.org/content/sustainability#:~:text=In%201987%2C%20the%20United%20Nations,development%20needs%2C%20but%20with%20the (accessed February 19, 2021).
49. The World Bank. e-Government. https://www.worldbank.org/en/topic/digitaldevelopment/brief/e-government#:~:text=%E2%80%9CE%2DGovernment%E2%80%9D%20refers%20to,and%20other%20arms%20of%20government.&text=Learn%20more%20about%20e%2DGovernment (accessed February 07, 2021).
50. Leydesdorff, L. 2013. Triple helix of university-industry-government relations. In *Encyclopedia of Creativity, Invention, Innovation and Entrepreneurship*,

ed. E. G. Carayannis, 1844–1851. New York: Springer New York. https://doi.org/10.1007/978-1-4614-3858-8_452

51. Kapoor, N. R., Baghel, A., Sharma, H., et al. 2017. Study of intelligent transportation systems in India. *International Journal of Advance Research in Science and Engineering* 6 (10):26–34.

52. Xia, F., Yang, L. T., and Wang, L., et al. 2012. Editorial internet of things. *International Journal of Communication System* 25: 1101–1102. https://doi.org/10.1002/dac.2417

53. Ashton, K. 2009. That 'internet of things' thing. *RFID Journal* 22 (7): 97–114.

54. ITU. 2021. New ITU standards define the internet of things and provide the blueprints for its development. https://www.itu.int/en/action/broadband/Documents/Harnessing-IoT-Global-Development.pdf (accessed January 31, 2021).

55. Sundmaeker, H., Guillemin, P., Friess, P., et al. 2010. Vision and challenge's for realizing the internet of things Cluster of European Research Projects on the Internet of Things 3:34–36. https://doi.org/10.2759/26127

56. Attwood, A., Merabti, M., Fergus, P., et al. 2011. Smart cities critical infrastructure response framework. *Proceedings of the 2011 Developments in E-Systems Engineering*, 460–464. https://doi.org/10.1109/DeSE.2011.112

57. Harper, R. 2003. Inside the smart home: Ideas, possibilities and methods. In *Inside the Smart Home*, ed. R. Harper, 1–13. London: Springer London. https://doi.org/10.1007/1-85233-854-7_1

58. Zielonka, A., Woźniak, M., Garg, S., et al. 2021. Smart homes: How much will they support us? A research on recent trends and advances. *IEEE Access* 9:26388–26419. https://doi.org/10.1109/ACCESS.2021.3054575

59. Balta-Ozkan, N., Rosemary, D., and Martha, B., et al. 2013. Social barriers to the adoption of smart homes. *Energy Policy* 63:363–374. https://doi.org/10.1016/j.enpol.2013.08.043

60. Giatec Scientific. An overview of the concrete maturity method Giatech. https://www.giatecscientific.com/strength-maturity/ (accessed February 04, 2021).

61. Kumar, A., and Navdeep, M. 2020. Prediction of accuracy of high-strength concrete using data mining technique: A review. *Proceedings of International Conference on IoT Inclusive Life (ICIIL 2019)*, 259–267 Singapore. https://doi.org/10.1007/978-981-15-3020-3_24

62. Meyers, R. J., Williams, E. D., and Matthews, H. S. 2010. Scoping the potential of monitoring and control technologies to reduce energy use in homes. *Energy and Buildings* 42 (5):563–569. https://doi.org/10.1016/j.enbuild.2009.10.026

63. Venkatesh, A., Erik, K., and Eric, C. S. 2003. The networked home: An analysis of current developments and future trends. *Cognition, Technology & Work* 5 (1):23–32. https://doi.org/10.1007/s10111-002-0113-8

64. Venkatesh, A. Smart home concepts: Currents trends. IT in home. https://escholarship.org/uc/item/6t16p6pf (assessed February 04, 2021).

65. GhaffarianHoseini, A., Nur, D. D., and Umberto, B., et al. 2013. Sustainable energy performances of green buildings: A review of current theories, implementations and challenges. *Renewable and Sustainable Energy Reviews* 25:1–17. https://doi.org/10.1016/j.rser.2013.01.010

66. Shan, M., and Bon-Gang, H. 2018. Green building rating systems: Global reviews of practices and research efforts. *Sustainable Cities and Society* 39:172–180. https://doi.org/10.1016/j.scs.2018.02.034

67. Psychoula, I., Chen, L., and Amf, O. 2020. Privacy risk awareness in wearables and the internet of things. *IEEE Computer Society* 19:60–66.
68. Meyers, R. J., Eric, D. W., and Scott Matthews, H. 2010. Scoping the potential of monitoring and control technologies to reduce energy use in homes. *Energy and Buildings* 42 (5):563–569. https://doi.org/10.1016/j.enbuild.2009.10.026
69. Ehrenhard, M., Bjorn, K., and Lambert, N. 2014. Market adoption barriers of multi-stakeholder technology: Smart homes for the aging population. *Technological Forecasting and Social Change* 89:306–315. https://doi.org/10.1016/j.techfore.2014.08.002
70. Edwards, W. K., and Grinter, R. E. 2001. At home with ubiquitous computing: Seven challenges. *International conference on ubiquitous computing: LNCS 2001*, Springer- Verlag, Berlin Heidelberg.
71. Li, W., Lee, Y.-H., Tsai, W.-T., et al. 2012. Service-oriented smart home applications: Composition, code generation, deployment, and execution. *Service Oriented Computing and Applications* 6 (1):65–79. https://doi.org/10.1007/s11761-011-0086-7
72. Jadhav, V., Kumar, K. N., and Rana, P. D. A., et al. 2017. Understanding the correlation among factors of cyber system's security for Internet of Things (IoT) in smart cities. *Journal of Accounting, Business and Management (JABM)* 24 (2):1–15. http://journal.stie-mce.ac.id/index.php/jabminternational/article/view/319 (accessed February 13, 2021).
73. Hosek, J., Masek, P., and Andreev, S., et al. 2017. A SyMPHOnY of integrated IOT businesses: Closing the gap between availability and adoption. *IEEE Communications Magazine* 55 (12):156–164.
74. Karakash, C. 1998. Intelligence: A nomadic concept. In *Intelligence and Artificial Intelligence: An Interdisciplinary Debate*, eds. U. Ratsch, M. M. Richter and I. O. Stamatescu, 22–40. Berlin, Heidelberg: Springer Berlin Heidelberg. http://doi.org/10.1007/978-3-662-03667-9.
75. Seising, R. 2012. John McCarthy, 1927–2011. *Artificial Intelligence in Medicine* 54 (3):151–154. https://doi.org/10.1016/j.artmed.2012.01.003
76. Gandhi, M., Ashok, K., and Rajasekar, E., et al. 2020. A review on shape-stabilized phase change materials for latent energy storage in buildings. *Sustainability* 12 (22):9481. https://doi.org/10.3390/su12229481
77. Mendula, M., Siavash, K., and Salih, S. B., et al. 2020. Interaction and behaviour evaluation for smart homes: Data collection and analytics in the Scaled Home Project. *Proceedings of the 23rd International ACM Conference on Modeling, Analysis and Simulation of Wireless and Mobile Systems*, Alicante, Spain. https://doi.org/10.1145/3416010.3423227
78. Ashril, N. A. N. M., Dahnil, D. P., and Abdullah, S. 2019. Wifi based smart home prototype development. *International Conference on Electrical Engineering and Informatics (ICEEI)*, 9–10 July 2019. doi: 10.1109/ICEEI47359.2019.8988871.
79. Reinisch, C., Kofler, M. J., and Kastner, W. 2010. ThinkHome: A smart home as digital ecosystem. *4th IEEE International Conference on Digital Ecosystems and Technologies*, 13–16 April 2010. doi: 10.1109/DEST.2010.5610636.
80. Cook, D. J., Youngblood, M., and Heierman, E. O., et al. 2003. MavHome: An agent-based smart home. *Proceedings of the First IEEE International Conference on Pervasive Computing and Communications, 2003 (PerCom 2003)*. 26–26 March 2003. doi: 10.1109/PERCOM.2003.1192783.

81. Helal, S., Mann, W., and El-Zabadani, H., et al. 2005. The gator tech smart house: A programmable pervasive space. *Computer* 38:50–60. https://www.cise.ufl.edu/~helal/projects/publications/helal_GTSH_IEEE_Computer_March_2005.pdf (accessed February 05, 2021).

82. Kidd, C. D., Robert, O., and Gregory, D. A., et al. 1999. The aware home: A living laboratory for ubiquitous computing research. *International Workshop on Cooperative Buildings*. Berlin, Heidelberg: Springer. https://doi.org/10.1007/10705432_17

83. Raaijen, T., and Daneva, M. 2017. Depicting the smarter cities of the future: A systematic literature review & field study. *2017 Smart City Symposium Prague (SCSP)*, 25–26 May 2017, Prague, Czech Republic.

84. Oktaria, D., and Kurniawan, N. B. Smart city services: A systematic literature review. *Proceedings of International Conference on Information Technology Systems and Innovation (ICITSI 2017)*. http://toc.proceedings.com/38032webtoc.pdf (accessed February 06, 2021).

85. Arroub, A., Zahi, B., Sabir, E., et al. 2016. A literature review on smart cities: Paradigms, opportunities and open problems. *International Conference on Wireless Networks and Mobile Communications (WINCOM)*, 26–29 October 2016. https://doi.org/10.1109/WINCOM.2016.7777211

86. Alamsyah, N., Susanto, T. D., and Chou, T. 2016. A comparison study of smart city in Taipei and Surabaya. *International Conference on ICT for Smart Society (ICISS)*, 20–21 July 2016. https://doi.org/10.1109/ICTSS.2016.7792859

87. Wikipedia. List of countries and dependencies by area. https://en.wikipedia.org/wiki/List_of_countries_and_dependencies_by_area (accessed February 11, 2021).

88. GuideMeSingapore. Weather and climate in Singapore. https://www.guidemesingapore.com/business-guides/immigration/get-to-know-singapore/weather-and-climate-in-singapore (accessed February 11, 2021).

89. Mahizhnan, A. 1999. Smart cities: The Singapore case. *Cities* 16 (1):13–18. https://doi.org/10.1016/S0264-2751(98)00050-X

90. IESE Cities in Motion Index. 2015. Center for globalization and strategy. https://media.iese.edu/research/pdfs/ST-0366-E.pdf (accessed February11, 2021).

91. CDC. Mosquito control centre for disease control and prevention. https://www.cdc.gov/zika/pdfs/BTI_Fact_Sheet.pdf (accessed February 12, 2021).

92. NEJM Catalyst. What Is Telehealth? https://catalyst.nejm.org/doi/full/10.1056/CAT.18.0268 (accessed February 13, 2021).

93. Kapoor, N. R., and Tegar, J. P. 2018. Human comfort indicators pertaining to indoor environmental quality parameters of residential buildings in Bhopal. *International Research Journal of Engineering and Technology* 5 (7):1744–1750.

94. Statista. The 30 largest countries in the world by total area. https://www.statista.com/statistics/262955/largest-countries-in-the-world/ (accessed February 16, 2021).

95. World Population. U.S. Census Bureau current population. https://www.census.gov/popclock/print.php?component=counter (accessed February 16, 2021).

96. Smart Cities. Smart cities mission, ministry of housing and urban affairs, GoI. http://smartcities.gov.in/content/innerpage/cities-profile-of-20-smart-cities.php (accessed February 17, 2021).

97. Smart Cities. Process of selection. Smart cities mission. Ministry of housing and urban affairs, GoI. http://smartcities.gov.in/content/innerpage/process-of-selection.php (accessed February 17, 2021).

98. Kapoor, N. R., Ashok, K., and Chandan, S. M., et al. 2021. A systematic review on indoor environmental quality in naturally ventilated school classrooms: A way forward. *Advances in Civil Engineering* 2021:8851685. https://doi.org/10.1155/2021/8851685

99. Haque, U. 2012. Surely there's a smarter approach to smart cities? http://www.wired.co.uk/article/potential-of-smarter-cities-beyond-ibm-and-cisco (accessed February 19, 2021).

2

A Smart City Analytical Framework: Evidence from Vietnam

Nguyen Viet Lam and Bui Huy Khoi

Industrial University of Ho Chi Minh City, Vietnam

CONTENTS

2.1 Introduction

In human history, we are living in a convergence of two big phenomena: the emergence of global urbanization, and the digital revolution. Currently, according to the United Nations, cities are home to 55% of the world's population, with that number predicted to rise to 68% by 2050. More than 2.5 billion people will live in cities in 30 years, mainly in Asia and Africa, based on current population growth rates [1]. Currently, Vietnamese cities are facing many challenges of urbanization pressure; population increment increase; traffic congestion; polluted environment; electricity, water, and traffic infrastructure are overloaded. To get through these challenges, urban paradigms with socio-economic, environmental, and transport infrastructure Smart management is an inevitable one. However, assess the current level of development toward Smart city access is a moderately new issue in Vietnam. Therefore, this research will mainly be about producing an analytical framework for the pillars of smart city asymptotic according to international and local practice measures to evaluate these pillars through selected contextual scales; at the same time, research proposes an analytical framework, and selected criteria will be applied to assess the asymptotic level of the model smart city in Vietnam. From there, we initially offer key suggestions policies to promote appropriate steps toward smart cities. Results will be a meaningful reference for researchers and policymakers when the smart urban model in the Vietnamese context is still quite new, but it is interested strong [2].

The paper's structure is as follows: literature reviews on smart cities and our proposed research model, smart city analytical framework, data in Vietnam, discussion, and conclusion.

2.2 Literature Review

The idea of a "smart city" was conceived in 1994. The smart city idea is focused on environmental sustainability, with the primary goal of using cutting-edge technology to reduce greenhouse gas emissions in urban areas. Many private and public investments in technology development and implementation have resulted from the increasing interest in the smart city concept, as well as the need to resolve urbanization's challenges [3]. There are many concepts of "Smart City," and this concept is used inconsistently with each other. Below, we present the idea of a "smart city" of previous researchers:Mitchell [4] argues that the origin of this concept lies in the "Connection City," promising the use of new telecommunications technologies that will provide an unprecedented amount of documents and data to families, and businesses by "information highways," resulting in a world dominated by the media. The digital city, a city characterized by technology that uses broadband infrastructure to enable electronic governance and is a "global world" for public transactions, is another forerunner of this Smart City. Giffinger et al. [5] argue that a smart city is a well-functioning city built on a clever combination of resources and the actions of self-determined, autonomous, and well-informed citizens.

According to Hollands [6], smart cities rely on the use of infrastructure networks to boost economic and political efficiency to ensure long-term growth. Caragliu et al. [7] conclude that a city would be wise if it invests in people and social capital in order to build a sustainable economy, enhance the quality of life and wealth, and manage nature effectively. Economic development, social sustainability, and quality of life, according to Alam et al. [8], are all dependent on human resources, social capital, and environmental resources. Aside from this definition, the authors want to demonstrate the importance of smart cities for long-term urban development. Many important problems associated with the current urbanization phase, such as traffic congestion, noise, and scarce natural resources, can be alleviated. Smart governance, smart people, smart economy, smart living, and smart environment, according to Petrolo et al. [9], are six factors that affect smart cities. The smart city model usually combines the city's economic, social, and environmental components in a way that improves the city's primary structures' performance over time. Elhoseny et al. [10] identified seven factors that are important to smart cities: smart government, smart living, smart economy, smart education, smart mobility, smart service, and smart society. Talari et al. [11] indicate the five elements affected are smart homes, smart buildings, attentive customers, smart energy, and smart grids. According to Kumar and Dahiya [12], the six key building blocks that make up a smart city structure are: (i) intelligent people; (ii) intelligent city economies; (iii) intelligent mobility; (iv) intelligent climate; (v) intelligent living; and (vi) intelligent governance.

Ngan and Khoi [13] identified policy analysts face a challenge in developing an effective quantitative model based on economic theory and empirical evidence. To find realistic evidence for the model, their research team gathered data on 362 people in Ho Chi Minh City, Vietnam. Their analysis results showed that the Smart city is influenced by six factors: Smart Economy (S-Econ), Smart Governance (S-Gov), Smart Environment (s-Env), Smart Citizens (S-Citi), Smart Traffic (S-Tra), and Smart living (S-Liv). Finally, six factors affect smart cities: Smart Economy (S-Econ), Smart Governance (S-Gov), Smart Environment (S-Env), Smart Citizens (S-Citi), Smart Traffic (S-Tra), and Smart living (S-Liv).

2.3 Smart City Analytical Framework

2.3.1 Smart Economy (S-Econ)

The characteristics of a smart economy take into account elements of competition such as creativity, entrepreneurship, branding, efficiency, labor market stability, and integration in both the domestic and foreign markets. A typical smart economy will always have e-commerce transactions (E-Business, E-Commerce) based on information and communication technology systems, to maximize efficiency in transactions, saving costs, encouraging innovation in products, services, and business models. Information and communication technology systems help to create smart Eco-Systems through the forms of business and digital entrepreneurship. Besides, the application of high technology in the economy also creates the flow of products, services, and knowledge both tangible and intangible over the network, thereby promoting the connectivity of local and global economic transactions [2].

2.3.2 Smart Governance (S-Gov)

Smart governance characteristics are evaluated based on the criteria of people's participation in the process of urban management, utility services, and administrative activities [5]. A municipality with smart governance is a city where residents have the right to contribute ideas to management activities to help the city operate more efficiently. To achieve these goals, information, and communication technologies play a key role in smart city development [14]. Information and communication technology systems include hard and soft infrastructures that help provide an open and transparent database through which people can connect or with the community such as monitoring urban activities. These electronic services and E-Government can be provided on mobile applications. Smart governance is the first important pillar that connects the remaining pillars of smart cities.

2.3.3 Smart Environment (S-Env)

The smart environment is assessed based on the criteria of habitat conditions, green area, pollution problem, and effectiveness of pollution treatment and remediation measures. Similar to the features mentioned above, a smart city applies high technology in environmental management, especially in the energy sector. According to Manville et al. [14], smart cities must use "smart energy" including reusable energy sources, and must apply information and communication technology systems in the Energy Grids when measuring, monitoring, and controlling pollution. Information and communication technology also needs to be applied to create Green Buildings, Green Urban Planning. Also, urban services such as streetlights, waste discharge systems, water supply, and drainage need to be monitored through the information and communication technology system to evaluate the effectiveness in environmental management.

2.3.4 Smart Citizens (S-Citi)

The Smart Citizen is not only based on urban residents' education but also the level of individual interaction with the community and the connection between people and people in society [5]. On the other hand, The Smart

Cities [15] assessed that cities with the characteristics of smart people are cities in which individuals must have the opportunity to learn for a lifetime, where they must find ways to increase social integration, improving the quality of people's lives, creating opportunities and motivations to enhance the creative spirit of the people, at the same time, people are guaranteed to have access to open data systems anytime, anywhere. Besides, Manville et al. [14] argued that smart people are individuals with E- skills and able to work in the application of information technology and information communication (ICT-Enabled Working).

2.3.5 Smart Traffic (S-Tra)

The smart traffic system in the city not only meets the normal traveling needs but also provides traffic information for people through applications, upgrading the existing transport system to modern and more sustainable. Specifically, the traffic network must be safe, clean, and especially flexible and effective, including trams, buses, trains, cars, bicycles, and pedestrians. People can easily change between transport modes to save the most time and money. In order to do this, the smart transport system must provide a source of information data based on the actual time of the vehicle so that the people can access the system and choose effective means of transportation. As a result, this system brings many benefits for both the people and the government, such as cost savings, reduced CO_2 emissions, and an improved feedback system from the connection of electronic citizens on the smart traffic system.

2.3.6 Smart Living (S-Liv)

This feature includes all aspects to assess people's quality of life such as health, safety, culture, housing, and tourism. Smart life is a life that applies ICT – Enabled Lifestyles and Behavior. The smart living must be healthy and safe with high-quality housing, social capital, and high levels of social cohesion [14]. In cities with a smart life, natural resources must be managed in a "smart" way; a sustainable urban environment should be created with "smart" plans for roads and spaces, public and facilities. Most significantly, community planning and management strategies can both work for the same aim of enhancing and evaluating urban residents' quality of life.

The six pillars above are often applied in the assessment of a smart city according to international practices; on the other hand, they are also the development goals of urban managers to be achieved. These six pillars are often built on three important foundations: (1) technology, (2) institutions, and (3) people [14]. In which, technology includes: physical infrastructure, smart technology, mobile technology, virtualization technology, digital network; institutions include: governance, policy, and regulations; and people (or human resources) includes: human resources and social capital. These three platforms act as vehicles and tools to help achieve the six pillars of smart cities: (1) S-Gov, (2) S-Econ, (3) S-Tra, (4) S-Env, (5) S-Citi, and (6) S-Liv.

2.4 Data in Vietnam

First, the research of Khoi and Ngan [16] showed that six factors influence the smart city. Their quantitative research was carried out with a sample of 314 people in Vietnam.

Challenges, according to the writers, could persuade people that Smart City is fair, in line with current trends, and that everyone should cooperate. Six factors influenced the Smart city, accounting for 73.3% of the total in Figure 2.1. Table 2.1 shows that the six hypotheses are accepted because their p-values are less than 0.05.

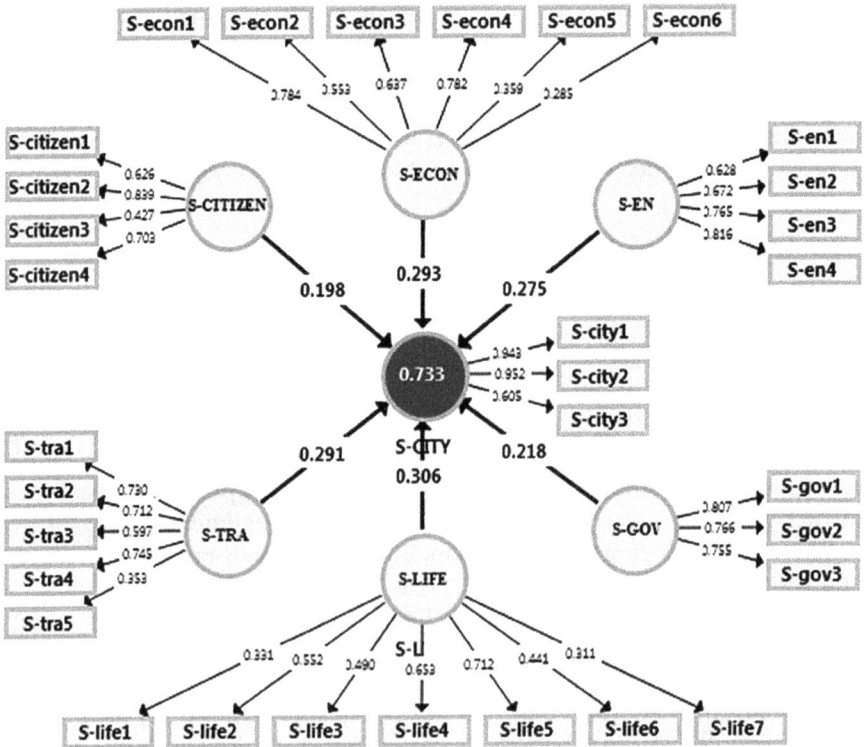

FIGURE 2.1
Smart city model [16].

TABLE 2.1

Hypothesis in the Model [16]

Hypothesis	Coefficients	Standard Error	t	p-Value	Decision
H1	0.29	0.05	5.53	0.00	Accepted
H2	0.19	0.042	4.73	0.00	Accepted
H3	0.21	0.04	4.87	0.00	Accepted
H4	0.30	0.04	6.79	0.00	Accepted
H5	0.27	0.04	7.19	0.00	Accepted
H6	0.29	0.04	6.76	0.00	Accepted

The second, based on the results of calculating the indexes of the six pillars of smart cities, each indicator is composed of 1 to 6 components, smart city index and six pillar indicators in Southeast Vietnam [2]. Overall, Ho Chi Minh City leads the way in the S-city approach, with the highest-ranked smart city index of 0.86, followed by Binh Duong, Dong Nai, Tay Ninh, Ba Ria – Vung Tau and Binh Phuoc with smart city indicators are 0.36, 0.07, −0.30, −0.41 and −0.90. Although Ho Chi Minh City leads in most of the indexes but ranks behind Binh Duong in the smart management index and only 4th in the smart life index (after Binh Duong, Tay Ninh, and Dong Nai). Another noteworthy point, out of six provinces/cities in the Southeast, Binh Phuoc is currently the province with the least potential for a smart urban approach. In general, the above figures show the superiority of Ho Chi Minh City in terms of pillars toward a smart urban model compared to others in Southeast Vietnam.

2.5 Conclusions

In short, to realize the goal of building a smart city to progress the quality of life of the people and toward sustainable economic development, Vietnam must first concentrate on three key tasks: enhancing infrastructure, human capital development, and evolving information and communication technologies, with a special emphasis on applying communications technology achievements to governance, economics, the environment, and transportation. Vietnam, in particular, must take advantage of favorable human, institutional, and technical conditions to play a leading role in the growth of smart city models. Vietnam's growth will serve as a model for achieving pervasiveness in Southeast Asia.

2.6 Limitations of the Research and the Next Researches

This study is still limited in terms of data and selection criteria for the proposed smart city framework, which will be improved in further studies when resources are available. Currently, Vietnam still only has the idea of a smart city, and there is no actual smart city that meets the standards or is recognized in the country or the region. Therefore, there is still a lot of debate about the criteria for the evaluation of local smart city proxy's efforts. Although the study tried to select the criteria that represent the pillars of smart cities from the previous researchers, these criteria are limited by existing data in Vietnam. Especially the observational data for the smart environment and smart life pillar.

Acknowledgments

This paper was funded by IUH (Industrial University of Ho Chi Minh City), Vietnam.

Conflicts of Interest

There are no conflicts of interest declared by the authors.

References

1. Bach, K. H. V., & Kim, S.-K. (2019). Developing smart city: Based on the assessment of smart projects in Medium-Size Cities, Vietnam. *American Scientific Research Journal for Engineering, Technology, and Sciences (ASRJETS)*, 56(1), 38–49.
2. Hoai, N. T., Dung, N. V., Duyen, T. T. P., & Vien, N. V. (2018). Smart urban analysis framework: Case study of Southeast Vietnam (Vietnamese). *Journal of Asian Business and Economic Studies*, 29(6), 5–26.
3. Ahvenniemi, H., Huovila, A., Pinto-Seppä, I., & Airaksinen, M. (2017). What are the differences between sustainable and smart cities? *Cities*, 60, 234–245.
4. Mitchell, W. J. (1999). Designing the digital city. Paper presented at *the Kyoto Workshop on Digital Cities*, Kyoto, Japan, September 1999.

5. Giffinger, R., Fertner, C., Kramar, H., & Meijers, E. (2007). City-ranking of European medium-sized cities. *Centre of Regional Science Vienna UT*, 1–12.
6. Hollands, R. G. (2008). Will the real smart city please stand up? Intelligent, progressive or entrepreneurial? *City*, *12*(3), 303–320.
7. Caragliu, A., Del Bo, C., & Nijkamp, P. (2011). Smart cities in Europe. *Journal of Urban Technology*, *18*(2), 65–82.
8. Alam, F., Mehmood, R., Katib, I., & Albeshri, A. (2016). Analysis of eight data mining algorithms for smarter Internet of Things (IoT). *Procedia Computer Science*, *98*, 437–442.
9. Petrolo, R., Loscri, V., & Mitton, N. (2017). Towards a smart city based on cloud of things, a survey on the smart city vision and paradigms. *Transactions on Emerging Telecommunications Technologies*, *28*(1), e2931. doi: 10.1002/ett.2931.
10. Elhoseny, H., Elhoseny, M., Riad, A., & Hassanien, A. E. (2018). A framework for big data analysis in smart cities. Paper presented at *the International Conference on Advanced Machine Learning Technologies and Applications*.
11. Talari, S., Shafie-Khah, M., Siano, P., Loia, V., Tommasetti, A., & Catalão, J. P. (2017). A review of smart cities based on the internet of things concept. *Energies*, *10*(4), 421.
12. Kumar, T. V., & Dahiya, B. (2017). *Smart economy in smart cities* (pp. 3–76): Springer.
13. Ngan, N., & Khoi, B. (2020). Determinants influencing to smart city. *Journal of Advanced Research in Dynamical and Control Systems*, *12*, 676–681.
14. Manville, C., Cochrane, G., Cave, J., Millard, J., Pederson, J. K., Thaarup, R. K., ... Kotterink, B. (2014). *Mapping smart cities in the EU*. Directorate General For Internal Policies, European Parliament's Committee on Industry, Research and Energy.
15. Smart Cities: Regional Perspectives (2015). The Government Summit Thought Leadership Series. United Nations.
16. Khoi, B. H., & Ngan, N. T. (2019). *Factors impacting to smart city in Vietnam with smartpls 3.0 software application* (Vol. 10, pp. 1–8): IIOAB.

3

The Contribution of Games in the Design of Smart Cities: A Look at Brazilian Slums

Daniel Oliveira Cruz and Sergio Nesteriuk

Anhembi Morumbi University, São Paulo, Brazil

CONTENTS

3.1 Introduction

The human being is a social being, and the social being tends to be an urban being. Even when we consider the nomadic beginning of the history of human beings, in general, humanity perseveres in collectivity, the maximum expression of groupings and cities.

As a result of the collective way of life, artifacts, customs, cultural practices, and social activities thrive, among which the games are inserted and
walk concurrently with the emergence of cities.

From its origins 5,000 years BC to the present, there is a multidimensional,
transversal, and interdisciplinary, quantitative and qualitative increase, in
a complexity of actors and processes that is equivalent to the information
flows required to understand the various dimensions of sustainability. Does
this logic apply only to the city? No! From clusters to a megalopolis, from
rudiment to the ubiquitous, both cities and games are clear testimonies of
humanity's leaps and setbacks.

Speaking specifically of cities, the complexity of the relationships existing
in these spaces involves joint studies of different areas of knowledge aiming at sustainable urban development. In this sense, the productivity and
competitiveness of cities (economic interests) grow concomitantly with the
promotion of social and environmental interests (Caragliu et al., 2009).

On the one hand, this scenario indicates that cities are spaces of problems
and challenges. On the other hand, cities are configured as stages of opportunities, innovation, knowledge, and creativity. According to market projection, the 600 largest cities in the world will generate 60% of the global GDP in
2025. For Capdevila and Zarlenga (2015) cities can be conceptualized as complex ecosystems, where different actors, with different interests, are obliged
to collaborate to guarantee a sustainable environment and an adequate quality of life.

The diversity of inputs and outputs that touch the management, nature,
and integration of all processes operating in a city, such as infrastructure,
agencies, services, the economy, and the wellbeing of citizens, requires solutions of an interconnected and synergistic nature, possible with the adoption
design strategies, Internet of Things (IoT) and in the so-called 4.0 generation
(Industry 4.0, Education 4.0, Health 4.0, among others). A Brazilian example
of these complexities is revealed in the growing number of Brazilians living
in slums, exceeding 6.5 million in 2000, 11.4 million in 2010, and 13.6 million
in 2020, distributed in 6,329 points in 323 municipalities, 88% of which are
concentrated in 20 large cities. On the other hand, its residents move R $ 119.8
billion per year, representing an income volume greater than twenty of the 27
units of the federation.

Such imperatives make pressing the bet on new models of urban development, as has been advocated by the so-called Smart Cities. Thus, a smart
city—among several other definitions that we will see in this work—can be
described as a city that uses Information and Communication Technologies
(ICT) to increase the quality of life of its inhabitants, also contributing to sustainable development (Capdevila and Zarlenga, 2015), through participatory
and transparent processes. This "new" taxonomy of cities elevates the citizen
to an active contribution role in facing local urban challenges and building
a more inclusive, sustainable, and humane city, emerging in participatory

bottom-up and joined-up processes, made possible by technology and augmented with participatory and instrumental methodologies in which gamification and the universe of games are inserted (Damiani et al., 2018; Aune, 2017; Vanolo, 2018; Olszewski et al., 2016).

In this context, this chapter develops a study of the contribution of games and gamification in the universe of a smart city design, with the focus or thematic focus on the socio-spatial inclusion of the so-called Brazilian subnormal territories, whose semantics and vernacular the term "favelas" takes on the present work.

3.2 Literature Review

Brazilian or international publications that simultaneously address the thematic triad of this research—Smart Cities, Favela, and Gamification—are very rare. Only one publication was found involving the thematic triad, addressed indirectly and peripherally to the central objective of the work.

The only work found that addresses the three themes, prepared by Ogie (2016), is entitled "Adoption of mechanisms to encourage large-scale participation in mobile crowdsensing: from the literature review to a conceptual framework" (Ogie 2016, p. 1) and focuses on the areas of Computer Science and Social Sciences. The study reviews the literature on incentive mechanisms for using crowdsensing to capture and map phenomena of common interest on a large scale in smart cities. The only mention of the favela (originally, the term slum was used) is related to privacy issues, as described below:

> In the context of Jakarta, privacy concerns exist, particularly for the urban poor living in informal settlements or slums situated along the watersides. These people may be reluctant to contribute geo-located tweets because of fear that they will be traced by the government and punished for living in illegal settlements. Though money can quickly incentivize people to forgo their privacy, the issues raised above with monetary incentives remain a concern in that context. Rather, it is argued that societal recognition of slum dwellers and their role in community building be strongly emphasized in PetaJakarta campaigns and endorsed by government authorities as an aspect of social incentive to motivate contributions from the millions of Jakarta's citizens residing in slums. Anonymity should also be encouraged as possibly permitted through the Twitter user account.
>
> (Ogie, 2016, p. 20)

Therefore, as detailed in the methodology, the initial search for works with an isolated focus on the thematic triad is oriented, to subsequently seek

NEIROTTI ET. AL (2014) ALBINO ET AL., (2015)
GIFFINGER ET AL. (2007)

SMART CITY

HALL ET. AL (2000)
KANTER & LITOW (2009)
CHEN (2010)   URSSI (2018)
SAKUMA (2014) INTELI (2012)
              TOPPETA (2010)
MARCUCCIA (2018)
              DAROS (2014)
LEMOS (2013)
              LEVY (1993)
HARRISON ET AL. (2010)
ARUP (2008)   GUIMARÃES (2018)
SOUSA E SILVA (2006)
WASHBURNET & SINDHU (2010)
SOARES (2012)   EGER (2009)
     CARAGLIU (2009)

DAMIANI (2018)
AUNE (2017)
KAZHAMIAKIN ET AL. (2015)
SPITZ ET AL. (2017)
ZICA ET AL. (2017)
VANOLO (2018)

OGIE (2016)

GAMIFICATION
GAMES

SANTAELLA,
NESTERIUK,
FAVA (2018)

HOSSE (2014)

BURKE (2015)

NORDMEDIA (2013)

ALVES (2014)

CHOU (2014)

FUCHS (2011)

HIMMELFARB (1984)   BRESCIANI (1984)
QUEIROZ FILHO (2015)   GONÇALVEZ (2011) OBSERVATÓRIO DA
QUEIROZ (2010)   FAVELA (2019)
SILVA ET AL. (2009) **BRAZILIAN**   LECLERC (1979)
PASTERNAK &   IBGE (2010) IBGE (2011) BARRET-DUCROCQ (1991)
D'OTTAVIANO (2016)   KLINTOWITZ & ROLNIK (2006)
          SLUMS
     VALLADARES (2000)   UN-HABITAT (2007) PAULINO (2007)

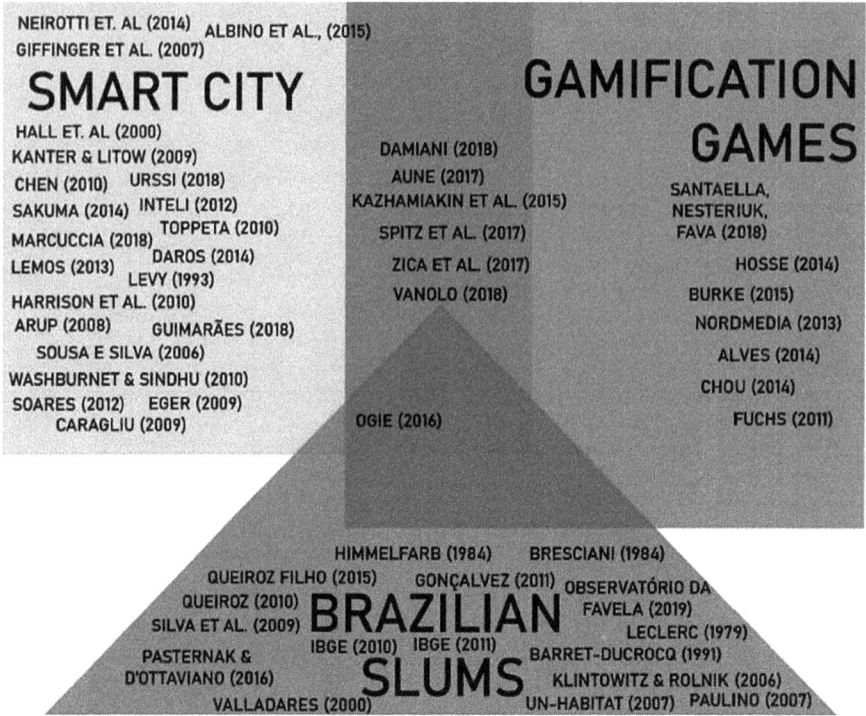

FIGURE 3.1
Venn diagram with the authors on the thematic triad.

correlations with two of the central themes in their study objects. Next, there is a graph called the Venn Diagram, which seeks to structure the authors of the research tripod, allowing them to be identified separately and together (Figure 3.1).

About the smart city, there are multiple approaches to work that develop from concepts and definitions, structure dimensions, frameworks, and proposals for Brazilian and international classifications, case studies, models, and implementation methods, represented by authors such as Hall et al. (2000), Giffinger et al. (2007), Arup (2008), Kanter and Ve Litow (2009), Caragliu et al. (2009), Giffinger and Haindl (2009), Chen (2010), Toppeta (2010), Washburn and Sindhu (2010), Harrison et al. (2016), Dutta (2011), Nam and Pardo (2011), Thite (2011), Neirotti et al. (2014), Albino et al., 2015, Guimarães (2018), Lemos and Bitencourt (2018), among others.

In the scope of the favela, works were raised whose approaches focus on its origin, historical evolution, the gradual process of construction or deconstruction of its conceptualization and characterization—both by governmental attempts and by citizens' initiatives—and its problematization, developed by authors such as V. Klintowitz and Rolnik (2006), Valladares (2000), Paulino

(2007), Pasternak and D'Ottaviano (2016),. As for gamification and games, there are historical approaches to the development of the theme in humanity, conceptual definitions, types of games, classifications and typologies, different applications, predictions and specificities in works by authors such as Kishimoto (1994), Hosse (2014), Nesteriuk and Fuchs (2018), Miri et al. (2019), Franco et al. (2015), Santaella et al. (2018), Robson et al. (2015), Hosse (2014), Thiel et al. (2016), Olszewski et al. (2016), Fernandes and Ribeiro (2018), among others.

The convergences between the themes of smart cities and slums are limited to the work of Ogie (2016) mentioned at the beginning of this review.

Thematic intersections between smart cities and gamification, despite being less than a decade old, are more common when compared to the previous attempt, having a great diversity of objectives and objects of study. Some works address the application of games and gamification in the urban planning process, survey of gamified application processes and tools for citizen engagement and participation in decision-making processes for the construction of smart cities, present in Reinart and Poplin (2014), Spitz et al. (2018), Vanolo (2018), Mueller et al. (2017), Damiani et al. (2018), Damiani et al. (2019), Virtanen et al. (2015), Pinos et al. (2020) and Olszewski et al. (2020).

As for the shading between gamification and slums, they only refer to the article already cited by Ogie (2016). Outside the scientific basis, a study was found that addresses the use of a game in the process of social inclusion through the qualification of urban spaces, the Block by Block (2018), which will be discussed later on.

Therefore, for each of the themes, bibliometric analysis and a systematic survey of elementary information are carried out for purposes of conceptualization and identification of the key research elements.

In the universe of smart cities, after the systematization and compilation of the definition of several authors, it is assumed as its definition: The smart city is an urban territory whose sustainable governance guarantees social equality, cultural, and political freedom, human development, economic maintenance, and environmental preservation, through a technological infrastructure—including ICTs—that allows transparent and participative management flows of all its population in its construction.

Still on this theme, when holistically scrutinized by multi, inter, and transdisciplinarity, one realizes the importance of establishing its dimensions so that none of its facets are forgotten. In this sense, despite the specialized literature defining between 3 and 18 dimensions, the number of 10 is assumed for identifying some shadings and overlaps of the items covered in the research, namely:

- Economy.
- Governance.
- Mobility.

- Environment.
- People.
- Education.
- Health.
- Safety.
- Technology.
- Infrastructure.

In the favela universe, after raising the main authors of the area, in addition to specialized Brazilian governmental organizations and institutions, it is assumed as its definition: as a historically privatized territory of basic infrastructure investments, with intense socio-spatial stigma, with self-management and self-construction predominantly and whose population spectrum has educational, economic, and environmental indicators below the average of its surroundings and the city.

Bringing the favela closer to the context of the discussion and definition of smart cities, favelas are territories that may even have ICT infrastructures, self and individually financed by their inhabitants—predominantly on cell phones—a little less on private computers and Wifi, in addition to the so-called illegal "gato velox" that spread band to all Brazilian cities. Such resources happen in a disjointed and fragmented way, self-implemented and without integration with an official public management system. But that, even before discussing ICTs, they are configured as territories that do not have the elementary requirements of an urbanized city and a provider of quality of life, due mainly to the lack of infrastructure and basic public services.

In the gamification universe, it is not intended to propose any innovation in terms of its definitions, using as a general concept and understanding such as: "the use of game mechanics and aesthetics with a view to increasing people's involvement and engagement, motivating actions and promoting learning and problem solving"(Kapp, 2012 and Olszewski et al., 2016).

3.2.1 Approaches between Smart Cities, Slumps, and Gamification

In the scope of the bibliographic review, a bibliometric analysis of the thematic triad is elaborated and the data found is integrated for the purpose of making possible relations possible, where historical and bibliometric intersections are shown, described below.

3.2.1.1 Historical Approaches

In the historical scope, the present chapter has as its theme the city and the games (from the board to the electronics) and as a general objective to understand and systematize the conceptual evolution and the temporal interactions between these two anthropic "entities," the city and the games.

The first cities arose between approximately 5,000 and 3,000 years BC, and began to perform administrative, artisanal, commercial, and security functions, in addition to agricultural functions, based on the need for interdependence and a network for exchanging services and products. In this period, 5000 BC there was the existence of the Senet Game (considered the father of backgammon) in the region of Egypt and, the Royal Game of Ur, dated from 4600 BC, originated in the region of Babylon (Syria, Iraq, Egypt, Iran, Crete, Cyprus, Sri Lanka).

Despite the different periods and locations in the world, the emergence of card games in the regions of Italy, Spain, and France has dated between approximately AD 1250 and 1500. In the universe of cities, this period is marked by the transition between the city of Feudal and the City of the Renaissance, with an emphasis on the European cities mentioned above.

From the eighteenth century onwards, the urbanization process intensified, with a greater offer of work in the cities and mechanization of work in the countryside, generating a one-way population movement, at the time being called the Industrial City. Although Chess was invented in 822 AD, it takes place in the years 1851 and 1886, the First International Chess Tournament and the Official World Chess Championship, respectively, in London and the United States.

In the mid-1970s, information technology created a cultural scenario, called cyberculture, configuring the era of the so-called "digital city" in the context of urbanity. At about the same time, there was the development of the first electronic games, with an emphasis on the game Tennis for Two, archaic, but disruptive to the opening of new possibilities in the face of technological development. As for analog games, the creation of the first city creation game called Supercity dates from this time.

With the percolation of ICT in companies, institutions, and residences, the city of information and knowledge was established in the 1980s and 1990s, with the emergence of a varied network of offers and services such as e-commerce, e-government, e-learning and e-bank. Despite the predominance of electronic games in this period, such as Super Mario (1985), SimCity (1990), Doom (1993) and Street Fighter II (1991), there are also analog board games, such as Banco Imobiliário, War, Naval Battle, Game of Life, Detective, among others that marked the 90s of the last millennium.

Beginning the new millennium, with the democratization of technologies and mobile data networks, access to consolidated information in the routine of cities allowed the emergence of new urbanities, known as Hybrid Spaces, which combine the physical and the digital in a social environment created by users connected by mobile communication devices. In this period, there is SimCity 4 (2003), SimCity DS and SimCity Societies (2007), Minecraft (2011), SimCity Reboot (2013).

Amid in its "technologization," with the transition from a rural to a predominantly urban world—in the inflection that occurred approximately in 2007 – the city began to influence decisively in human life and be influenced progressively and inevitably by ICTs. Around the first decade of 2010, there was the conformation of the connected city, of the ubiquitous city, in which resources and instruments are incorporated into the daily life of global urbanity, enabling the integration of data and the beginning of the horizontal participation of the population in construction of urbanity. This last scenario will take a "leap" from the next decade, with the 6 G – which will allow mobile devices with a connection speed of about 1 Tb/s (up to 8,000 times faster than the current 5 G).

From clusters to megalopolises, from its origin to the present, the city has been increasing multidimensionally, transversally, and with interdisciplinary, quantitatively and qualitatively, in a complexity of actors and processes that is equivalent to the information flows required to understand the various dimensions of sustainability. The complexity of the relationships existing in these spaces involves joint studies of different areas of knowledge aiming at sustainable urban development. In this sense, the productivity and competitiveness of cities (economic interests) grow concomitantly with the promotion of social and environmental interests.

Such imperatives make pressing the bet on new models of urban development (Inteli, 2012), as has been advocated by the so-called Smart Cities. Thus, a smart city can be described as a city that uses ICT to increase the quality of life of its inhabitants, also contributing to sustainable development (Capdevila and Zarlenga, 2015), through participatory processes and transparent. In parallel to games, there is the adoption of complex systems in games such as Supercity (2014), Cities Skylines (2015) and to a lesser extent Block'hood (2017), which demonstrate in their structures and mechanics, several dimensions recommended for smart cities.

Currently, if we were asked to cite two inseparable elements of the post-contemporary human life scenario, we would have the city as a backdrop and as an emerging peripheral member, technology, represented mainly by smartphones and cell phones. After the brief digression, the city reflects the human spirit, in which corporate and governmental arrangements are historically printed and reflected in urbanity, in the geography of the occupied and anthropized territory. And information technology, a portal for hybridization, the ubiquity of all human processes and relationships, of things and space, typical of the definitions of the Internet of Things (IoT) and the Internet of Everything.

This "new" taxonomy of cities elevates the citizen to an active role of contribution to face the local urban challenges and in the construction of a more inclusive, sustainable, and human city, emerging in participative processes bottom-up and joined-up, made possible with technology and augmented with participatory and instrumental methodologies in which gamification is inserted (mentioned for the first time in 2002 and consolidated in 2010) and the universe of games (Damiani et al., 2018; Aune, 2017; Vanolo, 2018; Olszewski et al., 2016). In this context, there is an emergence of the meaning of games and gamification in the context of cities. We move to a new level of game analysis as elements that contribute to the design of the cities of the present and the future, the so-called smart cities, now called Smart and Human Cities (Figure 3.2).

3.2.1.2 Bibliometric Approaches

Initially, the absence of publications that focus on the three main themes of the research Smart Cities, Favelas, Gamification, and Games is highlighted as the core of the study, considering the search engine—Scopus—employed. Even including others such as Capes and Science Direct journals, there were no positive results with the three themes grouped together. Given this observation, works are sought that include at least two common areas and—considering the low number of results—an encapsulated research and analysis of the themes is directed, that is, an isolated look at first, to weave combinations later.

In general, it can be said that Favelas, Smart Cities and Gamification are themes of increasing interest in national and international scientific literature, presenting a high trend line in the period evaluated, with some only small variations in systematized historical data, as seen in graphs below (Figure 3.3).

Among the three, the theme Favela presents itself as the oldest to be worked on in the scientific literature, with consistency in number and growth of publications since 2000 and with an average increase of 16% per year over the analyzed period (until 2019). Of the three themes surveyed, Favela appears as the oldest in publications of relevance and quantity, having already reached three digits in the early 2000s. Still, on this theme, it is worth highlighting the behavior of extrapolating the average in 2012, the year where Rio + 20 takes place, an important global sustainability event.

Regarding the analysis for the second theme, Smart City, in addition to observing the same growth trend as that of the previous theme, there is a more marked and well-marked acceleration from 2010 onwards, with an evolutionary average of more than 100% per year in the number of publications between 2010 and 2019. This period of marked advance combines with the period of technological diffusion of mobile telephony in Brazil and worldwide.

As for Gamification, it is clearly characterized as the most recent theme of the three surveyed, having its origins marked out at the beginning of 2000, but with more consistency of publications from 2011, with an even more significant increase and extrapolation of the historical average from 2015. Gamification and Smart City are the themes with the highest growth in publications in recent years, among the three surveyed. Such growth may

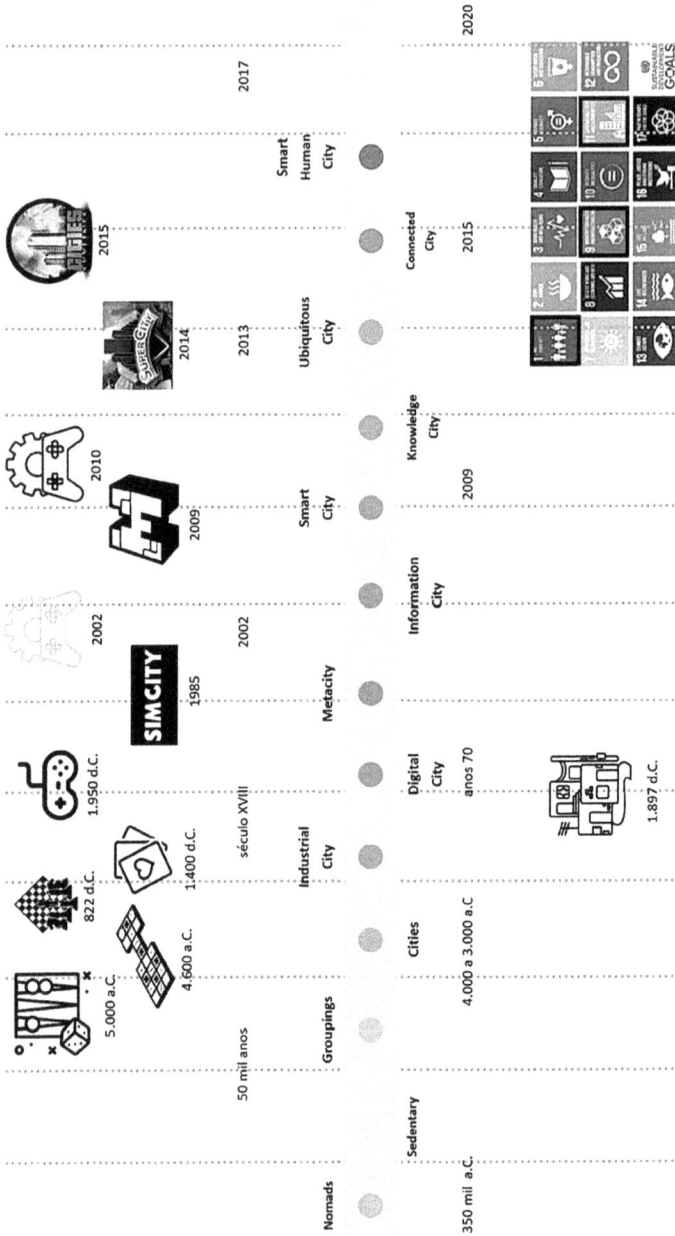

FIGURE 3.2
Timeline with historical integration of the Games and the Cities.

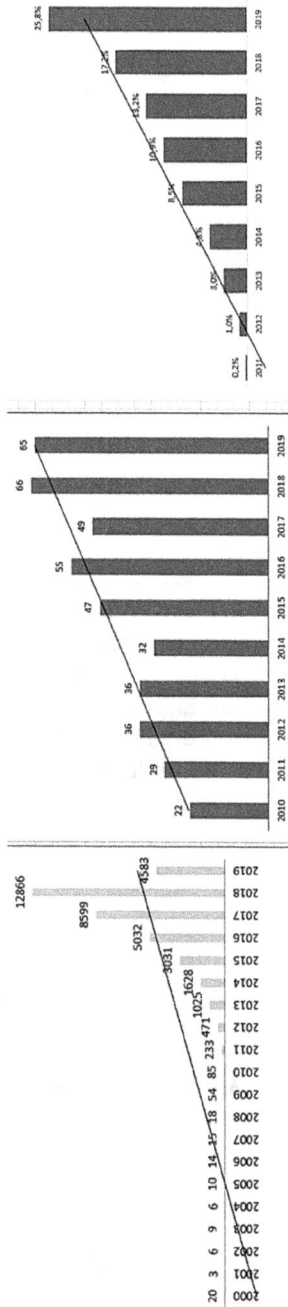

FIGURE 3.3
Comparison between YEAR of publication of key research topics: smart city, slum, and gamification.

be associated with the democratization and diffusion of mobile technologies (mainly smartphones) and the consequent development of urban and educational applications.

When the three themes are integrated and analyzed through the distribution of countries/territories, some points can be identified among the top 10, as listed in the following figure (Figure 3.4). United States, England, Germany, Australia, Italy, Canada are countries that appear in the three themes, with the first two having greater consistency in "leadership" and in the number of publications. The exception regarding leadership is made in the "Favela" theme, which, as already seen, has Brazil as the country that publishes the most, followed by the United States and England. In the only topic in which Brazil is not among the top 10, the country is in the fourteenth position.

A theme sequential to the previous result and consequent to international publication standards refers to the predominant language in publications, English, covering more than 90% of the articles produced in the themes and in the evaluated search engine.

As for the research areas—as seen in Figure 3.5—cutting out the five most productive areas in the themes, Smart City concentrates works in the area of Computer Science (35%) followed by Engineering (22%), Social Sciences (12%), Mathematics (8%) and Energy (6%). In the Favela theme, there are Social Sciences (49%), Arts and Humanities (15%), Medicine (9%), Engineering (5%) and Environmental Sciences (5%).

Evaluating the convergences and intersections common to the three themes, there are Social Sciences (26.6%) and Engineering (13%) appearing in the first positions among the five areas with the largest number of works, according to the classification and division methodology used by Scopus (Table 3.1). The Computer Science area (26%), is in second place in the overall sum and occupying first place in the theme Smart City and Gamification, it does not appear in the top five in the theme Favela.

Approaching the analysis of the keywords of the researches carried out in the Scopus database, it is not possible to establish correlations between the three themes at first (Figure 3.6). When evaluating the first six incidences, the terms are restricted to their areas, with the exception of gamification that demonstrates adjacencies with education and motivation, the latter being integrative with the areas of smart cities.

In a second step, when the evaluation expands to all incident items that can be analyzed and not just to the 6-word cut, some correlations can be made between the themes of the work. In the Favela theme, there are the keywords: Human, Controlled Study, Female, Brazil, Adult, Ubiquitous Computing, Healthcare, Temperature, Genetics, Adolescent, Aged, Child, and Poverty. The predominance of terms related to the Social Sciences Area is confirmed and it is noted the appearance of a term linked to the Smart City and Gamification area, Ubiquitous Computing. In addition, another concept related to the Favela is also found in Gamification, the Human (Figure 3.7).

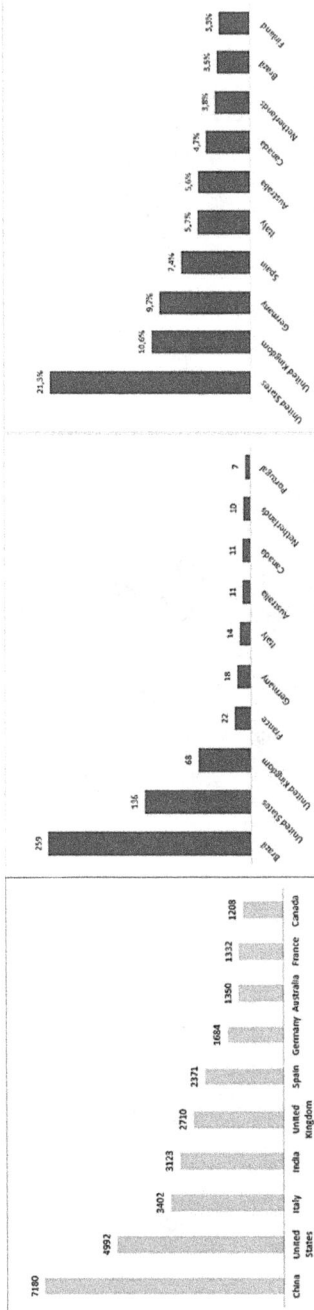

FIGURE 3.4
Comparison between the PLACE of publication of the key research topics: smart city, slum, and gamification.

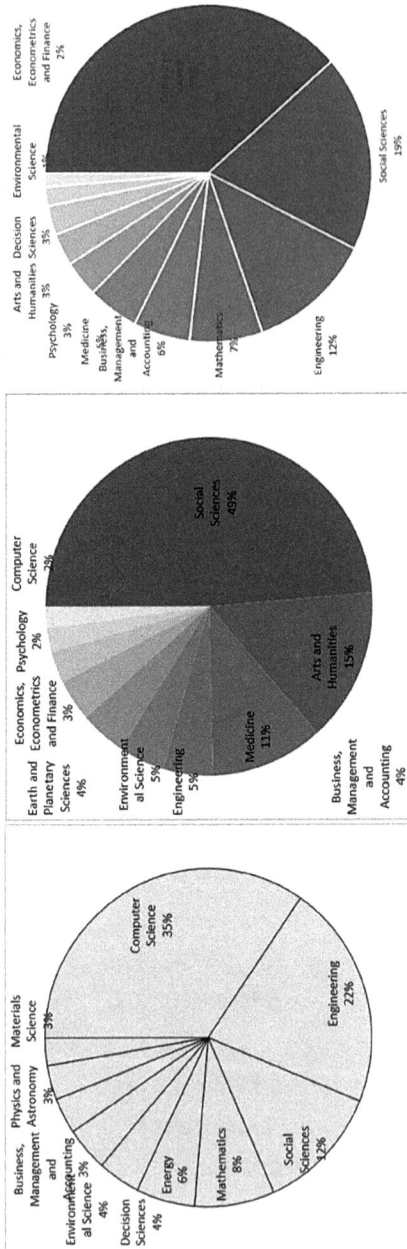

FIGURE 3.5

Comparison between RESEARCH AREAS of publications on key research topics: smart city, slum, and gamification.

TABLE 3.1

Comparison of the Incidence of Publications between the Research Areas

SMART CITY	FAVELA	GAMIFICATION
1 Computer Science	Social Sciences	Computer Science
2 Engineering	Arts and Humanities	Social Science
3 Social Sciences	Medicine	Engineering
4 Mathematics	Biochemistry	Business
5 Energy	Engineering	Medicine

The non-appearance of "human" or similar terms in Smart City confirms the criticism of specialized authors in the field of urban planning (Benites 2016), when they argue the lack of direction in the discussion of smarts cities for citizens, clinging fundamentally to the technological apparatus. At the same time, it confirms the concern of the authors who defend the recent concept of human Smart Cities, demanding this new look at the conceptual evolution of cities.

The Smart City theme includes the following keywords: smart city, internet of things, big data, smart cities, IoT, Energy Efficiency, Energy Utilization, cloud computing, information management, wireless sensor networks, IoT, Artificial Intelligence, Decision Making, Ubiquitous Computing, Sustainable Development, Digital Storage, and Automation. Among the words found, in Decision Making, adherence to the theme Gamification stands out, especially when considering the word "motivation," a key element for citizen engagement in decisions and participation in urban management.

When it comes to the topic of Gamification, expanded bibliometric analysis finds the following keywords: gamification, education, students, human-computer interaction, motivation, teaching, e-learning, human, serious games, virtual reality, computer-aided instruction, humans, design, engineering education, computer games, behavioral research, learning systems, article, networking, and game-based learning. It is interesting to note that one can find several and a greater number of words with adherence to other themes, such as motivation (often applied in the engagement of citizens, as previously mentioned), Serious Games (among the types of games, as we will see ahead, this is the one that comes closest to the heart of the problem of this research), Humans (approached by the theme Favela and ignored in the bibliometrics of Smart Cities, as mentioned earlier) and Computer Games (a word of great relevance among the case studies of this work). In the thirteenth place, there is the term, "design," which, although not occurring in bibliometrics, is closely related to the areas of Social Sciences and Arts and Humanities, and given its holistic meaning, it has integration with all areas.

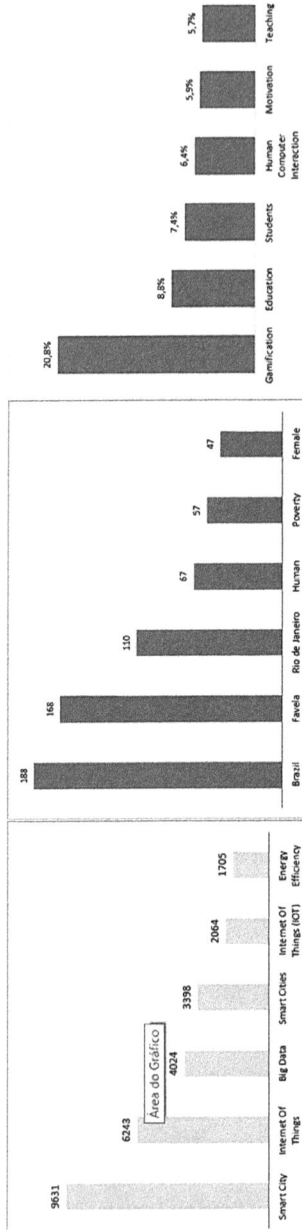

FIGURE 3.6
Comparison between Research Areas of publications on key research topics: smart city, slum, and gamification.

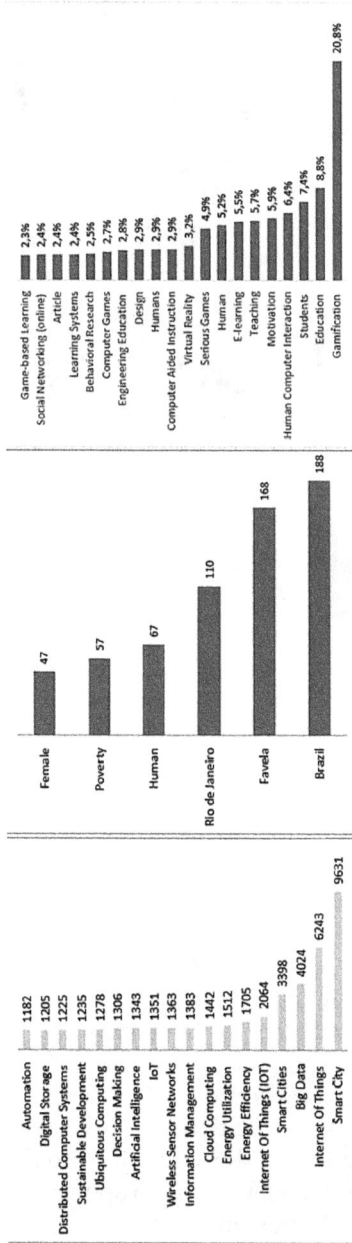

FIGURE 3.7
Comparison between RESEARCH AREAS of publications on key research topics: smart city, slum, and gamification.

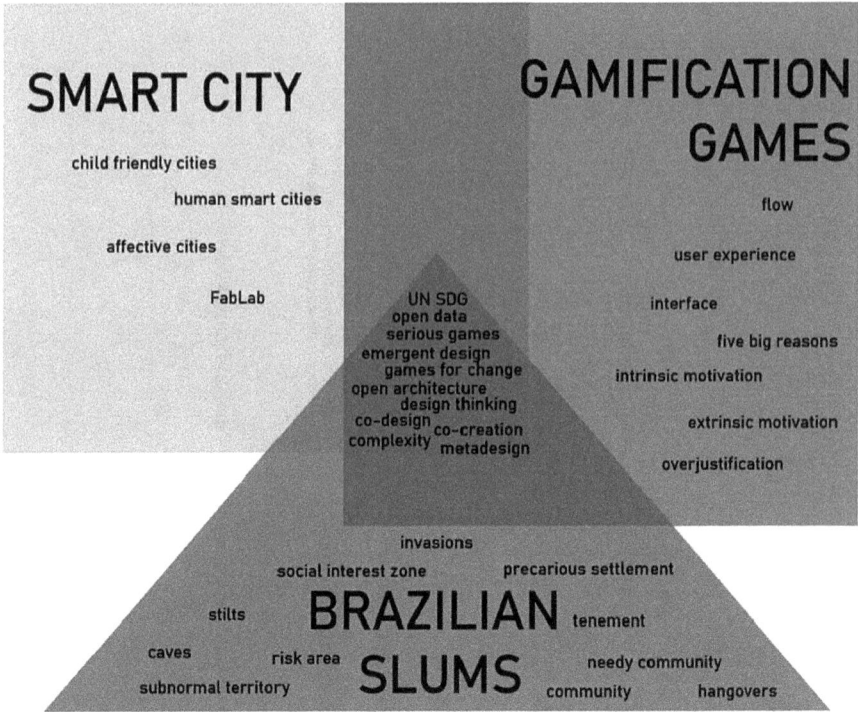

FIGURE 3.8
Search concepts and terms.

The intersection of the keywords found in bibliometrics, added to some others found in the deepening of the bibliographic review, results in the construction of a conceptual tree that confirms and indicate the need to review some steps and terms (and specific objectives) of the research of this dissertation (Figure 3.8).

Entering the thematic intersections and the specialized authors of the research tripod, authors stand out who sometimes contribute on time, sometimes work on concepts and discussions on more than one theme.

Gamification, like Serious Games and Games for Change, is used to stimulate and sustain the motivation of citizens to participate in the creation of the Intelligent Human City, one of the concepts unfolded in the universe of Intelligent Cities (and Aune, 2017), which also applies to the central theme of the work. Converges into this movement the statement originated from the gamification theme, by Fuchs (2014) and Santaella et al. (2018), that nothing will influence human life as much as mobility, social media, and gamification, having the latter, the greatest potential for impact.

Aune (2017) in the context of the development of smart cities, emphasizes the need to have, at local scales, innovation ecosystems—or Urban Living Labs—that bring together actors from different sectors of society in co-creating innovative solutions (technological or not) for urban issues as well as methodologies to ensure community involvement in the different work phases of the co-creation process, such as gamification. As a result, the interaction of digital technologies with smart cities can also be seen in the development of the so-called playful cities, in which the principle is to use the structure of the city, together with technology and gamification, to connect citizens and the urban space, aiming to minimize the exclusion of groups, as highlighted by Vanolo (2018) and Damiani et al. (2018).

It is in this universe that we approach the following examples of games and gamified applications that emerge from entertainment to solve urban problems, seeking to identify potential applications in slums.

3.3 Games and Urban Gamification

In the 1970s, Dupuy (2012) and Henriot and Molines (2019) identified approximately 100 simulation games that covered the theme of cities, which he called "urban games." Despite 50 years having passed, his work remains highly relevant (Henriot and Molines, 2019). The games industry is multifaceted and the focus on cities is very popular (Reinart and Poplin, 2014). Although many of these games are only intended to amuse the player, some are configured as serious games and include realistic visualizations of the city or district of the city (Reinart and Poplin, 2014). In this line, some games also allow the transfer of knowledge, which means that concrete information about a planning situation is provided (Reinart and Poplin, 2014).

The use of Serious Games and Games for Change for urban planning and especially, at the stages when public participation is necessary, has enormous potential (Poplin, 2011). Research in this area is considered to be a very promising, innovative field that is advancing in large steps, even though it was identified 10 years ago that it was walking in its initial stages (Poplin, 2011). The development and implementation of games in urban planning include an emerging research area. The games can show abstract and very specific planning processes in a fun and more readable way to ordinary citizens, unrelated to engineering and architecture.

Another evident aspect refers to the wide capacity of applications within the scope of the lifecycle of a city. It can be applied in the planning, design, management and maintenance, and remodeling stages of urban centers.

Such an application is based on a hypothesis that games can potentially enable easier and more joyful learning processes and bring playfulness to the public participation process, sometimes called playful public participation (Krek, 2008).

Research addressing all games applicable to the context of cities—or more specifically, so-called "intelligence"—can be vast and endless, extrapolating the material and immaterial resources available in a master's dissertation. For this reason, the search for applicable games focuses on scientific works on scientific bases such as Scopus, ScienceDirect, and Scielo, and on national and international festivals of games and games for change. The work identifies five categories of games: Analog or Traditional/Non-Digital (Reinart and Poplin, 2014), Digital, Diffused Games, Geolocative Games (no games were selected in this typology), and Gamified Urban Applications. Previous studies have stated that the first three covers most games and, considering the fourth, it is stated that—in the scope of serious games, games for change, and entertainment games—a large scope of games can be contemplated.

For the purposes of preliminary survey, SimCity games, Minecraft, PlastiCity, UrbanPlans, City Creator, Super City, and Cities: Skylines, found in scientific works and available in the games market, are listed as Block'hood games and Cidade em Jogo, award-winning games found in international and national game festivals, Gamesforchange.org and BIGfestival.com.br respectively. The City Statute Game, available at the Pólis Institute, unlike all others, includes an analog game without a digital version. And lastly, although it does not fit as a game itself, there is the Colab application, which makes use of gamification elements in its dynamics. For the purpose of systematizing the research, initial assessment points for games are proposed based on case studies and referential authors, as well as on the items raised in the chapter "Three Essential Focuses on Game Design," developed from Hosse (2014), in elements used in the article by (Reinart and Poplin, 2014) and in the comparative analysis by Damiani et al. (2018).

The points raised in each game address, in addition to the name, the presentation of the manufacturer, considering the official presentation of the game manufacturer; the objective of the game, considering whether the game has the objective of generating in the player reflection or behavior change; the topic, considering it to be a general political partner or to address a specific issue; social or political; the location and company, country and breeding company; the year and versions; a brief description of its functioning (Context, Physical System, Choices, Immediate Feedback, Collaboration/Competition, Cooperation, and Interaction between players); its evaluation in Metacritics or awards received (when applicable); and finally, case studies in which the respective games were used for urban planning, city management, or slum activities. The summary table with the evaluated and described points is shown below (Table 3.2).

TABLE 3.2

Analysis Points of the Games Raised

Name of the Game	
Manufacturer Presentation	Official presentation from the game's manufacturer.
Goal Game	The game aims to generate in the player the reflection or change of behavior in face of a social or political theme.
Theme	General or specific socio-political theme?
The Location/Company	Country and Company.
Year/Versions	Year of publication and versions.
Description Operation	Brief description of how the game works (Context, Physical system, Choices, Immediate feedback, Collaboration/ Competition, Cooperation, and Interaction between players).
Metacritic or Awards	Metacritcs scores and ratings and/or awards.
City Applications	Cases in which games were applied in urban contexts, in the practical and scientific scope.

Source: Prepared by the author based on Hosse (2014) and Reinart and Poplin (2014).

3.3.1 SimCity

SimCity is one of the most famous entertainment games related to urban planning and also one of the oldest available on the market, featuring a series whose player can create, build, and manage a city, the main objective of the game.

Some examples of the use of the game in urban planning are found on search engines and scientific databases. The oldest refers to the publication of (Pina and Leal, 2008), from the Federal University of Uberlândia, in the geography course, which aims to identify teaching methodological tools inserted in the context of urban geography, thus highlighting the great need to consider the study and planning of land use and occupation in a city.

Another example of the use of SimCity in the universe of urban planning comes from the application in the educational area, through the project "Virtual cities: use of games for the study of architecture and urbanism" by professor Pablo Lira, professor at the University of Vila Velha, Espírito Santo. The "Virtual Cities" project aims to encourage students to reflect on the contents exposed in classes in the search for solutions to complex problems in the field of Architecture and Urbanism. The project allows students to verify in a simulated virtual environment the applicability, and effectiveness of the theories, instruments, and mechanisms of urban planning, which is one of the winners of the 2015 Inova UVV Award.

Extrapolating the educational area and moving toward serious games, there is the use of SimCity for studies and urban predictions, present in the work of Bereitschaft (2016) entitled "Gods of the city? Reflecting on city-building games as an early introduction to urban systems"(Bereitschaft,

2016, p. 50, our translation). The work argues the importance of city-building games (simulation) and opposes a series of assumptions and prejudices produced by their developers, such as emphasizing personal transport over transit and public transport, self-centered development over mixed-use, and the simplified social dynamics in the face of the complexities of reality. But, he concludes that:

> No gaming simulation can be expected to reproduce urban systems in all their intricacy—nor would they necessarily aim to given that their primary purpose is to entertain.
>
> (Bereitschaft, 2016, p. 58)

3.3.2 Minecraft

Minecraft is an electronic survival sandbox game, considered to be the best-selling game of all time, with 200 million copies sold and over 126 million monthly active players.(.

Several studies and cases of application of the game in the universe of cities can be found. The first one refers to the UN-Habitat's Block by Block program, in partnership with Microsoft, and the Minecraft developer Mojang. The Block by Block program considers gambling as a great tool to engage people—especially young people, women, and slum dwellers—in planning sustainable, accessible, inclusive, and safe public spaces. On the website of the Block by Block program (2018), it is possible to access several initiatives carried out in various parts of the world (there are 30 projects in all), distributed between Africa, Arab States, Asia, and the Pacific, Europe, Latin America and the Caribbean, and North America.

Among the initiatives, a Brazilian woman is included, in a neighborhood of a poor community in São Paulo, in which a staircase was renovated with the participation of the Jardim Ângela community in the design and execution. The proposal, called "Olhe o Degrau," organized by "Cidade Ativa" aims to requalify the use of the stairs in the city of São Paulo, which are generally abandoned and degraded, seen as unsafe places by the population. With the support of HealthBridge Foundation of Canada and UN-Habitat, after choosing the intervention site, the Active City began work with surveys and interviews in April 2018, involving participation from the Oscar Pereira Machado State School through the interaction of the school community, teachers, and management.

To introduce the subject and present the project site to the students, the Corrida Amiga—an organization that addresses mobility on foot as a means of transport—led a Tram on Foot on a route of about 1 km between the school and the staircase. During the route, important elements that make up the mobility network on foot (sidewalks, crossings, accessibility, etc.) were highlighted, debates were raised on the rights and needs of all pedestrians and their different abilities: children, pregnant women, the elderly, and people with reduced mobility, and the senses are sharpened (aromas, noises, feelings). The activity ended with games for the children.

The next stage, still on the staircase—this time led by the Cidade Ativa team—invited the students to observe the conditions of the surrounding elements: steps, house facades, drainage, lighting, among others, and reflections were proposed for the transformation of the place and questions about possible wishes and solutions. Still on the first day of the workshop and back to school, the development of ideas and solutions designed for the staircase began using the Minecraft game. According to the organization, students quickly became familiar with the tool and modified the base model with several proposals, being developed until the end of the second day, when all groups were able to present their ideas and models of the renovated staircase. On the third and last day of the activity, a single consolidated model of the most common ideas among all the proposals was developed.

To add to the results of the Minecraft workshop, on the last day there was also a community engagement workshop, where important data about the users' profiles, the main daily displacements and suggestions for elements to improve the system were collected through interactive panels. Other recreational and artistic activities for children such as "Janela do Futuro" and "Arquitetando no Escadão" and picnic were developed to expand this engagement. The previous cleaning of the place was made possible by the Regional Municipality of M'Boi Mirim. Articulators and groups from the region, such as artists from the Social Cycle Art and the Nakamura family, as well as users and residents around the staircase were also present to express support for the action and give their suggestions for transforming the space.

The results of the different types of data collected during the three days of activity at Jardim Nakamura served as a basis for the development of the intervention project. Presented and approved by the local community, a workshop called "hands-on" was carried out, which brought about the transformations, providing a more pleasant and attractive space for all people. The community worked to create a park along the newly updated staircase, with paints, construction materials, and playful ideas, with color and imagination, developed by artists, children, and the community in general. Additional structures such as benches, a community library, and a slide offered new spaces for families and passers-by. As a result of the transformation and improved safety, the use of the ladder has increased dramatically.

3.3.3 Plasticity

Plasticity—contemplates a computer game based on the views of the British architect Will Alsop. Alsop's controversial vision for the redesign of Bradford's city center has led to media turmoil. It started intense discussions among the public about how far a "master plan" for urban reconstruction could reach (Poplin, 2011). Alsop's suggestion to replace two of the most important buildings in the geographic center of the city with a lake sounded like a joke to many of Bradford's inhabitants (Poplin, 2011).

In this scenario, the Plasticity game focuses on Bradford city center, where players can build, demolish, repaint, resize, and rotate buildings in this part of the city. The scenario takes place in a 3 D environment and supports complex interactions between the player and the game environment (Poplin, 2011). The game was created by Mathias Fuchs (senior professor at Salford University) and Steve Manthorp (special project manager, Bradford). The game's authors state that, led by Alsop's statement that the absence of joy is the greatest threat to society, we try to emphasize the playful and joyful elements in urban planning (Poplin, 2011).

The first phase of the project was carried out with the support of the Lightwave partnership: an organization based in Bradford, managed by the National Museum of Photography, Cinema and Television, Bradford University, the city of Bradford, and a regional development organization (Poplin, 2011). It is based on careful research by city planning institutions, such as the Municipal Planning Council, the city center master plan, and the wishes and demands of the local population. It included investigations into the present, past, and possible future of urban structures and, the history of visions of "cities of the future" related to visions provided by some well-known thinkers and architects, such as, for example, Charles Fourier and Le Corbusier (Poplin, 2011).

The Plasticity game includes realistic architectural models, basic gameplay, and a set of functions developed with the aim of creating and altering buildings (Poplin, 2011). A significant amount of time was spent taking pictures and modeling the main buildings. In the second phase of their development, the creators also included different possible urban planning strategies that can be used by players, and programmed some game elements specific to ethnicity and gender (Poplin, 2011). The game has no prize and is not listed in Metacritic for evaluation. As for the application in urban areas, the game was originally configured with such an objective.

3.3.4 Cities Skylines

Cities Skylines includes an electronic city-building game for one player, launched in 2015 for several platforms, including PC, XBOX, PS4, and Switch.

As for the scientific and practical studies using the Cities Skylines game directed to the urban area, we can mention those developed by Pinos et al. (2020), from the Czech Republic; by Bereitschaft (2016) from the United States; and by Olszewski et al. (2020), from Poland.

The article by Pinos et al. (2020) is entitled "Automatic methods of processing geographic data for visualizations of real-world cities in Cities Skylines" (Pinos et al., 2020, p. 1, our translation). According to the author, by simulating urban processes in visually attractive environments and 3 D, Cities: Skylines offers interesting possibilities for visualizing places in the real world. For these characteristics—despite the difficulty, the delay and, the imprecision of the geographic data—the game offers interesting possibilities for applications in the presentation of projects, in the social participation in urban policies and processes of preservation of buildings and, educational activities (Pinos et al., 2020).

Specifically, the article presents the methods and processes of a game modification (mod) called GeoSkylines, which allows to the creation of geographically accurate visualization of a location of cities or unoccupied spaces. In this sense, the work creates road and rail networks, wooded areas, river basins, planning zones, buildings, and urban services with playable models from the cities of Svit (Slovakia) and Olomouc (Czech Republic). Besides, the GeoSkylines game mod also describes methods for exporting game objects (roads, buildings, and zones) to a GIS (Geographic Information System) data format, allowing the game to be used as a data collection tool that can be used in city design and development projects.

Another work to address this game refers to the article by Bereitschaft (2016), entitled "Gods of the city? Reflecting on city-building games as an early introduction to urban systems" (Bereitschaft, 2016, p. 50, our translation). In addition to Cities Skylines, the work addresses SimCity in its analysis and relationships with real urban dynamics and process patterns. Despite the immense potential to shape players' understanding and expectations about the functioning of cities, Bereitschaft (2016) highlights some points that usually do not occur in real cities: unlimited concentration of power in the mayor/player, absence of mixed uses, historical heritage, the privilege of cars, social simplification, and utopian city management.

The first refers to the concentration of powers in the player, which in turn gives rise to the title of his article. Usually, cities are managed, and decisions are made according to a series of professionals and specialties, with the mayor having the role of manager of this knowledge, hierarchizing and articulating the integration of solutions to solve problems (Bereitschaft, 2016). Often, the Cities Skylines player is forced to perform additional tasks that would normally fall to city planners, transportation engineers, and planning and zoning boards, with the only fixed limits on territorial space and financial resources (Bereitschaft, 2016).

Having the player as a mayor, he then governs with more power than any mayor in history (Bereitschaft, 2016), being that "not only can they destroy an entire city and start over on a whim, they also make their decisions in a virtual politician vacuum, without elections" (Bereitschaft, 2016, p. 53).

Finally, the player's total ability in the game, which usually includes the ability to dramatically alter the landscape and unleash natural disasters, is more like a deity than a mayor. This top-down, one-sided, and virtually omnipotent form of urban development, while understandably designed to

maximize the entertainment value of the game, is a far cry from the constraints and complexities of reality. (Bereitschaft, 2016, p. 54, our translation)

As for the last statements of the top-down, unilateral and omnipotent government process present in the game, although it seems quite distant from
the reality of the place of origin of the article and the author when contextualized to the Brazilian reality, they are adherent to the recent rulers of the
parents. Another point highlighted by Bereitschaft (2016) refers to the lack
of mixed uses in the game. The use of mixed uses in cities (such as residential and commercial, or residential and corporate, or residential and services)
is seen with very good eyes in the universe of sustainable urban development (Jabareen, 2006 and Bereitschaft, 2016). This statement is justified by
the reduction in automotive dependence, the reduction of travel, the possibility of using alternative and active transport, such as walking, bicycles,
scooters, among many others. Potential benefits include improved health
due to increased physical activity, better air quality, and lower CO_2 emissions, reduced urban sprawl and associated losses in agricultural, and natural areas and a more equitable and vibrant urban environment (Parker 1994;
American Planning Association, 1998 and Bereitschaft, 2016).

"In Cities: Skylines, players are also strongly discouraged from implementing mixed horizontal use. Commercial buildings, along with industry,
roads, and power plants, produce noise pollution that can reduce land values
and decrease health and happiness in adjacent residential areas. Therefore,
players are encouraged to separate potentially compatible commercial and
residential areas into separate districts, aggravating automotive congestion."
(Bereitschaft, 2016, p. 54). Several users ended up commenting on this rigidity
and, in response to the comments, Colossal Order stated that mixed-use zoning can be implemented in future versions of the game (Paradox Interactive
2015b and Bereitschaft, 2016).

The next point commented by the author refers to issues of preservation
of historical heritage. The most recent Cities: Skylines have only an architectural style, a kind of generic Euro-American contemporary mix (Bereitschaft,
2016). The buildings in the game have a more dynamic and generic aspect,
being able to undergo drastic improvements without necessarily preserving
the original elements of their formation, without having an identity or vernacular characteristic of their culture (Bereitschaft, 2016).

As for urban mobility, despite the various types of transport modes existing in Cities: Skylines (considered to be one of the most multimodal games
to date) such as buses, passenger and cargo trains, subways, cruise and cargo
ships, airplanes, and trucks with several options of roads and highways, cars
still reign (Bereitschaft, 2016). On the other hand, the game includes pedestrian paths, although it excludes bike lanes (ibid.). However, it can be said
that Cities: Skylines facilitates dependence on automotive transport, despite
the availability of other public transport options (Bereitschaft, 2016). Such
a statement is justified by the fact that any congestion can be minimized by

increasing the number of lanes or highways, which does not happen in the real world. Induced demand, in which more lanes stimulate more demand and, ultimately, more traffic, is not a consideration (Bereitschaft, 2016).

As for social simplification, from the possibility of using the camera in the first person, it can be noted that citizens apparently have only one race and ethnicity: Caucasians (Bereitschaft, 2016). The incorporation of unique racial and ethnic identities in the simulated cities would increase their complexity and realism, although perhaps not necessarily their entertainment value (Bereitschaft, 2016). The potential implication, however, is that players may underestimate or disregard the important socio-spatial patterns and inequalities that often manifest along racial and ethnic lines. As for social simplification, from the possibility of using the camera in the first person, it can be noted that citizens have only one race and ethnicity: Caucasians (Bereitschaft, 2016). The incorporation of unique racial and ethnic identities in the simulated cities would increase their complexity and realism, although perhaps not necessarily their entertainment value (Bereitschaft, 2016). The potential implication, however, is that players may underestimate or disregard the important socio-spatial patterns and inequalities that often manifest along racial and ethnic lines.

Olszewski et al. (2020) in the article "Developing a serious game that supports the resolution of social and ecological problems in the toolkit environment of Cities: Skylines" (Olszewski et al., 2020, p. 1) also addresses the game in an applied context. The authors' objective contemplates the use of serious games in the engagement and in the population's participatory process in solving social and environmental problems, through the project "Increasing the participation of the residents of Zuromin in the management, environmental monitoring, and creation of a vision for the development of the city, stimulating social participation," (Olszewski et al., 2020, p. 5). Both the purpose of the article and the project are the result of a demand from the Ministry of Investments and Economic Development at the University of Warsaw, linked to the "Smart and Human City" program. In the publication, the authors present an unprecedented concept of using the Cities: Skylines environment associated with the C# programming language to automate the process of importing official topographic data into the game engine. In addition, they develop a prototype of a serious game to support the solution of social and ecological problems in the city of Zuromin, a city in central Poland, located 100 km from Warsaw (Olszewski et al., 2020).

Zuromin's main social and environmental problem concerns the high density of poultry farms (20 million head) and pig farms (600,000) in an area of around 135 km 2, with about 15,000 inhabitants (Olszewski et al., 2020). This situation negatively and intensely influences the lives of residents, through over-fertilization and consequent degradation of the soil; damage to roads by heavy transport; and, above all, the emission of an extremely offensive odor (Olszewski et al., 2020). Economic and social analyzes carried out showed that the construction of a biogas plant must provide a comprehensive solution to the problem (Olszewski et al., 2020).

This particular serious game means convincing users (residents, municipal authorities and, livestock producers) that building a biogas plant is a win-win solution (Olszewski et al., 2020). The purpose of the developed concept of this serious game is to show the benefits of the cooperation of the inhabitants of a smart city with their authorities and local companies and to visualize the beneficial environmental changes related to the construction of a biogas plant (Olszewski et al., 2020, p. 17).

3.3.5 Block'hood

Block'hood includes a neighborhood building simulation game that allows for the diversity and experimentation of cities and those of unique ecosystems inserted within them. "Block'hood is a city-building simulation video game that focuses on ideas of ecology, interdependence, and decay."

"The game is an educational and research initiative that explores the connection between games and architecture, contributing to a form of digital infrastructure for the ecological and systemic thinking needed in contemporary urbanism." (Plethora Project statement in 2016. The main objective of the game is to build a neighborhood as a productive network of resources that works sustainably, maintaining itself over time. The game has an educational character (partially), incorporating the mechanics in which all the built blocks must be accessible by stairs and corridors, to guarantee the flow of the processes of a city. Buildings must also be architecturally structured; for example, corridors must be supported before they can be built.

As for the scientific studies using the Block'hood game aimed at the urban area, two developed by Sanchez (2019), called "Massive re-standardization of the urban landscape" and "Architecture for the commons: participatory systems in the age of platforms" respectively. A third, published by Bullivant (2017), is entitled "The hyperlocal: less intelligent city, more shared social value" (ibid.). The articles cited above were not fully accessed for financial reasons, with only the science of the abstracts published by Sanchez, and Bullivant's article does not have an abstract published openly.

The summary of the first article by Sanchez (2016) follows below:

> Gambling is usually seen as a way to escape from reality; but could your open-source tools be used to shape the physical environment? Game architect, programmer, and designer Jose Sanchez created the Block'hood video game as an urban simulation, with users building urban blocks to design a community. Emerging standards are tested against ecological and energy concerns. The results increase the public's understanding of these issues and present the possibility of an alternative based on dialogue to top-down planning.
>
> Sanchez, 2016

The second article by Sanchez (2019) is as follows:

> Does Discrete offer a path to fairer architecture? Jose Sanchez, director of the Plethora Project's research and learning initiative and an assistant professor at the School of Architecture at the University of Southern California (USC) in Los Angeles, laments the exclusivity of parametric design and argues why Discrete could be the answer. Two combinatorial assembly projects and a gaming platform that he was involved in creating serve as practical illustrations of the possibilities for social engagement that this new paradigm presents.
>
> Sanchez, 2019

3.3.6 Cidade Em Jogo

Cidade em Jogo is a Brazilian game developed by the Brava Foundation in partnership with the Woodrow Wilson Center. The Game—which can be considered a serious game or a game for change—aims to foster critical thinking and social protagonism in students to become more engaged citizens in the context of their city. The Game proposes to show, inside the classroom, how the management of a municipality works in the best way: getting hands-on and learning in practice. The game allows students to be mayors for a day, and decide which solutions are best for their city. In this process, students are encouraged to think critically and to discuss with different colleagues their different points of view about their city.

The game is online and free, includes a simple method, and with elements of fun themes related to citizenship, public management, and democracy, the player acts as the mayor of a c-ity. The game was launched in 2017 and has since been recognized as the British School of Creative Arts (EBAC) award for best game and second place in the Games category of the Festival. In 2019, he received the award at BIGfestival.com.br in the BIG IMPACT: EDUCATIONAL category, with the jury's justification for its relevance to political education, standing out by demanding from the player skills such as reflecting on the challenges and prioritizing investments in the city move. It has no mention on the Metacritic website.

As for the applications of the game in the real world, the website of the creators of Cidade em Jogo discloses four initiatives carried out in Brazil, specifically within schools. The first is the Mayor for a Day Project in Jundiaí/ SP. Through a partnership of the Brava Foundation with the São Paulo State Department of Education and the Jundiaí City Hall, a pilot project was implemented in three schools in the city. After conducting a few rounds with the game Cidade em Jogo, students from each school were challenged to develop projects that aimed to identify problems in their respective neighborhoods, analyze them, and raise difficulties and present proposals to solve them. Those responsible for the best projects in each school had the opportunity to go to the city hall to present their proposals to the mayor, secretaries and, managers linked to the measures suggested by the students. Several practical improvement measures suggested by the students were followed by the municipal government.

In E.E. Dom Joaquim Justino Carreira, the teacher created a three-class mini-course that talked about elections, state structure and, political education. The winning project talked about the lack of leisure options in the neighborhood, which generated violence and made young people approach crime and evade school. In E.E. Maria de Almeida Schledorn, after using the game Cidade em Jogo, the students developed a project about the main challenges related to basic sanitation and the floods that some streets present during the rainy season in the Jardim das Tulipas neighborhood.

In E.E. Professor Deolinda Copelli de Souza Lima, the teachers developed an interdisciplinary project of Mathematics and Sociology that used Cidade em Jogo as the guiding thread of the discipline, used at different times. The winning project dealt with urban janitorial and public lighting of a staircase with important access to the school, which increased the feeling of insecurity and the occurrence of crimes in the place. The second initiative took place at Colégio Humboldt in São Paulo/SP, with an initiative that motivated students to reflect on citizenship and public policies. Teachers in the humanities field developed a didactic sequence with a problematization and used the experience of the game as a stimulus for discussion. After the experience with Cidade em Jogo, the students took their impressions of what they experienced during the rounds with the game for a broader discussion, drawing parallels with real-life issues. Thus, it was possible to demonstrate limits to the action of municipal public management, to promote reflection on political issues that were initially less tangible, and to simulate how individual decision-making impacts the results obtained within a more general context.

At Colégio Dante Alighieri in São Paulo/SP, classes from the 1st year of high school participated in an interdisciplinary project called "Eu Cidade," which was started in the computer lab using SimCity. The goal was for them to have a sense of what it would take to create a city, learn about the entire infrastructure and how management takes place. Then, they started the experiment with the game Cidade em Jogo, where they could see how the issue of public policies works, putting learning into practice and, discussing how each group dealt with specific issues, in addition to being able to feel the importance of their choices. From there, students were challenged to make social impact business plans to solve urban problems in São Paulo, culminating in the presentation of proposals and prototypes for a bank. The project aimed to make students feel protagonists, to know their importance as citizens and, that they have the right to question and actively participate in the process of social change.

The fourth and last example occurred in Recife/PE, at Colégio Lubienska. We worked with the game Cidade em Jogo as a trigger tool for holding debates and reflections over a week, whose central theme was "citizenship and participation." An electoral process was initiated in which the three candidate slates made videos defending their government platforms, based on the priorities they chose when playing the game. The other students formed four groups, each representing a different legislative group and interest group. Due to active participation, 9th-grade students formed a 5th group, representing organized civil society.

The elected mayor and vice mayor then played Cidade em Jogo again, presenting their proposals for the application of public policies to the benches, which should deliberate for approval or rejection, until all rounds were completed. Altogether, 80 students actively participated in the same round of the game, simulating the functioning of the powers.

3.3.7 O Jogo Do Estatuto Das Cidades

The City Statute Game is configured as an analog game, not digital or traditional, also called an RPG (Role-Playing Game) of Urbanism. It was created in 2002 (with the second edition launched in 2005) by Instituto Pólis, a nongovernmental organization that works to build cities that are more together, sustainable, and democratic. According to the creator's definition:

> A possibility to deal with the different audiences involved in the daily construction of the city playfully: at the same time that discussions of conflicting urban situations are proposed, they are alternatives are presented using the new instruments contained in the Statute. The proposal is to awaken the various actors who live and build the urban space, through the role play, of the interest in the knowledge of the urban regulation instruments available by the Statute, stimulating reflection on the innumerable possibilities of solutions to the issues that affect the cities.
>
> (Santos et al., 2005a, p. 3)

The City Statute Game contemplates a set of roles and scenarios, the objective of which is to familiarize society with the contents and instruments of the City Statute, whose participants are challenged to use them in hypothetical situations (Santos et al., 2005b).

As for the application of the game, according to a publication by Folha de São Paulo, in the various tests carried out with the material, three São Paulo cities stood out: Guarulhos, Limeira, and Caraguatatuba.

3.3.8 Colab

Colab does not fit as a game, but a gamified application of urban janitorial, collaborative, created to receive contributions from any citizen in the management and maintenance of the urban quality of the city in which he lives (Colab, 2013). The software's slogan is:

> Making the city better is in your hand.
> Colab (2013)

Colab was created in Recife/PE in 2013 by partners Gustavo Maia and Paulo Pandolfi and aims to encourage and disseminate collaborative, transparent and, efficient public management by the citizen himself, connecting them to the responsible bodies (city halls) to solve problems found in cities (Colab, 2013).

In general, the application has a lean appearance, easy usability and navigation, without difficulty in using it.

3.4 Results and Reflections between the Thematic Tripod

Currently, the exploration of games in the context of smart cities represents a great challenge, both because we are not aware of the "limitations" of smart cities and because we do not adequately understand the full range of possibilities that technology offers (Cavada and Rogers, 2019). As long as participants are trained, literate, and willing to share personal information—the latter point of resistance being highlighted by the Ogie study (2016)—technology can provide a foundation for the smart city as a sharing platform that supports local governance as well as for citizens (Kamel Boulos et al., 2017).

After the survey and analysis of the case studies, it is possible to outline groupings and categorizations, as well as to make considerations regarding games and applications directed to urban areas. The first distinction to be made refers to the existence of Analog and Digital resources and options. Analog resources—represented here by the City Statute Game (B. J. dos Santos et al., 2005a)—are those that develop without the use of ICTs, using boards and cards. Digital resources, present in greater number and represented by SimCity, Minecraft, PlastiCity, UrbanPlans, City Creator, Super City, Cities: Skylines, Block'hood, Cidade em Jogo and Colab, demand the use of electronic devices for execution; mobile or not. The Analog Game (City Statute Game) has a wide application capacity in disadvantaged territories since it does not demand the existence of technological devices, for its execution. On the other hand, overcoming the barrier of the absence of hardware (more related to desktops and laptops, since the percolation of mobile devices, breaks social barriers), digital resources have greater visual resources and attractions and, consequently, they can have a more positive result. significant in the scope of engagement.

A second look understands the existence of Games and Gamified Applications. The games are represented by SimCity, Minecraft, PlastiCity, UrbanPlans, City Creator, Super City, Cities: Skylines, Block'hood, Cidade em Jogo, and Jogo da Cidade. As for gamified applications, there is Colab that loads in its operation, dynamics, and mechanisms of gamification, without necessarily being a game in itself (Poplin, 2011; Reinart and Poplin, 2014; Olszewski et al., 2016; Bereitschaft, 2016; Pinos et al., 2020; Olszewski et al., 2020).

A third consideration to make refers to the type of game. In the simulation category, there are SimCity, UrbanPlans, City Creator, Cities: Skylines and, Block'hood. In the sandbox entertainment category (or non-linear game), there is Minecraft. The PlastiCity game was born as a serious game with a focus on political issues. The Jogo da Cidade and Cidade em Jogo are inserted in an educational and political context. Colab, as mentioned earlier, cannot be considered a game but a gamified urban janitorial app.

A fourth distinction refers to the types of games and the possibilities of transmutation of their original typology, that is, of emergency. As described by and Ferreira (2017) as what happens when two or more things come together and interact intensively in a specific way, to constitute a system in which their attempts at combination and self-organization generate new behaviors.

A fifth point to be highlighted refers to the possibility of personalizing the game within the geographical aspects of existing territories and cities. The SimCity, Minecraft, and Cities: Skylines games with a lesser or greater degree of difficulty, allow the insertion of georeferenced data and, consequently, simulate with great precision several dimensions and processes existing in a given region, such as environment, mobility, society, among several other elements (Olszewski et al., 2016; Bereitschaft, 2016; Pinos et al., 2020; Olszewski et al., 2020).

The sixth consideration, which became was not studied in this dissertation, refers to Locative Mobile Games, which became popular with the spread of the internet on mobile devices with GPS. In these games, the boundaries between virtual and physical almost disappear. According to and Reinart and Poplin (2014, p 7.):

> The family of broadcast games is diverse, including individual games, ranging from simple single-player mobile games to artistic and politically ambitious mixed reality events.

The penultimate one emphasizes the intention and the expected effect of using games in non-game contexts, that is, the application as serious games or serious games. In addition to the educational intention, the role of games in engagement, in encouraging the participation of local society in urban issues delimited by each case study, is highlighted (Poplin, 2011; Reinart and

Poplin, 2014; Olszewski et al., 2016; Bereitschaft, 2016; Damiani et al., 2018; Pinos et al., 2020; Olszewski et al., 2020).

Finally, there is a complement to engagement, facilitating the understanding of proposals and projects in the field of architecture, urbanism, and even strategic planning of cities, highlighting here its inclusive bias. Usually in the urban field, the proposals made for society are presented through projects full of technical drawings and architectural planks, which are difficult for the ordinary citizen to understand and understand. Some proposals are accompanied by 3 D representations and perspectives that contribute to their understanding. And, as an evolution of this instrument, there are the game technologies that allow the visualization and the accelerated evolution of the implementation of these proposals, democratizing and providing an opportunity for their understanding by a good part of society.

Once again, Cavada and Rogers (2019) highlight that Virtual Reality (VR) provides a powerful tool for visually influencing people and for being able to bring about changes in the way citizens think and act. The use of VR is in line with the technical requirements of technocratic stakeholders such as planners, engineers, and local authorities, but is usually not leveraged in a way that is integrated into a system that supports social ties and real-time communication and is available for use of individuals (Cavada and Rogers, 2019). However, it contemplates a great resource to be applied in the "serious game" as an artifact to assist in communicating with citizens and in making complex decisions (Cavada and Rogers, 2019).

There is a summary table containing the results of the analysis of the universe of the application of games in the slums, in the context of smart cities, shown below (Table 3.3).

3.5 Final Considerations

The last chapter aims to establish conclusions regarding the subjects discussed in the work and to suggest some recommendations for future research, related to the main theme.

This research aimed to establish approximations between smart cities, Brazilian slums, games, and gamification, having the field of design as a critical means for seams and integrations. Additionally, it ended up moving toward studying the contribution of games and gamification in the design of smart cities, with a focus on urban improvements and the socio-spatial inclusion of favelas in the territory of the official city, with a focus on the territorial universe of Brazilian favelas.

In this context, this objective is considered to have been achieved, since even though it is not a highly studied or applied subject, the research

TABLE 3.3

Results of the Analysis of the Universe of the Application of Games in the Slums, in the Context of Smart Cities

Games					Criteria						
	SimCity	Minecraft	Plastic City	Urban Plans	City Creator	Supercity	Cities Skylines	Block'hood	Cidade Em Jogo	Jogo Da Cidade	Colab
Area	[icon]	[icon]	[icon]	[icon]	[icon]	[icon]	[icon]	[icon]	[icon]	[icon]	[icon]
Objective (what for)	[icon]	[icon]	[icon]	[icon]	[icon]	[icon]	[icon]	[icon]	[icon]	[icon]	[icon]
Smart city dimensions	●	◐	◐	◐	◐	◐	●	◐	●	◐	●
Was there NA urban application?	↻	↻	↻	⊘	⊘	⊘	↻	↻	⊘	↻	↻
Emergency	⊘	↻	⊘	⊘	⊘	⊘	↻	⊘	⊘	⊘	⊘
New area	—	[icon]	—	—	—	—	[icon]	—	—	—	—
Was there application in slum?	⊘	↻	⊘	⊘	⊘	⊘	↻	⊘	⊘	⊘	⊘
Result	—	Urban requalification	—	—	—	—	—	—	—	—	—
Potential for slum application?	↻	↻	⊘	⊘	⊘	⊘	↻	⊘	↻	↻	↻

(Continued)

TABLE 3.3 (Continued)

Results of the Analysis of the Universe of the Application of Games in the Slums, in the Context of Smart Cities

						Criteria					
Games	SimCity	Minecraft	Plastic City	Urban Plans	City Creator	Supercity	Cities Skylines	Block'hood	Cidade Em Jogo	Jogo Da Cidade	Colab
Authors	Poplin (2011), Pina and Leal (2008), Borba (2016), and Bereitschaft (2016)	Blockbyblock (2018)	Alenka (2011) and Poplin (2011)	QWERD (2001) and Poplin (2011)	Wiltone and Henderson (2002) and Poplin (2011)	Playkot (2009) and Poplin (2011)	Pinos, Vozenilek and Pavlis (2020), Olszewski et al. (2020), and Bereitschaft (2016)	PLETHORA PROJECT LLC (2016), Sanchez (2016), Sanchez (2019), and Bullivant (2017)	Brava (2017)	Duran (2002) and Santos et al. (2005)	Colab (2013, 2020)

Symbols:

Simulation; Sandbox; Serious games; Educational games; Gamified application; Amusement; Educational games; Urban management; All dimensions; Some dimensions; Yes; No

correlates with each tripod theme in addition to finding a case of application of games and gamified processes in the development of participatory solutions for urban improvements and socio-spatial inclusion in Brazilian slums.

Even if in most of the cases raised the object of study of the applications of games or gamified processes are not favelas or in Brazilian favelas, it is clear that the use of games, the methods, and instruments adopted can and should be applied in participatory processes in the Brazilian urban context. Such an application, considering the dimensions studied and the joined-up philosophy addressed by Aune (2017) in the participation of society in the construction of public policies and more humane and intelligent cities, has unlimited potential in the transformation of cities and, consequently, the slums in inclusive and intelligent territories.

Regarding the work developed, the author reassures its existence, considering the unprecedented and impermanent tone, the legitimacy and, the great importance—vernacular and even globalized—of thematic integration: smart city, favela, and games and gamification. The shortness of the regular period of a master's degree and the unthinkable pandemic could be used as a justification for certain limitations, but they were used as fuels for its development.

Finally, given the young age of thematic integration and the exponential growth and demand for the application of games and gamification for urban management purposes, the research effectively demonstrates that it is necessary for its continuity. There are several desires to deepen the studies in the inclusion of a quantitative character and to respond to the application of the questions raised in a favela and with personas and background contextualized to the Brazilian reality, a subject to be further developed in further research.

References

Albino, V.; Berardi, U.; Dangelico, R. M. (2015) "Smart Cities: Definitions, Dimensions, Performance, and Initiatives," *Journal of Urban Technology*, v. 22, n. 1, pp. 3–21. Disponível Em: <10.1080/10630732.2014.942092>.

Arup (2008), "Smart Cities—Transforming the 21st Century City via the Creative Use of Technology," *Journal of Communications Software and Systems*, v. 1, n. 1, p. 32.

Aune, A. (2017), "Human Smart Cities—O Cenário Brasileiro E A Importância Da Abordagem Joined-Up Na Definição De Cidade Inteligente," Disponível Em: <https://www.maxwell.vrac.puc-rio.br/32955/32955.pdf>

Benites, A. J. (2016), "Análise Das Cidades Inteligentes Sob A Perspectiva Da Sustentabilidade: O Caso Do Centro De Operações Do Rio De Janeiro," p. 224. Disponível Em: <http://repositorio.unicamp.br/jspui/bitstream/reposip/321541/1/benites_anajane_m.pdf>

Bereitschaft, B (2016), "Gods of the City? Reflecting on City Building Games as an Early Introduction to Urban Systems," *Journal of Geography*, v. 115, n. 2, pp. 51–60.

Bullivant, L. (2017). "The Hyperlocal: Less Smart City, More Shared Social Value," *Architectural Design*, v. 87, n. 1, pp. 6–15. https://doi.org/10.1002/ad.2126

Capdevila, I.; Zarlenga, M. I. (2015), "Smart City or Smart Citizens? The Barcelona Case," *Journal of Strategy and Management*, v. 8, n. 3, p. 266–282.

Caragliu, A.; Del Bo, C.; Nijkamp, P. (2009), "Smart Cities in Europe 2009," *3rd Central European Conference in Regional Science—Cers*, pp. 45–59.

Cavada, M.; Rogers, C. (2019), "Serious Gaming as a means of Facilitating Truly Smart Cities: A Narrative Review," *Behaviour & Information Technology*, v. 39, n. 6, p. 20.

Chen, T. (2010), "Smart Grids, Smart Cities Need Better Networks," *Ieee Network*, v. 24, n. 2, pp. 2–3.

Covas, A. (2020), "Da "Smart City" À Cidade Inteligente E Criativa _ Opinião _ Público," Disponível Em: <https://www.publico.pt/2020/03/09/opiniao/opiniao/smart-city-cidade-inteligente-criativa-1906861>. Acesso Em: 23 July.

Cruz, D. O. et al. (2016), "Cidades Inteligentes, Uma Visão Geral A Partir Da Produção Científica," In: Edufes (Ed.). Comunidades Urbanas Energeticamente Eficientes. 1 Edição Ed. Vitória/ES: Edufes, pp. 43–54.

Cunha, R. R. (2019), "Rankings E Indicadores Para Smart Cities: Uma Proposta De Cidades Inteligentes Autopoiéticas," p. 132.

Damiani, C.; Machado, G. M.; Gasparini, I. (2019), "Personalização E Gamificação No Contexto De Cidades Inteligentes," v. 10.

Damiani, C. et al. (2018), "Análise Da Gamificação No Contexto De Cidades Inteligentes E Sustentáveis," *Anais Do Workshop De Aspectos Da Interação Humano-Computador Na Web Social*, v. 9, pp. 71–82. Disponível Em: <http://Portaldeconteudo.Sbc.Org.Br/Index.Php/Waihcws/Article/View/3897/3849&Hl=Pt-Br&Sa=X&D=5086299682410631382&Scisig=Aagbfm2xlj3w5-Bq-Bmp47q6lzn13rw4dg&Nossl=1&Oi=Scholaralrt&Hist=Qlgyuxaaaaaj:4507962020913692037:Aagbfm0ockwnd6-V7qej8ry3xkvaqmmgqw>

Daros, C. and Kistmann, V. B. (2016), "Gestão De Design E Cidades Inteligentes," *Strategic Design Research Journal*, v. 9, n. 1, pp. 14–26.

Davis, M. (2006), "Planeta Favela," *Pós. Revista Do Programa De Pós-Graduação Em Arquitetura E Urbanismo Da Fauusp*, v. 26, n. 25, p. 357.

Dupuy, G. (2012), "Les Jeux Urbains," *L'actualité Économique*, v. 48, n. 1, p. 85.

Dutta, S. (2011), "The Global Innovation Index 2011: Accelerating Growth And Development," [S.L: S.N.].

Eck, R. Van. (2006), "Digital Game-Based Learning: It's Not Just the Digital Natives Who are Restless ….," *Educause Review*, v. 41, n. 2, pp. 1–16,.

Eco, U. (1977), "Como Se Faz Uma Tese," Disponível Em: <www.editoraperspectiva.com.br>.

Fernandes, C. W. R.; Ribeiro, E. L. P. (2018), "Games, Gamificação E O Cenário Educacional Brasileiro," Ciet: Enped, pp. 1–22. Disponível Em: <http://cietenped.ufscar.br/submissao/index.php/2018/article/view/344>

Ferreira, N. B. (2017), "Design De Emergência Em Games," Universidade Anhembi Morumbi,. Disponível Em: <http://download12.docslide.com.br/uploads/check_up03/202015/547ee252b4af9f5e228b475d.pdf>

Franco, P. M.; Ferreira, R. K. Dos R.; Batista, S. C. F. (2015) "Gamificação Na Educação: Considerações Sobre O Uso Pedagógico De Estratégias De Games," In: Congresso Integrado Da Tecnologia Da Informação, Rio De Janeiro.

Freitas, E. De. (2020), "Urbanização No Mundo," Disponível Em: <https://brasile-scola.uol.com.br/geografia/urbanizacao-no-mundo.htm>. Acesso Em: 18 Mar. 2020.

Fuchs, M. (2014), "Predigital Precursors of Gamification, Rethinking Gamification," pp. 119–140.

Giffinger, R.; Haindl, G. (2009), "Smart Cities Ranking: An Effective Instrument for the Positioning of Cities?," pp. 703–714.

Giffinger, R. et al. (2007), "Smart Cities: Ranking of European Mid-Sized Cities," Digital Agenda for Europe, p. 28.

Guimarães, J. G. (2018), "De A. Cidades Inteligentes: Proposta De Um Modelo Brasileiro Multi- Ranking De Classificação," p. 278.

Hall, R. E. et al. (2000), "The Vision of a Smart City." *2nd International Life* v. 28, p. 7. The Vision of A Smart City. 2nd International Life Extension Technology Workshop, Paris, France, September 28, 2000.

Hamari, J.; Koivisto, J. (2015), "Why Do People Use Gamification Services?," *International Journal of Information Management*, v. 35, n. 4, pp. 419–431. Disponível Em: <http://dx.doi.org/10.1016/j.ijinfomgt.2015.04.006>

Harrison, C. et al. (2016), "Foundations for Smarter Cities," *IBM Journal of Research and Development*, v. 54, n. 4, pp. 1–16.

Henriot, C.; Molines, N. (2019), "Urban Serious Games and Digital Technology," Netcom.

Höjer, M.; Wangel, J. (2015) "Smart Sustainable Cities: Definition and Challenges," Advances in Intelligent Systems and Computing, v. 310, Springer, Cham.

Hosse, I. R. (2014). *O design de games for change*. São Paulo: Universidade Anhembi Morumbi.

Jordão, K. C. P. (2016), "Cidades Inteligentes: Uma Proposta Viabilizadora Para A Transformação Das Cidades Brasileiras", p. 307. Disponível Em: <http://tede.bibliotecadigital.puc-campinas.edu.br:8080/jspui/handle/tede/900>

Kamel Boulos, M. N. et al. (2017), "From Urban Planning and Emergency Training to Pokémon Go: Applications of Virtual Reality Gis (Vrgis) and Augmented Reality Gis (Argis) in Personal, Public and Environmental Health," *International Journal of Health Geographics*, v. 16, n. 1, p. 1–11.

Kanter, R. M., Ve Litow, S. S. (2009), "Informed And Interconnected: A Manifesto For Smarter Cities," Harvard Business School General Management Unit Working Paper, pp. 09–141.

Kishimoto, T. M. (1994), "O Jogo E A Educação Infantil," *Perspectiva*, v. 22, n. 1, pp. 105–128.

Klintowitz, D.; Rolnik, R. (2006), "Cidade Ou Favela?," *Oculum Ensaios: Revista De Arquitetura E Urbanismo*, n. 5, p. 1.

Krek, A. (2008), "Games In Urban Planning: The Power Of A Playful Public Participation", *Real Corp 2008: Mobility Nodes As Innovation Hubs Verkehrsknoten Als Innovations- Und Wissensdrehscheiben*, v. 2, Krek 2005, pp. 683–691, Disponível Em: <http://www.corp.at/archive/corp2008_45.pdf>

Lemos, A. L. M.; Bitencourt, E. C. (2018), "Sensibilidade Performativa E Comunicação Das Coisas," *Matrizes*, v. 12, n. 3, pp. 165–188.

Miri, D. H. et al. (2019), "Gamificação: Uma Análise Bibliométrica Sobre Artigos Científicos De 2008 À 2018," February, p. 1–13.

Mueller, C.; Klein, U.; Hof, A. (2017), "An Easy-To-Use Spatial Simulation For Urban Planning In Smaller Municipalities," *Computers, Environment And Urban Systems*, v. 71, December, pp. 109–119, Disponível Em: <Https://Doi.Org/10.1016/J.Compenvurbsys.2018.05.002>

Nam, T.; Pardo, T. A. (2011), "Conceptualizing Smart City with Dimensions of Technology, People, and Institutions," *Acm International Conference Proceeding Series*, pp. 282–291.

Neirotti, P. et al. (2014), "Current Trends In Smart City Initiatives: Some Stylised Facts," *Cities*, v. 38, September, pp. 25–36. Disponível Em: <Http://Dx.Doi.Org/10.1016/J.Cities.2013.12.010>

Nesteriuk, S.; Fuchs, M. (2018), "Precursores Pré-Digitais Da Gamificação," (Transl.) Sergio Nesteriuk. October.

Ogie, R. I. (2016), "Adopting Incentive Mechanisms For Large-Scale Participation In Mobile Crowdsensing: From Literature Review To A Conceptual Framework," [S.L.] Springer Berlin Heidelberg, v. 6.

Olszewski, R.; Turek, A.; Laczyński, M. (2016), "Urban Gamification as A Source of Information For Spatial Data Analysis And Predictive Participatory Modelling of A City's Development," *Data 2016 – Proceedings of the 5th International Conference on Data Management Technologies and Applications*, Data, pp. 176–181.

Olszewski, R. et al. (2020), "Developing A Serious Game That Supports The Resolution Of Social And Ecological Problems In The Toolset Environment Of Cities: Skylines," *Isprs International Journal of Geo-Information*, v. 9, n. 2, p. 118.

Pasternak, S.; D'ottaviano, C. (2016), "Favelas No Brasil E Em São Paulo: Avanços Nas Análises A Partir Da Leitura Territorial Do Censo De 2010," *Cadernos Metrópole*, v. 18, n. 35, pp. 75–100.

Paulino, J. (2007), "O Pensamento Sobre A Favela Em São Paulo: Uma Historia Concisa Das Favelas Paulistanas," USP, Disponível Em: <Https://Www.Teses.Usp.Br/Teses/Disponiveis/16/16137/Tde-17052010-111743/Publico/Jorge01.Pdf>

Pina, J. H. A.; Leal, P. C. B. (2008), "Aplicação Do SimCity Como Ferramenta Para Um Melhor Entendimento Sobre O Planejamento Do Uso E Ocupação Do Solo Em Uma Cidade," In: 4 Semana Do Servidos E 5 Semana Academica, Uberlancia. Anais... Uberlancia. Disponível Em: <Http://Www.Seer.Ufu.Br/Index.Php/Caminhosdegeografia/Article/View/15812>

Pinos, J.; Vozenilek, V.; Pavlis, O. (2020), "Automatic Geodata Processing Methods For Real-World City Visualizations In Cities: Skylines," *Isprs International Journal Of Geo-Information*, v. 9, n. 1, p. 17.

Poplin, A. (2011), "Games and Serious Games In Urban Planning: Study Cases," In: *Computational Science and Its Applications—Iccsa 2011*, May, Spain. Anais... Spain, Disponível Em: <Https://Doi.Org/10.1007/978-3-642-21887-3>

Reinart, B.; Poplin, A. (2014), "Games in Urban Planning—A Comparative Study," *Society*, v. 2, pp. 49–56.

Robson, K. et al. (2015), "Is It All a Game? Understanding the Principles of Gamification," *Business Horizons*, v. 58, n. 4, pp. 411–420, Disponível Em: <http://dx.doi.org/10.1016/j.bushor.2015.03.006>

Sanches, Jose (2016) https://archive.curbed.com/2016/5/17/11672254/architecture-video-game-blockhood-jose-sanchez

Sanchez, Jose (2019). "Common'Hood, 2019," In: Gilles Retsin, Manuel Jimenez, Mollie Claypool and Vicente Soler (Eds.). *Robotic Building: Architecture in the Age of Automation.* München: DETAIL, pp. 104–106. https://doi.org/10.11129/9783955534257-022

Santaella, L.; Nesteriuk, S.; Fava, F. (2018), Gamificação Em Debate, 1st Ed. São Paulo: Blucher.

Santos, B. J. Dos et al. (2005a), Jogo Do Estatuto Da Cidade—Manual De Instruções—Santo Expedito, 2nd Ed. São Paulo: Instituto Pólis.

Santos, B. J. Dos et al. (2005b), Jogo Do Estatuto Da Cidade—Manual De Instruções—Rurópolis, 2nd Ed. São Paulo: Instituto Pólis.

Spitz, R. et al. (2018), "Gamification, Citizen Science, and Civic Technologies: In Search of the Common Good," *Strategic Design Research Journal*, v. 11, n. 3, pp. 263–273.

Thiel, S.-K. et al. (2016), "Playing (with) Democracy: A Review of Gamified Participation Approaches," *Jedem—Ejournal Of Edemocracy And Open Government*, v. 8, n. 3, pp. 32–60.

Thite, M. (2011), "Smart Cities: Implications of Urban Planning For Human Resource Development," *Human Resource Development International*, v. 14, n. 5, pp. 623–631.

Toppeta, D. (2010), "How Innovation and Ict the Smart City Vision: How Innovation and Ict Can Build Smart, Liveable, Sustainable Cities," *Think Report*, v. 5, pp. 1–9.

Valladares, L. (2000), "A Gênese Da Favela Carioca. A Produção Anterior Às Ciências Sociais," *Revista Brasileira De Ciências Sociais*, v. 15, n. 44, p. 05–34.

Vanolo, A. (2018), "Cities and the Politics of Gamification," *Cities*, v. 74, pp. 320–326.

Virtanen, J. P. et al. (2015), "Intelligent Open Data 3 d Maps In A Collaborative Virtual World," *Isprs International Journal Of Geo-Information*, v. 4, n. 2, pp. 837–857.

Washburn, D.; Sindhu, U. (2010), "Helping Cios Understand "Smart City" Initiatives," *Growth*, v. 17, pp. 1–17.

4

An IoT-Based Framework toward a Feasible Safe and Smart City Using Drone Surveillance

Nayyar Ali Usmani

Hanu Software Solutions, Noida, India

Tasneem Ahmed and Mohammad Faisal

Integral University, Lucknow, India

CONTENTS

4.1 Introduction

There has been an immense increase in urbanization across the globe over the past few years. Most of the worlds' population started moving toward the cities in 2009. With the boom in technology and growing urbanization, there is a need for smart solutions for the environment, governance, quality of living, and transportation, etc. This proposed the idea of "smart cities," and [1] in 1994 launched the concept of "smart city," and with the arrival of smart city schemes and funding from the European Unions (EU); the number of publications on the subject has risen significantly since 2010 [2]. A smart city policy, in general, seeks to use technologies to improve the quality of living in metropolitan areas, including enhancing environmental quality and providing more services to residents. The most important instruments for promoting smart city policies are ICT, and as a result, the terms "internet city" and "smart city" are often interchanged. Other terms are used as well, but the most common is "digital city." However, it is unclear if these two terms – smart and modern – mean the same thing or whether they refer to various communities, strategies, and technology [1]. The relationships and distinctions between digital cities and smart cities were shown by Dameri and Cocchia [1], and they can be used to guide municipal and central governments in developing strategies for urban innovation, as well as to assess and compare the outcomes for public policy and people in improving the quality of life in much bigger and more diverse cities.

A sophisticated and ubiquitous application was offered by the "Internet of Things (IoT)" for smart cities. The Internet of Things, which is a crucial enabler for smart cities, is made up of sensing sensors and actuators, as well as networking and network equipment. With the rapid growth in smart cities, there has been a drastic increment in IoT devices and applications. By 2021, according to an Ericsson Mobility study published in June 2016, there will be up to 16 billion IoT-enabled connected devices in the world [3]. This has given rise to many environmental as well as energy management issues because of which the smart city is considered non-sustainable. Some Indian companies like Aeris and Sterlite Technologies Ltd. are working to develop network infrastructure in smart cities to allow IoT technologies to link information and intelligence to devices that help to achieve the main objective of making smart cities. Also, various IoT applications such as smart buildings, smart hospitals, smart waste management, smart streetlights, smart parking, and secure cities may be very helpful to provide the quality of life by using the service efficiency for city residents (ibid.). More data on productivity, threats, and opportunities for infrastructure and environmental growth would be available as IoT becomes more embedded into the urban landscape and our everyday lives. IoT, out of all the emerging technology, has the potential to have the biggest impact on the urban growth of cities. When comparing the past to the current, most cities have created whole IoT environments, providing people with numerous benefits such as accessibility, convenience, healthcare, and increased productivity. The world is shifting, and we are sure that the next industrial revolution is just around the corner. New technologies will arise as a result of technical advancements, making urban communities more sustainable, convenient, and healthier for their residents [4]. By 2025, 66% of the world's population would be residing in cities as per the United Nations predictions, [5], posing significant problems in terms of traffic congestion, smart waste disposal, air quality, and human wellbeing [6]. Since the United Nations and European Union determined energy goals and climate standards for the futures, smart ways to address the challenges of urbanization are urgently needed [5, 7].

Assessing susceptibilities and refining plans are difficult tasks for smart cities. Given the fact that security is a costly task that necessitates a sufficient budget, processing in the public sector takes longer. Hence, Security concerns in smart cities are a serious problem that must be addressed for smart cities to continue to thrive. Security and privacy are key issues, particularly in "smart city" technology and systems, which have become increasingly critical for enhancing cities and raising living standards. It's important to keep in mind, though, that smart cities have provided enormous benefits to consumers, users are worried about the protection of their data, which is transmitted via wireless networks. Therefore, to extend stable support for moving data, particularly over wireless networks, a secure communication channel is required [8]. The main challenge of smart cities is to provide people with unrivaled quality service to improve their lives and provide a standard of living. As a consequence, it is important to consider anonymity, security procedures, and threats to protect citizens' privacy. Residents will constantly be using vulnerable Wifi networks to access their e-banking services, emails, and other internet resources; thus, they might expose themselves to the cracking man in the middle (MITM) attacks and the cyberattacks like denial of service (DoS). But from the other side, smart cities' infrastructure is vulnerable and susceptible to threats which would result in serious denial of service to smart cities and smart production industries, as well as hinder the distribution of other utilities [9]. Consequently, smart cities will find it costly to rebound from these kinds of cyberattacks. As a result, government agencies must take the appropriate measures to promote cybersecurity in order to build more protected smart cities. To build cyber-safe smart cities, government agencies should implement some security features like identification of vulnerabilities, development of regulations, utilization of two-factor authentication, encryption of confidential data, and education of the citizens [10]. A smart city can be protected with these basic security measures. On the other hand, these kinds of security services are not freely available. Cost is the main concern, while many cities and governments do not have a dedicated cybersecurity budget at this time. Hence, a smart city is just as intelligent as the people in charge of it. You're protecting yourself and your fellow people by safeguarding a city's smart infrastructure. Because IoT devices are used in smart applications in the city, hackers can unknowingly access the details of a citizen's day-to-day activities without the user or administrator's knowledge or can make devices unsafe. Therefore, a stacking classification-based approach has been presented in [11] to prevent cyberattacks in smart cities, which goes beyond scientific inputs and has economic and social consequences.

A plethora of applications is offered in IoT on a daily basis. However, the heterogeneity of this paradigm makes it very complex, and the implementation of this paradigm is also hindered due to a lack of clear and widely accepted models [12]. This heterogeneity has also led to different problems arising in distinct sectors such as safety in smart cities, scalability, communication issues, and a lot more. Drones have recently gained popularity in research and a variety of applications due to their versatility and ability to be used in a variety of applications, including protection, regulation, surveillance, and exploration of terrestrial areas that are otherwise difficult to access quickly. Drone data collection using wearable sensors in smart cities or at festivals is the most appealing use of drones. A drone can connect between smart objects on a field that has had to interconnect across vast distances and can transmit data to the desired target to make a definitive assessment and take action quickly [13]. Changes in how we interact with the large majority of objects linked to the Internet would benefit from the advancement of those networking technologies. Drones are expected to play an important role in future connected smart cities, given their growing popularity. They'll transport things and products, act as portable hotspots for internet connectivity and keep smart cities safe and secure [14]. Although drones can be beneficial to humanity, and they may be used by hostile organizations to carry out cyber and physical -attacks and pose a threat to society [15]. In the proposed work, we consider some of the major issues that need to be worked upon in the near future in the smart city. Also, this chapter addresses some measures taken to attain a better form of a "Sustainable smart city". Along with highlighting major issues and their cures, an IoT-based system has been proposed in this chapter for the betterment of society and achieving a more profound utilization of IoT applications in numerous sectors of the city.

This chapter is organized in multiple sections where Section II provides the area of challenges in smart city development, Section III consists of the measures taken in favor of the issues, Section IV consists of the proposed work and Section V includes results and conclusion.

4.2 Theoretical Background

4.2.1 Application of IoT-Enabled Smart Cities

IoT refers to the increasingly increasing number of digital devices, now in the billions, which can communicate and interact with one another over a global network and be monitored and managed centrally [16]. IoT has the ability to alleviate urbanization's pressures, provide new experiences for city people, and make daily life more enjoyable and safer. IoT-enabled smart cities can be used for a variety of applications, but five main applications are smart infrastructure, road traffic, smart parking, public transport, street lighting, and waste management, which are considered mainly for the development of the fully IoT-enabled smart city.

4.2.1.1 Smart Infrastructure

Cities must provide the environment for long-term growth: emerging innovations are becoming more critical, and buildings and urban infrastructure must be designed effectively and sustainably. Emissions of CO_2 gas will be kept to a minimum by investing in hybrid vehicles and self-driving cars. Smart cities make use of smart technology to build infrastructure that is both energy-efficient and environmentally sustainable. To conserve money, smart lighting can only switch on when someone steps by it, such as by setting brightness thresholds and monitoring everyday consumption [16].

4.2.1.2 Road Traffic

Smart cities make sure that people can move from point A to point B most securely and effectively as feasible. To achieve this goal, municipalities will use IoT architecture and smart traffic technology. To measure the number, position, and speed of vehicles, smart traffic solutions employ a range of devices as well as drivers' smartphones GPS data for location monitoring. Simultaneously, smart traffic lights linked to a cloud management network consent for the tracking of green light timings and the automated alteration of lights depending on current traffic conditions to avoid congestion. Smart traffic control solutions can also forecast where traffic will go based on historical data and take steps to avoid future congestion [17].

4.2.1.3 Smart Parking

Using GPS data from drivers' smartphones (or road-surface sensors located in the ground on parking spots), smart parking solutions determine whether parking spaces are occupied or open and produce a real-time parking map. Drivers get a message when the nearest parking space becomes open, and instead of wandering around aimlessly, they use their phone's tracker to find a parking space faster and more conveniently [17].

4.2.1.4 Public Transport

Data from IoT sensors may be used to uncover trends in how people use public transportation. Public transportation services may use this data to enhance the traveling environment, as well as increase protection and reliability. To do more complex analysis, smart public transit networks should combine multiple sources, such as ticket sales and traffic statistics [17].

4.2.1.5 Street Lighting

In smart cities that rely on the Internet of Things, streetlamp operation, and maintenance has been made simpler and more cost-effective. Sensors attached to streetlights and linked to a cloud management system aid in the adaptation of lighting schedules to lighting areas. To refine lighting schedules, smart lighting systems gather data on luminance, persons, and vehicle activity and combine it with historical and temporal data (for example, public transit schedules, special events, time of day and year). As a consequence, a smart lighting solution "tells" a streetlight to dim, brighten, turn on, or turn off the lights depending on the season [17].

4.2.1.6 Smart Waste Management

When it comes to emptying bins, the bulk of waste management companies work to set targets. This is not a safe solution because it results in the waste bins being used inefficiently and waste collection vehicles consuming excess fuel. Through monitoring waste levels and delivering path optimization and organizational analytics, Waste disposal schedules can be improved with IoT-enabled smart city technology. Each trash can is equipped with an instrument which collects information on the number of discarded items in the bin. When a sensor report reaches a confident level, it processed through the waste management solution and a message sent to the truck driver's smartphone app. Hence, the truck driver avoids emptying the half-containers and empties the full container [17].

4.3 Challenges In IoT-Enabled Smart Cities

Smart city development, within time, has come across a lot of challenges in different sectors causing a great concern worldwide. Though the smart city is technologically sound there is a need for sustainability that is also environmentally friendly. Some of the issues are listed below.

4.3.1 Sustainability

Digital or smart cities play a critical role in combating climate change, and the emerging innovations that are being implemented are seen as a vital factor in reducing greenhouse gas pollution while simultaneously improving city energy quality. Cities have been pushing for smart city targets rather than environmental priorities in recent years [18]. From the European Union's perspective, since its key goal is to reduce greenhouse gas emissions in urban cities through the implementation of advanced technologies, the smart city model promotes the principle of environmental sustainability [19]. Overall, a city that aspires to be a smart sustainable city must be appealing, sustainable, inclusive, and balanced for the people who live, work, and travel in the city [20]. The term "smart sustainable city" refers to a city that is enabled by ubiquitous computing and widespread use of advanced ICT, as well as different urban structures and realms and how they intricately interrelate and are organized, and allows the city to manage available capital securely, sustainably, and efficiently to improve societal and economic outcomes [21]. The growing popularity of the smart city idea, as well as the pressing need to address the problems it poses, has resulted in several private and public developments in technological creation and implementation.

4.3.2 Energy Management

Due to continuous increases in usage needs, there is an urgent need for energy conservation all over the world, and future generations are at risk from global warming and air pollution. Energy management is generally categorized into two types: energy harvesting operations and energy-efficient solutions. Scheduling planning, lightweight protocols, energy-efficient technologies, a cloud-based methodology, and a cognitive management system statistical for IoT-enabled smart cities provide modeling for energy consumption and low-power transceivers. IoT systems may use energy harvesting to get power from both ambient and dedicated RF sources. Energy recycling is used to extend the life of Internet of Things appliances [22]. Because of the continuous growth in consumer demands, there is a pressing need for energy management all over the world. Future generations face grave dangers from global warming and air pollution. This is due to higher emissions of fumes as energy consumption rises. According to Cisco, more than 50 billion IoT computers will be wired to the internet by 2020 [22]. Because of the proliferation of smartphones, there is a critical need to control resources for IoT devices so that digital cities can attain sustainability more efficiently.

4.3.3 Security and Safety

Smart cities are based on data and required sensors. Sensors are needed to hear, smell, see, taste, and feel on behalf of humans. This is what technology is all about: reducing human efforts. Thus, IoT applications are used in many areas, including knowing weather/climate changes, checking air quality, getting to know infrastructure integrity (e.g. if a bridge/building that has been constructed is safe). As a result, different steps are taken to ensure that residents in smart cities are healthy and secure. But there are a lot of sectors still left that need development in the IoT sector for ensuring a safe and secure environment to live in. There is not only a need for a smart city but also a safe and secure smart city to live in Hamid et al. [8].

4.4 Measures Taken for IoT-Enabled Smart Cities

As the proverb goes, "Necessity is the mother of invention," there were various measures were taken to overcome the challenges posed in smart city development. A holistic solution is required to better address the relationships between the different aspects of the city. A vast variety of evaluation mechanisms and frameworks have been built for the construction industry to aid government policymaking and make sure that the environment and transportation sectors are working toward sustainability targets [23]. Instead of single houses, the appraisal priorities have recently shifted to cities and neighborhoods, allowing for simultaneous evaluation of the urban environment, public transit, and utilities, among other things. [24]. Some of the steps taken are as follows:

4.4.1 The Livable City Concept

A city can be described as smart and sustainable "if its condition of production does not destroy over time the conditions of its reproduction" [25], according to the original concept of sustainable development [26].

At least two well-ranked media firms: Monocle's most livable city [27] ranking and International livings' Quality of Index [28] have promoted the "livable city concept." There is the development of various tools and indicators that measure the sustainability of the cities and lead toward a better living.

4.4.2 IoT-Based Safe City

Safe city projects are in vogue and city surveillance is the new buzzword for the security industry. City surveillance is being a growing concern nowadays and needs immediate attention to get a secure smart city. The residents are provided with safety and security in almost all the cities. In a safe city project, the command center acts as the brain [3]. Various IoT devices including ultrahigh resolution cameras, NVRs (video management solutions) network-attached storage solutions are brought into use to achieve better security within cities.

4.5 Model Development

City surveillance is one of the toughest tasks to handle, and with the growing rate of increase in criminal activities, it is becoming more difficult for the government bodies to keep a check on the safety of the citizens, particularly in metro cities. One of the main problems which lead to the trouble is the inaccessible or congested areas of the city with a dense population, as police patrolling is not easy and fully efficient in such areas. Other factors that make the whole city surveillance difficult is the dependency on the human resource which is an expensive affair (deployment of more staff in administration) and is not fully accurate. Here, IoT technology can be implemented by the means of UAVs (Unmanned Aerial Vehicles) otherwise called "Drones" along with a real-time crowd analysis and mood analysis sensor attached to the device. In this approach, every police station in the city will be provided with drones that are designed in a manner to cover a specific range of areas (e.g. covering 15–20 km of the area around the station it is deployed). Drones with surveillance cameras are automated to move within the demarcated regions but at the same time, their movement can also be controlled from their control centers (such as police stations) if and when required.

The people in the area are analyzed on their moods (by using the mood analysis sensor) and their identity is shown on the screen installed in the stations for live monitoring with the help of a real-time crowd analysis sensor. A database of information of all the residents in the region under

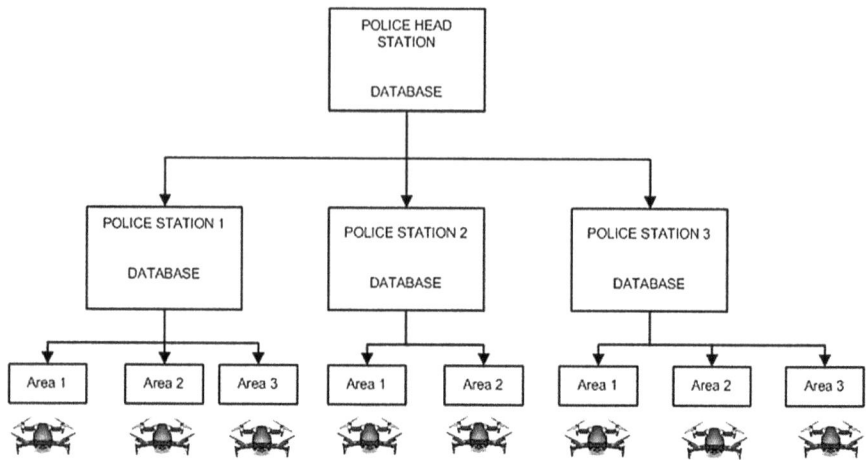

FIGURE 4.1
The framework of drone surveillance in smart city.

surveillance is stored in the stations and these databases from all the stations are connected. This enables the administration to keep an eye on every resident of the city, and also any unknown person found can be instantly detected and recognized. Any suspicious activity or violence is immediately spotted and analyzed, hence optimizing the surveillance and governing within the city. The live footage is recorded and maintained on a daily basis, and can be used as evidence of the crime activity against the criminal. The framework of drone surveillance is shown in Figure 4.1.

The proposed prototype will execute as follows:

a) The drone flies over the region allotted to it while the people's activities under the surveillance are monitored and recorded by the next-generation surveillance cameras.

b) The real-time crowd analysis sensor provides the details such as resident's name, age, address (as per the requirement), which is displayed on the screen as the crowd is monitored. Unknown identity is also highlighted and worked upon the databases linked to getting the information of the respective person.

c) The mood analysis sensor displays the mood of every person under observation on a scale of blue to red where blue shows calm, and red indicates anger while other mood states are indicated with different shades that lie within the range of it (e.g. purple indicates panic state).

d) The staff monitoring the drones inform the police officers as soon as they see any suspicious activity, and immediate actions can be taken.

e) The coverage by the drones is recorded and saved on a minute basis and a database is maintained.

As a result, the implementation of drone-based wireless networking will protect lives and habitats by assisting community security agencies in effectively responding to threats and managing emergencies. The smart city solution aims to increase residents' and tourists' quality of life by using IT technology and novel connectivity technologies by doing drone-based surveillance as shown in Figure 4.2.

A smart city is a prototype of productivity, sustainability, and easy access to a variety of resources. Smart tourism, smart climate, smart economy, smart governance, smart living, and smart mobility are all axes which can be used to determine a city's smartness. Though, the widespread usage of drones in future smart cities raises a number of technological and social issues and problems that must be tackled, like anonymity, cybersecurity, and public safety. Although drones can be beneficial to humanity, they may use by some of the hostile groups to carry out cyber- and physical attacks and can pose a threat to society.

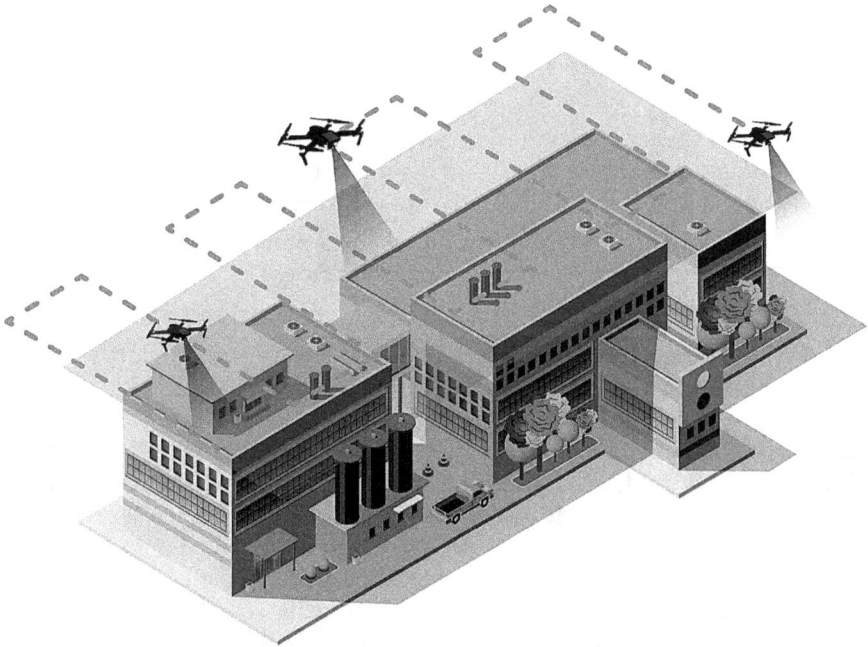

FIGURE 4.2
Drone surveillance in the smart city [29].

4.6 Results and Discussions

In terms of growth, transforming every city into a smart city is the latest global trend. To create a linked and resilient Smart City, which will use emerging technology like the Internet of Things and Cloud computing. This provides enough space for the use of UAVs to aid in the achievement of the aim. UAVs (Drone) may offer a variety of programs and resources to smart cities. Drones are being strongly considered as a defense enhancement tool. In the near future, there will be a significant increase in the use of drones for defense and surveillance. Smart police services will be the trend, alongside smart cities. The police services would be armed with cutting-edge technology, allowing them to quickly address challenging security issues using drones. Consider how motivated the police would be if a drone was deployed on their behalf in a dangerous or inaccessible location. Drones and the Internet of Things will work together to deter crime. They may be used to conduct inquiries successfully. A well-trained and well-connected swarm of drones on the lookout for criminals can be extremely successful. This will also affect how justice is delivered. We should anticipate fast sentences and, hopefully, lower crime rates.

The results achieved from the proposed work are as follows:

a) The safety of the city is ensured and achieved at a higher rate with accurate observations.

b) Helps in saving human resources as well as energy resources since the investment in a large number of patrolling cars and people deployed for the patrolling activities are minimized considerably.

c) Pre-hand safety measures can be taken to prevent future chaos by analyzing the moods and activities of the crowd.

d) Terrorist activities can be controlled and stopped on a broader scale.

e) Provisioning of the shreds of evidence of the incident against the criminal, thus, leading to better enforcement of the law.

This work commits to improving public safety in emerging cities by combining smart wearable devices and drone technology. By 2050, urbanization is projected to rise by up to 90%, and technology will be the best way to exploit current infrastructure to satisfy the demand. To deal with the looming conflict, smart city programs have been implemented all over the world. Drones and the smart city are a fantastic way to alleviate traffic congestion, smart resource and assets management, public safety, and security to any type of anonymous threat. As policing is an important service in a smart city provided by the state to its citizens, the safety of citizens, the maintenance of public order, the investigation of criminal cases, information security, etc.

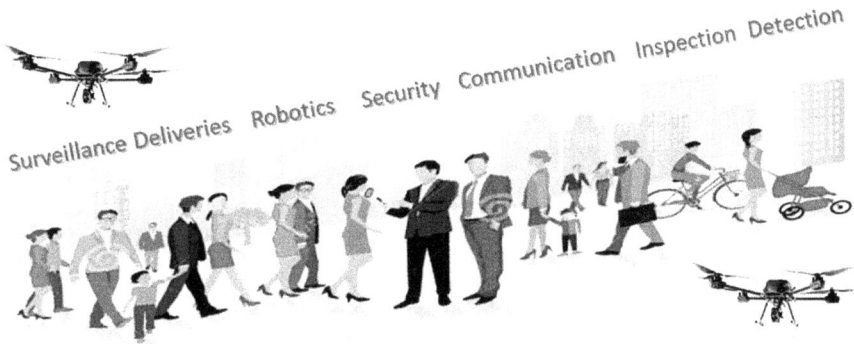

FIGURE 4.3
Detection of drone-based surveillance anomalous behavior in the smart city [30].

police-related services are provided; it also required harmonization with the smart city concepts listed above. The working of drone-based surveillance of smart city is shown in Figure 4.3.

Unlike other citizen services and their distribution to the public, however, policing in a smart ecosystem faces some problems, all of which derive from the innovations that are so central to smart city processes and networks.

In a smart city, the aforementioned policing template would necessitate a mix of policing infrastructure for data capture and collation, data aggregation, data analytics, performance data distribution, response configuration, and input capture/collection, among other things. An R&D setup, consisting of data scientists, academic fellows, and an incubation center, must be in place to test and implement emerging technology for police positions. As a result, the following are some of the major components of a smart city policing set up like CCTV cameras and ground sensors (RFID, gunshot, etc.) are used in a citywide surveillance and tracking system, installation of enterprise-level intelligent database integration for all processing smart city systems and databases for data sharing and smart imaging links, a database analytics engine that provides real-time inputs that process multiple databases and direct data from multiple channels, usage of machine learning, deep learning, and AI to make data analytics and decision algorithms self-improving, central management, control, and coordination center to bring everything components under one roof for better functional coordination, especially in response systems, and a smart data network that has been overstated to transmit (input/output) information received and transmitted in connection with law enforcement processes within a smart city is required. The dedicated network is of particular importance in the work of police officers in ensuring the security of confidential information of detainees and ensuring the stability of the system, emergency response functions, and maintaining public order.

4.7 Conclusions

Drone incorporation in smart cities is rapidly approaching. All personal and commercial drone operations may have significant advantages for cities, and rather than completely banning them, cities should consider them. In this chapter, the major issues arising in smart city development were discussed and analyzed. Also, solutions being proposed or available for the implementation of the IoT technology in smart cities to dissolve the issues were discussed. The mentioned technologies and models that are similar to them have been already implemented on an experimental basis and people in this industry are showing active participation in the production of these devices to take advantage of the technologies for enabling the applications of interests highlighted in Section III, whereas Section IV introduces the idea with which we can effectively achieve a safer smart city. The work includes applying the concepts of IoT for security features in a smart city and analyzing its efficiency. The desire of various stakeholders to rethink how this new technology tool will better represent the diverse interests of people living in cities will determine the potential success of commercial drones. In the future, the proposed framework will be helpful to build feasible, safe, and smart cities in India.

References

1. Dameri, R. P., and A. Cocchia. 2013. 'Smart City and Digital City: Twenty Years of Terminology Evolution'. *X Conference of the Italian Chapter of AIS, ITAIS 2013*, 1–8.
2. Jucevičius, R., I. Patašienė, and M. Patašius. 2021. 'Digital Dimension of Smart City: Critical Analysis'. *Procedia-Social and Behavioral Sciences* 156: 146–50. doi:10.1016/j.sbspro.2014.11.137.
3. Bhasin, S., T. Choudhury, S. C. Gupta, and P. Kumar. 2017. 'Smart City Implementation Model Based on IoT'. *International Conference on Big Data Analytics and Computational Intelligence (ICBDAC)*, 211–16. Chirala, Andhra Pradesh, India.
4. 'How IoT Is Enhancing the Development of Smart Cities'. 2021. Accessed March 18. https://lvivity.com/iot-and-smart-cities.
5. Sataloff, Robert T., Michael M. Johns, and Karen M. Kost. n.d. *World Urbanization Prospects. The 2014 Revision*.
6. Marchal, Virginie, Rob Dellink, Detlef Van Vuuren, Christa Clapp, Jean Château, Eliza Lanzi, Bertrand Magné, and Jasper Van Vliet. 2011. 'OECD Environmental Outlook to 2050 Chapter 3: Climate Change'. November: 90. doi:10.1787/9789264122246-en.

7. 'Climate Action | European Commission'. 2021. Accessed March 14. https://ec.europa.eu/info/topics/climate-action_en

8. Hamid, Bushra, N. Z. Jhanjhi, Mamoona Humayun, Azeem Khan, and Ahmed Alsayat. 2019. 'Cyber Security Issues and Challenges for Smart Cities: A Survey'. *MACS 2019 – 13th International Conference on Mathematics, Actuarial Science, Computer Science and Statistics, Proceedings, IEEE*, Karachi, Pakistan. doi:10.1109/MACS48846.2019.9024768.

9. Khatoun, Rida, and Sherali Zeadally. 2017. 'Cybersecurity and Privacy Solutions in Smart Cities'. *IEEE Communications Magazine* 55(3): 51–9. doi:10.1109/MCOM.2017.1600297CM.

10. '5 Steps to Protect Smart Cities from Cybersecurity Threats'. 2021. Accessed March 21. https://www.allerin.com/blog/5-steps-to-protect-smart-cities-from-cybersecurity-threats

11. Rashid, Md Mamunur, Joarder Kamruzzaman, Mohammad Mehedi Hassan, Tasadduq Imam, and Steven Gordon. 2020. 'Cyberattacks Detection in IoT-Based Smart City Applications Using Machine Learning Techniques'. *International Journal of Environmental Research and Public Health* 17(24): 1–21. doi:10.3390/ijerph17249347.

12. Laya, Andres, Vlad-Ioan Bratu, and Jan Markendahl. 2013. 'Who Is Investing in Machine-to-Machine Communications?' *24th European Regional ITS Conference,* Florence, Italy, 20–23 October 2013.

13. Alsamhi, Saeed H., Ou Ma, M. Samar Ansari, and Sachin Kumar Gupta. 2019. 'Collaboration of Drone and Internet of Public Safety Things in Smart Cities: An Overview of Qos and Network Performance Optimization'. *Drones* 3(1): 1–18. doi:10.3390/drones3010013.

14. Khan, Muhammad Asghar, Engr Alamgir Safi, Inam Ullah Khan, and Bilal Ahmed Alvi. 2018. 'Drones for Good in Smart Cities: A Review'. *International Conference on Electrical, Electronics, Computers, Communication, Mechanical and Computing (EECCMC)*, 1–6.

15. Vattapparamban, Edwin, Ismail Güvenç, Ali I. Yurekli, Kemal Akkaya, and Selçuk Uluağaç. 2021. *Drones for Smart Cities: Issues in Cybersecurity, Privacy, and Public Safety*. Ieeexplore.Ieee.Org. Accessed March 25. https://busybox.net/about.html

16. 'What Is the Role of IoT in Smart Cities?' 2021. Accessed March 24. https://www.finextra.com/blogposting/17931/what-is-the-role-of-iot-in-smart-cities

17. 'IoT for Smart Cities: Use Cases and Implementation Strategies'. 2021. Accessed March 24. https://www.scnsoft.com/blog/iot-for-smart-city-use-cases-approaches-outcomes

18. Marsal-Llacuna, M. L., J. Colomer-Llinàs, and J. Meléndez-Frigola. 2015. 'Lessons in Urban Monitoring Taken from Sustainable and Livable Cities to Better Address the Smart Cities Initiative'. *Technological Forecasting and Social Change* 90: 611–22. Accessed March 14. https://www.sciencedirect.com/science/article/pii/S0040162514000456

19. Bg, E. L. E. N. 2011. *Градовете На Cities of Тоv Tomorrow Πόλεις Αύριο Бъдещето.* doi:10.2776/41803.

20. Belli, Laura, Antonio Cilfone, Luca Davoli, Gianluigi Ferrari, Paolo Adorni, Francesco Di Nocera, Alessandro Dall'Olio, Cristina Pellegrini, Marco Mordacci, and Enzo Bertolotti. 2020. 'IoT-Enabled Smart Sustainable Cities: Challenges and Approaches'. *Smart Cities* 3(3): 1039–71. doi:10.3390/smartcities3030052.

21. Bibri, Simon Elias. 2018. 'The IoT for Smart Sustainable Cities of the Future: An Analytical Framework for Sensor-Based Big Data Applications for Environmental Sustainability'. *Sustainable Cities and Society* 38: 230–53. doi:10.1016/j.scs.2017.12.034.

22. Ejaz, W., M. Naeem, A. Shahid, A. Anpalagan, and M. Jo. 2017. 'Efficient Energy Management for the Internet of Things in Smart Cities'. *IEEE Communications Magazine* 55(1): 84–91 doi:10.1109/MCOM.2017.1600218CM.

23. Ahvenniemi, Hannele, Aapo Huovila, Isabel Pinto-Seppä, and Miimu Airaksinen. 2017. 'What Are the Differences between Sustainable and Smart Cities?' *Cities* 60: 234–45. doi:10.1016/j.cities.2016.09.009.

24. Haapio, Appu. 2012. 'Towards Sustainable Urban Communities'. *Environmental Impact Assessment Review* 32(1): 165–9. doi:10.1016/j.eiar.2011.08.002.

25. Castells, Manuel. 2000. 'Urban Sustainability in the Information Age'. *City* 4(1): 118–22. doi:10.1080/713656995.

26. Keeble, Brian R. 1988. 'The Brundtland Report: "Our Common Future"'. *Medicine and War* 4(1): 17–25. doi:10.1080/07488008808408783.

27. 'Quality of Life Survey III (2013/14) | GCRO'. 2021. Accessed March 14. https://www.gcro.ac.za/research/project/detail/quality-of-life-survey-iii-2013/

28. 'Best Places to Live in the World in 2021 – International Living'. 2021. Accessed March 14. https://internationalliving.com/world-rankings/.

29. 'Drone Automation Solution for Security & Surveillance – FlytBase'. 2021. Accessed March 25. https://flytbase.com/drone-security-solution/.

30. 'Drones and The Smart City'. 2021. Accessed March 25. https://www.airborne-drones.co/smart-city/

5

An End-to-End Framework for Dynamic
Crime Profiling of Places

Shailendra Kumar Gupta, Shreyanshu Shekhar, Neeraj Goel, and Mukesh Saini

Indian Institute of Technology Ropar, Punjab, India

CONTENTS

5.1 Introduction 114
    5.1.1 Background and Related Work 114
    5.1.2 Experimental Setup and Dataset 116
    5.1.3 Methodology 116
    5.1.4 Chapter Organization.. 117
5.2 Preprocessing... 117
    5.2.1 Crime Article Detection 118
        5.2.1.1 Baseline Method: Crime Word-Based
                Classification.. 118
        5.2.1.2 Improved Method: Ambiguity Score Based
                Classification.. 119
    5.2.2 Duplicate Article Detection 120
5.3 Building Crime Database... 122
    5.3.1 Entity Extraction 122
    5.3.2 Classification of Entities as Location or Non-Location 123
    5.3.3 Classification of Extracted Locations as Crime
        or Non-Crime Location 124
5.4 Crime Score Calculation 124
    5.4.1 Crime Class... 124
    5.4.2 Marginal Crime Score.. 125
    5.4.3 Final Crime Score 126
5.5 Auxiliary Database and APIs 127
5.6 Evaluation 127
5.7 Conclusion 130
Acknowledgments 131
References... 131

5.1 Introduction

The information on the probability of meeting a crime at a place is essential for travelers, investors, and private companies. Police and a few government agencies have this information in bits and pieces [1]. However, it is generally not available in the public domain. Besides, the information is mostly static, and the changes are incorporated manually much later after the crime events have occurred. Consequently, to make the safety information easily available to the public, there have been a number of safety apps [2–5]. Yet, these are not helpful, as most of them depend on users' feedback/rating about a place, which may not be available in a timely manner [3]. In addition, many of us prefer not to give feedback to such apps. This chapter proposes a framework to measure and dynamically calculate the crime score of a given location. The core idea of this approach is to exploit information present in the news articles published in different newspapers on the Internet. It extracts the date, place, and type of crime from each article, and models an aggregated crime score of a location. The aggregated score of a location depends on the number of crimes, their severity, and date. The resultant score is dynamic; that is, it changes with time. In this way, this chapter presents a real-time crime rate/score of different locations. As a case study, it focuses on some major cities of India. This knowledge will help travelers to plan their trips accordingly. Also, the crime profile can be used by the routing apps to find the safest path. This can also provide sufficient information to police forces so that they can accordingly disperse themselves in crime-prone areas. Several challenges need to be addressed to extract crime information from news articles. First, it has to separate crime-related news articles from other types. Once it has a crime article, it needs to identify the location and time of the crime. Extracting data, place, and type of crime from news articles is challenging because the information in news articles is context-specific and unstructured [6]. Further, multiple meanings of the same word, multiple news articles on the same incident, and ambiguity between places, names, and organizations, make the extraction more challenging. This chapter provides solutions to these individual problems and finally integrates these solutions to build an end-to-end framework to calculate crime density. This framework requires two databases, the first one is to store all the crawled articles (raw data) and the second one is to store all the processed data like locations, geo-coordinates, crime score, crime type, etc. It analyzes over 345,448 news articles published by six daily newspapers on the Internet over approximately two years to demonstrate the efficacy of the proposed solution, and the results are promising.

5.1.1 Background and Related Work

A lot of research attempts have been made in the past to extract relevant information from online available news articles. Arulanandam et al. [7] propose a method to detect crime location in theft-related news articles. After detecting the location using NER, the authors use conditional random fields (CRF) to classify whether the detected location is related to the crime or not. The method is evaluated only on 70 theft-related articles in New Zealand. Jayaweera et al. [8] analyze news articles published in Sri Lanka to extract crime information. The articles are classified as a crime or non-crime using an SVM classifier trained on TF-IDF representation of the articles. Location is extracted using NER. To find duplicate articles, the authors calculate the SimHash value of the entities extracted from the article. The overall goal of the work is to build a database handler that supports viewing crime statistics on a map.

The crime information is also available with various government organizations. Tayal et al. [9] use KNN based data mining technique to cluster data according to crime type and detect criminal attributes. Although the information received from government organizations is more reliable, it is not as up to date as the information extracted from the news articles. Joshi et al. [10] take a supervised approach to detect different types of crimes like thefts, homicide, and various drug offenses in a crime dataset of the North Wales region, Australia. Yadav et al. [11] also implement a system to derive state-wise crime statistics by analyzing government records of 14 years.

Researchers have explored crime analysis in various other safety-related applications as well. Sharma et al. [12] transform an article into a lower-dimensional space and apply KNN to detect crime-related articles. KNN requires a large number of labeled documents, also, it is not scalable with the number of articles. The authors employ NER based methods to detect locations in the document and calculate crime scores as the crime count. This crime score is used to find a safe path between two points. Similarly, Goel et al. [3] use the crowd-sourced safety score to find safe routes. Hassan et al. [13] find crime articles using an SVM classifier and then cluster these articles into groups according to the crime. Documents in each cluster are used to build a crime story. NER is used to find the location term. The crime locations are found by classifying the whole sentence as crime or no-crime, using an SVM classifier.

Further, Rollo et al. in [14] focus on crime event visualization using news articles. They divided their whole process into seven phases with each focusing on small tasks. Their method focused on 11 types of crimes and used around 13,000 news articles. They only considered one location per article hence not counting the effect of a single article on several locations which is a commonly observed scenario for news articles. They applied their approach in the Modena province(Italy). Mukherjee et al. in [15], focus on developing a system that uses online documents for estimation and faster visualization of real-time district-wise crime scenarios of a state. Paper [16, 17] uses the count of crime articles per location to estimate the crime rates, hence losing the crime severity information in the final results.

Table 5.1 shows a summary of the related work. It shows that there has been only piecemeal work on crime density calculation. This chapter integrates

TABLE 5.1

Comparison with the Related Work

Work	Focus	Crime Article	Crime Type	Duplicate Detection	Location Detection	Location Classification	Crime Score
[7]	Theft detection	No	No	No	Yes	No	No
[8]	Database handler	Yes	No	Yes	Yes	No	No
[9]	Data mining	No	No	No	No	No	No
[12]	Safe navigation	Yes	No	No	No	No	Yes
[13]	Crime story	Yes	No	Yes	Yes	Yes	No
[10]	Crime classification	No	Yes	No	No	No	No
[18]	Crime mapping and prediction	Yes	No	Yes	Yes	No	No
[14]	Crime event visualization	Yes	Yes	Yes	Yes	No	No
[15]	Visualization of real-time crime scenarios	Yes	No	No	Yes	No	No
[16]	Crime register	Yes	No	No	Yes	No	No
[17]	Crime analysis	Yes	Yes	Yes	Yes	No	No
This work	Crime density	Yes	Yes	Yes	Yes	Yes	Yes

and customizes these works to develop an end-to-end framework, which is currently running. In terms of individual components, it has improved document classification and location extraction; in terms of novelty, this is the first to consider spatio-temporal fusion to calculate crime score.

5.1.2 Experimental Setup and Dataset

It uses six popular English news websites in India, given in Table 5.2. The crawler has been running daily to collect new articles from these websites since May 2018. The title, body, date, time, and URL of each article, are stored in a database. The database has around 345,448 news articles (both crime and non-crime) collected over 2 years. The break-up of these articles, used for analysis, is shown in Table 5.2. The collected data was unlabeled. For verification and accuracy studies, a part of data was labeled. Around 50% of the labeled articles were crime related. The remaining unlabeled data are used for verification of the framework. More details discussed in the Evaluation section.

5.1.3 Methodology

The objective of this chapter is to create an automated framework that can estimate a reliable crime score for all possible locations by extracting information from online newspaper articles.

TABLE 5.2

Popular News Websites

Source	#Articles	Time Duration
TOI – Times of India	123,623	Sep 2018–Nov 2019
Hindustan Times	35,562	Jan 2018–Nov 2019
India Today	33,960	Jan 2018–Nov 2019
The Hindu	85,035	Dec 2018–Nov 2019
News18	36,232	Dec 2018–Nov 2019
NDTV	30,036	March 2018–Nov 2019

The proposed method consists of three main steps. The first step is preprocessing. In this step, it crawls a news article from the Internet, determines
whether or not it is a crime-related article, and checks for existing duplicate
articles. From each original crime article, in the second step, it extracts location and crime type information, and builds a crime database. Finally, in the
third step, it calculates the final crime score using the spatio-temporal information. Finally, it creates a heatmap of crime score to visualize the crime density across India. Also, an interactive web interface is designed using which
the real-time crime score of a location can be viewed.

5.1.4 Chapter Organization

The preprocessing step is discussed in Section 5.2. The preprocessed articles
are analyzed to build an article database in Section 5.3. Section 5.4 calculates the overall crime score in terms of the article entities in the database.
The proposed work is evaluated in Section 5.5, and conclusions are stated in
Section 5.6.

It performs two tasks. First, it detects whether the crawled article is crime
related or not. If it is a crime-related article, it compares the article further
with other existing articles in the crime dataset to find whether it is an original article or a follow-up article.

5.2 Preprocessing

This starts with the collection of news articles from online news websites.
Each article is classified as crime or non-crime based on the body and title.
There can be multiple articles related to the same crime. Hence, crime articles
are further classified as original or duplicate. Only original articles are added
to the crime database. Duplicate articles are rejected. The overview of the
preprocessing steps is given in Figure 5.1.

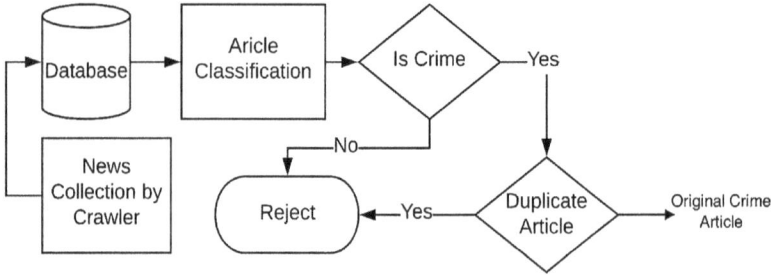

FIGURE 5.1
The preprocessing steps.

5.2.1 Crime Article Detection

Intuitively, if crime-related words appear in an article, they can be classified as a crime article. However, a word can have multiple meanings depending on the context. For example, one can hit another person to commit a crime, or hit the floor to dance. Due to such context-dependent multiple meanings of the same word, this process is very difficult. First, the baseline method is explained, and then an improved version being used in the system.

5.2.1.1 Baseline Method: Crime Word-Based Classification

In the baseline method, the crime score is calculated based mainly on the presence of the crime words. Let w^c be a crime word from the initial crime word list. Not all crimes are equally severe; murder is more severe than theft [19]. These severity scores are obtained through crowdsourcing (Section 5.5). Let $\psi(w^c)$ be a function that maps a crime word (w^c) to the corresponding severity score. The list of initial crime words is relatively small, as the severity had to be crowd-sourced. To enrich the crime word list, it obtains synonyms of all crime words based on the similarity score. Let w_j^c be jth synonym of w^c. The similarities between words are calculated using the method by Wu-Palmer [20]. Let $S(w^1, w^2)$ be a function that returns the similarity between w^1 and w^2. The severity score of a synonym word is calculated as follows:

$$\psi\left(w_{ij}^c\right) = \psi\left(w_i^c\right) * S\left(w_i^c, w_{ij}^c\right)$$

Once the severity scores are calculated for all the crime-related words (initial words and their synonyms), it looks for these words in article A and calculate the article severity score as follows:

$$\psi(A) = \sum_{\forall w \in A} \psi(w)$$

TABLE 5.3

Results of Crime Classification Models

Methods	Naive	Title Only	Body Only	Improved
Accuracy	0.749	0.810	0.669	0.845
Precision	0.652	0.822	0.584	0.794
Recall	0.982	0.752	0.994	0.942

To classify an article as crime or no-crime, a suitable threshold is obtained through experiments. For an experimentally selected optimal threshold, it runs this algorithm on the labeled dataset, and the results are shown in Table 5.3. The results show that the baseline algorithm has a very high recall, but accuracy and precision are low. By analyzing the results it is found that the low accuracy was due to the ambiguity of words since one word can be used for several meanings. Further, it is observed that treating the title as a separate entity could enhance accuracy. Most related works suffer from this limitation [8].

5.2.1.2 Improved Method: Ambiguity Score Based Classification

To improve the baseline methods, it makes two important changes. First, it considers the title and body as two separate entities of a news article. Second, it introduces an ambiguity score to mitigate the effect of contextual ambiguity. The title of a news article is an abstract description of the complete article. If there is a crime word present in the title, it classifies it as a crime article. Results in Table 5.3 show that if it uses only the title for classification, accuracy is good, but the recall is not effective. Similarly, if it classifies based on the article's body alone, then recall is 99.6%, but precision and accuracy suffer. Therefore, it treats title and body as separate features, and the overall article score is determined as a combination of both.

Contextual ambiguity is a property of a word [21]. Therefore, it counts the number of contexts in which a word can be used. Its ambiguity score is calculated as the number of crime contexts possible divided by the total number of contexts for that word. For example, "hit" can be used in two contexts, "Hit the road," or "Hit a man." Thus, the ambiguity score for "hit" is $\psi(\text{hit})/2$. Based on the ambiguity score, it calculates the overall score of the article. In this improved approach, first classifies based on the title and body of the article. If both title and body give the same classification result, the resulting classification is given as output. However, if there is a contradiction in classification, then it uses the ambiguity score of crime words and treats both body and title as a single entity and then classifies the news article. The idea is that at least one of the classifiers should be over-confident about the class

TABLE 5.4

Confusion Matrix of the Duplicate Detection Method

	Predicted	
Actual	Duplicate	Not Duplicate
Duplicate	188	15
Not duplicate	90	537

of the article in case of contradiction. The results (Table 5.4) show significant improvement in accuracy and precision with a slight degradation in recall in comparison to all other approaches.

5.2.2 Duplicate Article Detection

It is common for a single incident to be covered by more than one news website. On a single website new articles are added over time to give updates on the incident. Such duplicate articles may inflate the crime score for a location if each is considered a new crime. To avoid this, it needs to detect and discard all duplicate articles. If an article is a duplicate, similar words will appear in both the original and duplicate articles. Hence, it calculates the similarity between articles using Bag-of-Words (BoW) [22] representation of the news articles using TF-IDF metric [23] and applies a threshold on the similarity score to find duplicates. However, this method performs poorly because it decides the importance of words based on frequency, whereas in this case the importance of words does not depend on frequency. Instead, article entities are more important. Main article entities are very likely to be the same in duplicate articles, e.g. both would have the same crime type, persons (victim, accused, and/or investigators) and crime location.

Based on the above observations, it also matches the entities to determine the similarity of two news articles using SimHash technique [8]. Hence, it has two scores, one based on the TF-IDF vectors, and another based on entities. The final score is the weighted sum of both the scores.

There is a limitation of the entity score. The entities are only available in large documents. Therefore, the entity score is used only when both the documents are large. If even one document is smaller than a threshold, it uses the TF-IDF based score. Also, TF-IDF uses

normalization which avoids biasing toward shorter or longer documents. The algorithm is tested by creating a new dataset. It enumerated two sets of news articles, A and B. Both sets contain 20 news articles. All articles of Set A are based on the same story, whereas all articles of Set B are from different stories. Finally, combining all 40 articles it randomly generated 770 pairs,

TABLE 5.5

Results of Duplicate Detection by Fixing the Time Span for Comparison as X Days, Where X Is 15, 30, 60, and 90 Days Respectively

15 Days		30 Days		60 Days		90 Days	
ID	Dup ID	ID	Dup ID	ID	Dup ID	ID	Dup ID
1001	None	1001	28402	1001	28402	1001	28402
1002	26961	1002	26961	1002	26961	1002	26961
1013	12948	1013	12948	1013	12948	1013	12948
1021	6710	1021	6710	1021	6710	1021	6710
1031	6663	1031	6663	1031	6663	1031	6663
1035	2327	1035	2327	1035	2327	1035	2327
1050	9503	1050	9503	1050	9503	1050	9503
1062	None	1062	None	1062	8586	1062	8586
1078	None	1078	None	1078	7852	1078	7852
1088	None	1088	7852	1088	7852	1088	7852

ID refers to Article ID and Dup ID refers to respective Duplicate Article ID.

which are used to produce results of this algorithm as shown in Table 5.4. It is able to detect duplicates with an accuracy of 94.14%.

For the framework to detect whether an incoming news article is duplicate or not, it needs to compare this news article with all other news articles present in the database. Initially, this cost of comparison will be very low due to fewer articles in the database, however, as the database increases, the number of comparisons will also shoot up, which will slow down the system. To address this problem, it uses article information to prune out some unnecessary comparisons. First, it fixes a period for comparison, i.e. articles published today are compared only with articles published in the last N months. The value of N is flexible and can be changed as per requirement. Second, there is no need to compare articles from two different locations. Hence, the overall complexity of duplicate detection will be reduced to the number of articles published at the given location for the past N months. Table 5.5 shows duplicate detection results for ten articles. The result is calculated for four different periods. Table 5.6 shows the time taken by the system to execute the duplicate detection algorithm for 50 articles. The first column shows the time required for comparing articles with all articles published in the past 30 days from the current article's date. The second column shows the time required for comparing articles with only those articles published at the same location as the current article in the past 30 days. It can be observed that it significantly saves time by restricting N and using the location information.

TABLE 5.6

Time Taken by the System to Run a Duplicate Detection Algorithm over 50 Articles

Days	Without Location (mins)	With Location (mins)
15	67.11	21.74
30	104.99	28.69
60	146.57	37.18
90	171.50	44.87

Location means comparing only those articles which have the same crime location. Days indicates that the current article will be compared to articles that are published within X days before the current article.

5.3 Building Crime Database

The overview of building the crime database process is given in Figure 5.2. Location, crime type, date, and named entities of each article are extracted and stored in the database. Extracting location information from an article is very tricky. In many cases, location names match precisely with a person's or organization's name. To address this challenge, the location extraction task is divided into three different parts: (1) named entity extraction, (2) identifying potential location entities, and (3) classifying these potential locations as crime or non-crime locations. The details of these three steps are given in the following subsections.

5.3.1 Entity Extraction

The first step is to extract all named entities in the article with tags ORGANIZATION, LOCATION/GPE, or PERSON. This chapter uses NLTK's NER [24] and Stanford NER tagger [25] for named entity tagging. NLTK's NER tagger is a supervised Maximum Entropy classifier trained on Automatic Content Extraction (ACE) data. Stanford NER tagger is computationally more expensive than NLTK's NER tagger as it uses the advanced

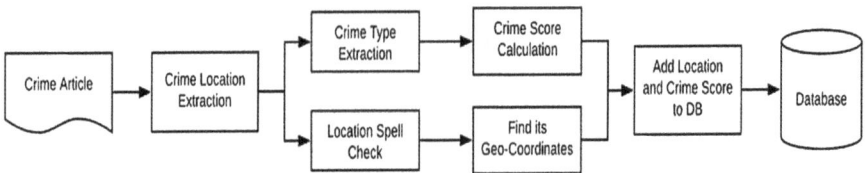

FIGURE 5.2
Important steps to build a crime database from the original crime articles.

TABLE 5.7

Accuracy Improvement Results for Location Separation from All Entities by Performing the Check, Presence of Common Words in Entities

Method	Without Check (%)	With Check (%)
NLTK	52.15	63.21
Stanford Tagger	78.96	82.77

statistical algorithms, Conditional Random Field (CRF) model. It is observed that, in both of these taggers, some names were classified as locations and vice versa. In the next subsection, the chapter discusses an algorithm to separate names and locations.

5.3.2 Classification of Entities as Location or Non-Location

Here is one example in which both NLTK's NER tagger and Stanford NER tagger go wrong: "A man found murdered in Anand Vihar, Delhi." In this sentence, NLTK's NER predicts Anand Vihar as location, whereas it predicts Delhi as a person entity. On the other hand, Stanford NER Tagger predicts Anand Vihar as a person entity and Delhi as a location entity. To mitigate such ambiguities, an extra country-specific step is added. If you go through the location names in India, you can observe some commonly used words in location names such as Vihar, Nagar, Pradesh, Chowk. Using this knowledge, a list of words, commonly used in Indian location names, is created and referred to as "common words," and a check is performed that ensures no person entity contains any tags from this list. It is relatively less frequent that a person's name contains such tags. A tagged set of news articles is used to further analyze the efficacy of this proposed algorithm for both NLTK and Stanford NER Tagger. There were 577 locations in the tagged set of news articles. After adding this check along with NLTK's NER tagger for separating location entities from all other tagged entities an increase of 11.06% in the accuracy of NLTK's NER tagger and a 3.81% increase in the accuracy of Stanford NER Tagger is observed as shown in Table 5.7.

Using the ideas from the above paragraph, an algorithm is developed to classify an entity as location or non-location. First, entities are extracted with their respective tags using NLTK or Stanford tagger. Next, all the entities containing any word from the common word list are classified as locations along with the entities which are marked as LOCATION by the tagger. Finally, this combined list is returned as the set of locations from a given article. Although this solution is specific to India, changing common words according to the applicable country is sufficient for the algorithm to work for any country.

5.3.3 Classification of Extracted Locations as Crime or Non-Crime Location

The previous step only gives us a set of locations. Not all locations found in a news article are related to the crime. The following observations made from news articles helped to identify potential crime locations: (1) distance of location word from the beginning of the article; and (2) distance of location word from the crime word. It is observed that in most cases, the writer mentions the name of the crime location at the start of the article. For example, sentences "SadarRaikot police on Wednesday booked two persons for the murder of an electrician after a fight over the phone in village Brahampura late on Tuesday evening." and "A 36-year-old man was found murdered in Anand Vihar, Delhi." are the first sentences of two news articles. Similarly, it can be observed that the crime locations are mentioned closer to crime words. For example, in the sentence "Ramesh, a resident of Sultanpura, was arrested in the case of murder in Anand Vihar, Delhi." It can be observed that the location "Anand Vihar," where the victim was murdered, is closer to crime word "murder" compared to the location "Sultanpura", which is the place of residence of accused.

The algorithm that is used to classify locations as crime or non-crime takes a set of locations, returned from the previous algorithm, and article text as input and returns a list of crime locations. First, a SentDistScore is assigned to each sentence based on its distance from the beginning of the article. Next, each location entity gets a CrimeWordDist score, the sum of the distance of all crime words from the location entity in a sentence. Here only those sentences are considered which contain that location entity. Finally, a total score is assigned to each location which is calculated as the sum of reciprocal of SentDistScore and reciprocal of CrimeWordDist. Finally, a threshold is decided empirically for separating crime locations from non-crime locations.

5.4 Crime Score Calculation

After the processing of an article is done, all the information of the article is updated into the DB. In this way, the DB will be ready with spatio-temporal information of crime-related incidents. The goal is to fuse this extracted information and obtain an aggregated crime score. This section of the chapter explains how the extracted information is used to calculate the final crime score of a location which reveals the crime severity of that location.

5.4.1 Crime Class

Unlike many other proposed works that consider the severity of each crime type equally, the methods discussed in this chapter have divided the crimes into ten distinct crime classes and determined their severity as discussed

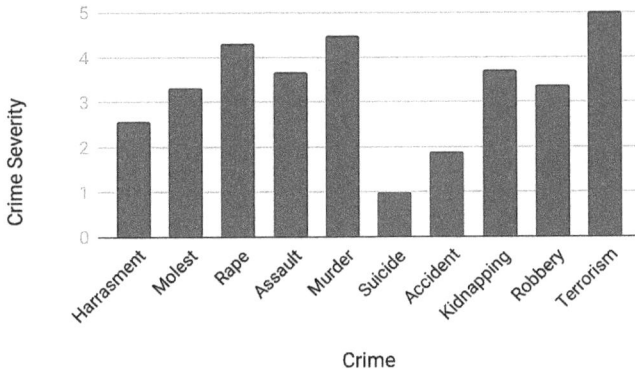

FIGURE 5.3
Crime severity score.

further. These classes are shown as X-axis labels in Figure 5.3. Each crime class may consist of different sub-crimes. For example, 'robbery' includes 'theft,' 'snatching,' and 'stealing.' Also, one word can be associated with more than one crime class. For example, death is associated with murder, suicide, accident, and terrorism. Intuitively, each class has a different crime severity. For example, murder is more severe than robbery, so its severity score should be higher. However, the judgment of the severity score is subjective. It varies from person to person. Therefore, a survey was conducted to record users' opinions about the severity of different crimes. Users were asked how unsafe they would feel, on a scale of 1 to 10 [1 – very safe, and 10 – very unsafe], if a particular crime out of these 10 (crime class) has occurred at a place they are visiting. The response to the survey was post-processed and is normalized to a scale of 1 to 5. The survey was filled by 186 people. The severity score of each crime class is shown in Figure 5.3. It can be observed that suicide got the minimum severity and terrorism got the maximum severity.

5.4.2 Marginal Crime Score

The marginal crime score is an intermediate crime score associated with each location extracted from news articles and updated into the DB. This subsection explains how the marginal crime score is calculated. This marginal crime score of a location is updated whenever an article containing that location is processed. While processing crime articles, a crime class (mentioned in Figure 5.3) is associated with each article. Based on the crime class severity, an article score $[\gamma^{art}]$ is assigned to each article. Once the article score is available, the marginal crime score of all the locations associated with this article is calculated and updated into the DB. Since the effect of a crime diminishes with time, the database is also updated with time information.

Further half-life concept is used to estimate the current marginal crime score of a location $[\gamma_c^m]$ as follows:

$$\gamma_c^m = \gamma^{art} + \gamma_p^m \times e^{-(\lambda \times (t_c - t_p))}$$

Where:

$$\lambda = \frac{\ln 2}{T}$$

T is half-life constant, t_c is the current time, t_p is the time when the marginal crime score of a location was updated last and γ_p^m is the previous crime score; all times are in the number of days. Thus, if T is 180 days, the marginal crime score of a location will be half after every 180 days.

5.4.3 Final Crime Score

One can use this marginal score to understand the crime severity of a location or to compare the crime severity of different locations. However, this marginal score of a location is purely dependent on that location information only, whereas it is generally observed that the crime density/rate (by implication crime score) of a location does not depend only on the crime events occurring at that location but also on the crime events occurring in the neighboring areas. Furthermore, the locations which are not present in DB (as the method has not seen any crime news article covering that location) will not be assigned any marginal crime score. To overcome the above-mentioned issues, the effect of nearby crime events is incorporated in the total crime score of a location, which is called the final crime score (or crime score) of a location. Intuitively, the impact of an incident decays with distance from the incident location [26]. This decay is assumed to be Gaussian in nature. The variance is assumed to be common for all locations. Now, whenever the crime score for a location is queried, a Gaussian is initiated centered at that location. All the locations lying within the 2.5 σ radius of queried location L are taken into account. Their marginal crime scores (γ_i^m) are added to estimate the final crime score of the queried location (γ_L^f), as follows:

$$\gamma_L^f = \frac{1}{\sqrt{2\pi\sigma^2}} \sum_{\forall i \in Range(2.5\sigma)} e^{-\frac{d(i,L)}{2\sigma^2}} \times \gamma_i^m$$

where $d(i, L)$ gives the distance between location i and L.

Before returning the crime score, it is normalized using the max marginal crime score to make sure that the returned crime score always lies between 0

TABLE 5.8

Accuracy Results for Location Extraction

Method	Potential Locations (%)	Crime Locations (%)
NLTK	63.21	60.08
Stanford Tagger	82.77	79.24

and 1 so that any user can understand the relative safety of a location. Note that this normalized score can be greater than 1 when the queried location is having a marginal score equal to max marginal score due to the addition of scores from nearby places, in those scenarios the final crime score is rounded off to 1 to ensure that the final crime score is ≤ 1. Also, note that the proposed method of providing a crime score between 0 and 1 is more effective in comparing crime at two locations rather than giving an absolute crime index.

5.5 Auxiliary Database and APIs

While going through different newspapers, it is observed that there are different spellings—and even different words—used by reporters for a location. These can be due to mistakes while writing, or historical/cultural reasons. For example, reporters from Mumbai use Bombay, Mumbai, to talk about Mumbai while writing articles. To make sure that the crime score is trustworthy even in such a diverse scenario, a separate database is created to map these words to a canonical form. Also, for the cases of wrong spelling, Bing Spell Check API is used to correct the spellings of important words (location name, entity name, etc.) before feeding them into other algorithms. Additionally, LocationIQ API is utilized to get the geo-coordinates of a location. The geo-coordinates are used to calculate the distance between nearby locations when calculating the final crime score (Table 5.8).

5.6 Evaluation

This section evaluates the working of the complete framework. The above framework is tested on a dataset crawled from various news websites, mentioned in Table 5.2. As already specified in Section 5.2.2, the dataset is divided into two parts. The first part (tagged manually) is used to verify various methods of the framework, and the second part is used to demonstrate the efficacy of the system. Unfortunately, there is no direct way to quantitatively measure or compare the performance of the system. To indirectly assess the quality of the final crime score, the city crime ratings are compared with the ratings published online as discussed in the latter part of this section.

The following are the details of the final framework. The best crime classification accuracy obtained after tuning all parameters is 85.4%, with 79.4% precision and 94.2% recall. There is more focus on recall to avoid missing any crime article. Crime location extraction was performed with both NLTK and Stanford tagger. Using the NLTK tagger alone, an accuracy of 75.92% is achieved in extracting all potential locations and 69.88% in separating all possible crime locations. For Stanford Tagger, an accuracy of 81.02% and 78.24% is observed respectively for both tasks. Finally, the overall framework is executed on the second part of the dataset to build a database that contains marginal crime scores and other details of crime locations.

To prove that the results are adequate, the final crime scores are cross-verified with crime reports available on some well-known websites. The results are compared with three online sources:

- **The Economic Times**: Seven things you didn't know about urban crime in India.
- **Numbeo**: Crime Index by City.
- **Statista**: Crime rate in major cities across India.

The crime index/rate was collected from all these online sources and plotted into charts, as shown in Figure 5.4, to show the similarity between the order of their crime rating and this framework's rating. Furthermore, Spearman's Rank-Order Correlation, a non-parametric measure of rank correlation that helps in understanding the strength and direction of association between two ranked variables, is used to understand the correlation between online source crime rating and this framework's crime rating results. Some of the major locations are used for comparison which was rated by online sources as well. These major locations are ranked based on their crime score/index/rate. Next, Spearman's correlation coefficient ρ is calculated to understand the correlation between crime rating of locations from different sources with this framework results. The results are shown in Table 5.9. The number of samples is different in each case due to the unavailability of crime score for some locations either in online sources or in this framework. From the values, it can be seen that there is a good correlation between this framework's results and online results. Spearman's correlation coefficient value of 0.661538 is observed with the Economic Times ranking which shows a good correlation. Though the correlation is not very strong, being in line with Economic Times and other online source rankings and a good correlation with them shows that the crime score/severity estimated by this framework can be trusted.

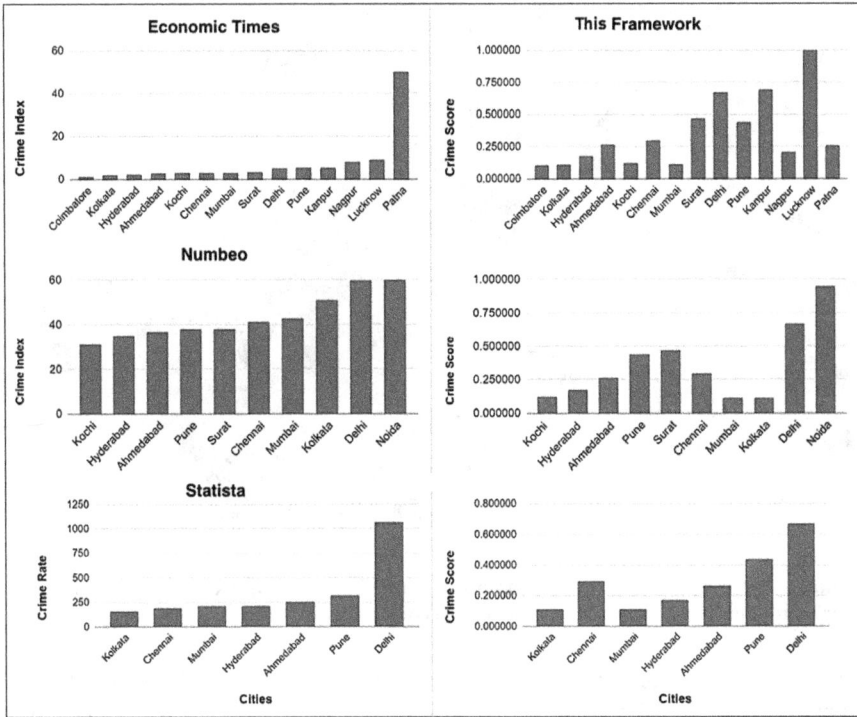

FIGURE 5.4
Chart showing the crime rating order of Top row: Economic times vs. this Framework; Middle row: Numbeo vs. this Framework; Bottom row: Statista vs. this Framework.

TABLE 5.9
Spearman's Correlation Coefficient ρ, Showing a Rank Correlation between this Framework and Online Source

ρ	Economics Times	Numbeo	Statista
This Framework	0.661538	0.369697	0.785714

Furthermore, the crime score results obtained are represented on an interactive map of India. To show the effectiveness of the framework at a smaller level, the heatmap of New Delhi is shown in Figure 5.5; the blue color represents the colder regions, i.e. low-crime occurring areas, and the red color represents the hotter regions, i.e. high-crime occurring areas. This map shows only locations available in the database. To make this framework available for the end-user, a web interface is developed, where the user can query the crime score of a location by providing the location name. The web interface is currently available only on the intranet of IIT Ropar. The plan is to make

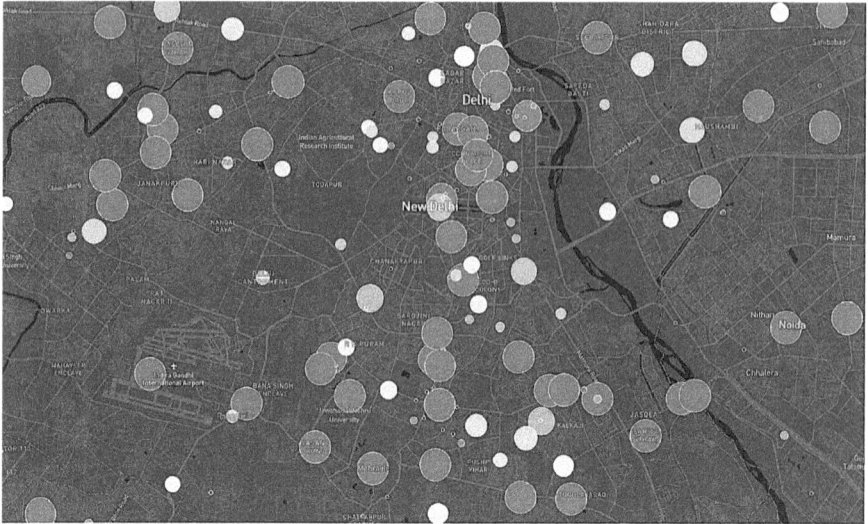

FIGURE 5.5
Heatmap of New Delhi and nearby locality generated using marginal crime score. The blue color represents a low crime score while the red color represents a high crime score.

it publicly and freely available to the users. The source code of this framework is publicly available at https://github.com/shreyanshu007/Crime-Analysis-BTP git repository.

5.7 Conclusion

Crime information of a location is very useful from a safety perspective. It is difficult for a normal person to get this information. This chapter of the book discusses a dynamic end-to-end framework to calculate the crime score of any given location. In the discussed framework, news articles are collected from online websites, classified into crime or non-crime, further identifying the duplicate articles and identifying crime locations, and finally, building a model for dynamic crime score for a given location. Experiments show that the crime classification accuracy is 84.5%, crime location extraction accuracy is 79.24%, and crime class identification accuracy is 76.12%. The methods can be used for crime profiling of any region as long as there is news coverage. The final crime score obtained thus is in agreement with the manual crime reports available on various online platforms. The framework is general and flexible as it can be integrated with other applications, for example, the safest route-finding application.

Acknowledgments

This work was supported by the grant received from the Department of Science & Technology, Government of India, for the Technology Innovation Hub at the Indian Institute of Technology Ropar in the framework of National Mission on Interdisciplinary Cyber-Physical Systems (NM – ICPS).

References

1. Ku, C.H., Leroy, G.: A decision support system: Automated crime report analysis and classification for e-government. *Government Information Quarterly* **31**(4), 534–544 (2014).
2. Deshmukh, A., Banka, S., Dcruz, S.B., Shaikh, S., Tripathy, A.K.: Safety app: Crime prediction using GIS. In: *2020 3rd International Conference on Communication System, Computing and IT Applications (CSCITA)*, pp. 120–124. IEEE (2020).
3. Goel, N., Sharma, R., Nikhil, N., Mahanoor, S., Saini, M.: A crowd-sourced adaptive safe navigation for smart cities. In: *2017 IEEE International Symposium on Multimedia (ISM)*, pp. 382–387. IEEE (2017).
4. Shah, S., Bao, F., Lu, C.T., Chen, I.R.: Crowdsafe: Crowdsourcing of crime incidents and safe routing on mobile devices. In: *Proceedings of the 19th ACM SIGSPATIAL International Conference on Advances in Geographic Information Systems*, pp. 521–524 (2011).
5. Utamima, A., Djunaidy, A.: Be-safe travel, a web-based geographic application to explore safe-route in an area. In: *AIP Conference Proceedings*, vol. 1867, p. 020023. AIP Publishing LLC (2017).
6. Kim, S.M., Hovy, E.: Extracting opinions, opinion holders, and topics expressed in an online news media text. In: *Proceedings of the Workshop on Sentiment and Subjectivity in Text*, pp. 1–8 (2006).
7. Arulanandam, R., Savarimuthu, B.T.R., Purvis, M.A.: Extracting crime information from online newspaper articles. In: *Proceedings of the Second Australasian Web Conference*, vol. 155, pp. 31–38 (2014).
8. Jayaweera, I., Sajeewa, C., Liyanage, S., Wijewardane, T., Perera, I., Wijayasiri, A.: Crime analytics: Analysis of crimes through newspaper articles. In: *2015 Moratuwa Engineering Research Conference (MERCon)*, pp. 277–282. IEEE (2015).
9. Tayal, D.K., Jain, A., Arora, S., Agarwal, S., Gupta, T., Tyagi, N.: Crime detection and criminal identification in India using data mining techniques. *AI & Society* **30**(1), 117–127 (2015).
10. Joshi, A., Sabitha, A.S., Choudhury, T.: Crime analysis using k-means clustering. In: *2017 3rd International Conference on Computational Intelligence and Networks (CINE)*, pp. 33–39. IEEE (2017).
11. Yadav, S., Timbadia, M., Yadav, A., Vishwakarma, R., Yadav, N.: Crime pattern detection, analysis & prediction. In: *2017 International Conference of Electronics, Communication and Aerospace Technology (ICECA)*, vol. 1, pp. 225–230. IEEE (2017).

12. Sharma, V., Kulshreshtha, R., Singh, P., Agrawal, N., Kumar, A.: Analyzing newspaper crime reports for identification of safe transit paths. In: *Proceedings of the 2015 Conference of the North American Chapter of the Association for Computational Linguistics: Student Research Workshop*, pp. 17–24 (2015).

13. Hassan, M., Rahman, M.Z.: Crime news analysis: Location and story detection. In: *2017 20th International Conference of Computer and Information Technology (ICCIT)*, pp. 1–6. IEEE (2017).

14. Rollo, F., Po, L.: Crime event localization and deduplication. In: *International Semantic Web Conference*, pp. 361–377. Springer (2020).

15. Mukherjee, S., Sarkar, K.: Analyzing large news corpus using text mining techniques for recognizing high crime-prone areas. In: *2020 IEEE Calcutta Conference (CALCON)*, pp. 444–450. IEEE (2020).

16. Dasgupta, T., Naskar, A., Saha, R., Dey, L.: Crime profiler: Crime information extraction and visualization from news media. In: *Proceedings of the International Conference on Web Intelligence*, pp. 541–549 (2017).

17. Margagliotti, G., Bollé, T., Rossy, Q.: Worldwide analysis of crimes by the traces of their online media coverage: The case of jewellery store robberies. *Digital Investigation* 31, 200889 (2019).

18. Rohini, D.V., Isakki, P.: Crime analysis and mapping through online newspapers: A survey. In: 2016 International Conference on Computing Technologies and Intelligent Data Engineering (ICCTIDE'16), pp. 1–4. IEEE (2016).

19. Wolfgang, M.E.: *The national survey of crime severity*. US Department of Justice, Bureau of Justice Statistics (1985).

20. Wu, Z., Palmer, M.: Verbs semantics and lexical selection. In: *Proceedings of the 32nd Annual Meeting on Association for Computational Linguistics*, pp. 133–138. Association for Computational Linguistics (1994).

21. Borowsky, R., Masson, M.E.: Semantic ambiguity effects in word identification. *Journal of Experimental Psychology: Learning, Memory, and Cognition* **22**(1), 63 (1996).

22. Zhang, Y., Jin, R., Zhou, Z.H.: Understanding bag-of-words model: A statistical framework. *International Journal of Machine Learning and Cybernetics* **1**(1–4), 43–52 (2010).

23. Ramos, J. Using TF-IDF to determine word relevance in document queries. In: *Proceedings of the First Instructional Conference on Machine Learning*, vol. 242, pp. 133–142. Piscataway, NJ (2003).

24. Johnson, R.M.: Lifting the hood on NLTK's NE chunker. URL https://mattshomepage.com/articles/2016/May/23/nltk_nec/. Last Accessed 2019.

25. Manning, C., Surdeanu, M., Bauer, J., Finkel, J., Bethard, S., McClosky, D.: Stanford named entity recognizer (NER). URL https://nlp.stanford.edu/software/CRF-NER.html. Last Accessed 2019.

26. Kumari, P., Singh, M., Saini, M.: Multimodal drunk density estimation for safety assessment. In: *2018 15th IEEE International Conference on Advanced Video and Signal Based Surveillance (AVSS)*, pp. 1–6. IEEE (2018).

6

Intelligent Transport Systems and Traffic Management

R. Shyam Shankaran and Logesh Rajendran

L&T Smart World, Chennai, India

CONTENTS

6.1 Introduction

Congestion of road traffic is an ongoing worldwide issue. The issue is experienced in almost every major city with a rapidly expanding economy. This is mainly because of space and cost restrictions; infrastructure development is slow compared to growth in the number of vehicles. Communication and innovations in information technology are the key drivers behind some of the most impressive developments in the smart cities. Communication technology using smart phones have changed our lives in the last two decades by encouraging information to be exchanged anywhere and every day. The advances in technology allow one to observe that vehicles increasingly have integrated applications and equipment such as cameras, sensors, processors, and communication tools. These features have an influence on existing networks by sending and processing information in real-time, enabling data collection, and accountability to help users and devices take necessary course of actions. Vehicles have become a valuable source for smart cities because they are able to feel and react accurately to artifacts and activities in the environment; they not only help to control traffic, but are also an instrument for the capture of relevant information used in resource management in real-time. A smart city can be identified among its current meanings as an intelligent environment which integrates ICT and interactive solutions. These networks introduce connectivity to the real world, to address the problems of residents in their urban conglomerations.

Intelligent Transportation Systems (ITS) is one of the methods for resolving or at least reducing the traffic congestion [1]. ITS applies to all modes of transportation which includes railways, roadways, water, and airways, as well as the various components of each mode, such as automobiles, utilities, connectivity, and software platforms. Different nations have developed methods and approaches to incorporate the various components into an interconnected system based on the spatial, ethnic, economic, and ecological backgrounds. In addition, every ITS implementation employs a Traffic Management Center (TMC), which collects, analyses, and combines data to solve complex transportation issues. Generally, a network of traffic operation centers is used to share the administration of transportation infrastructure among different agencies. Data and information are often distributed locally, and various standards are used by different centers to achieve traffic control objectives. This interconnected autonomy in services and prediction is critical due to the variation of trend and performance characteristics of interconnected sub-systems.

In this sense of intelligent networks, ITSs therefore include the combination of many innovations aimed to offer the urban mobility of the region comprehensively and provide greater protection for drivers as well as convenience and passenger entertainment. These technologies are focused on coordination between elements embedded in urban and transport networks (sensors, mobile devices, vehicles), in order to incorporate environmental knowledge in real-time. The adequate combination of these considerations leads to the sensing and collection of data by a control system for the assessment and eventual execution of adequate responses. The ITS consists of a range of innovations and software intended to enhance the safety and mobility of travel, maximize the productivity of people and reduce the detrimental effects of traffic. In the twentieth century, scientific researchers in the United States (US) proposed the original ITS description in the aim of improving the traffic conditions and improving the sustainability of the transport sector [2]. However, the ITS is now very much attracted by industry and academia because such technologies not only enhance the road state of the vehicle but also increasingly improve the safety, sustainability, and efficiency of the transport sector to reduce traffic congestion and climate impacts. We define the main ITS architectures and components in the upcoming sections and explain the variations between each model proposed. Furthermore, a few initiatives are presented to build an ITS and the challenges of ITSs.

6.2 Overview of Intelligent Transport System

An intelligent transportation system is a cutting-edge technology that seeks to deliver customized solutions related to various modes of transportation and traffic management, as well as to enable users to be better aware and use transportation networks in a smoother, more organized, and smarter manner [3]. As shown in Figure 6.1, the overview of the National ITS Architecture can be depicted as an illustration [4]. New developments in traffic management are being driven by cutting-edge data acquisition, video surveillance, assessment technology, digital mapping, communication networks, sensors, and variable message signs around the world. The integration of data collection, interpretation, assessment, and dissemination aids in the development of an all-encompassing traffic management system that allows managers and users to share information. The National ITS Architecture encompassing the essential features is depicted below in the Figure 6.1. The National ITS Architecture is an approach for analyzing system functionality, knowledge sharing, and component interoperability, as well as planning for them. The architecture directs the construction of new ITS designs by planners and engineers. It accomplishes this by providing a structure for ITS systems and subsystems to communicate. Only a small portion of this system can be used in small projects. Many of its features can

Travelers	Centers				
Remote Traveler Support	Traffic Management	Emergency Management	Payment Administration	Commercial Vehicle Administration	Maintenance and Construction Management
Personal Information Access	Information Service Provider	Emissions Management	Transit Management	Fleet and Freight Management	Archived Data Management

Wide Area Wireless (Mobile) Communications — Fixed Point-to-Fixed Point Communications

Vehicle-to-Vehicle Communications	Vehicle	Roadway	Field-to-Vehicle Communications
	Emergency Vehicle	Security Monitoring	
	Commercial Vehicle	Roadway Payment	
	Transit Vehicle	Parking Management	
	Maintenance and Construction Vehicle	Commercial Vehicle Check	
	Vehicles	Field	

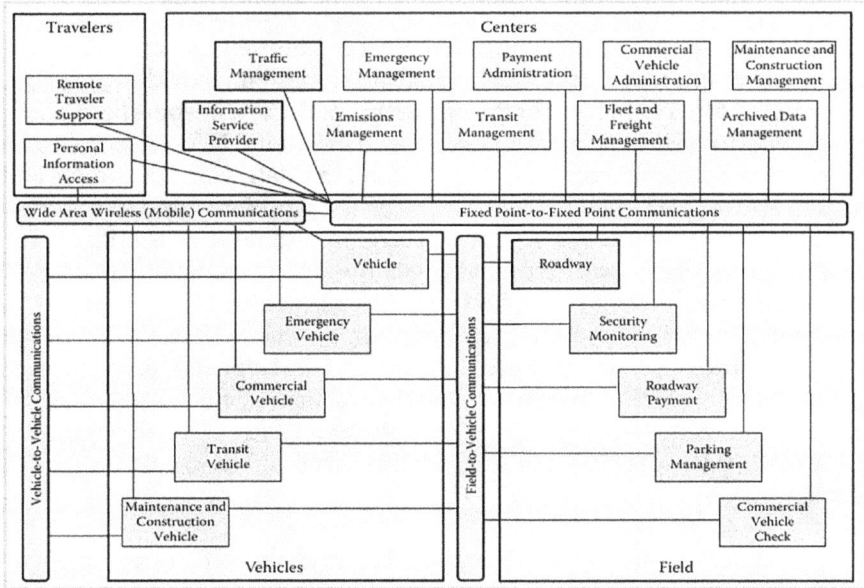

FIGURE 6.1
National ITS Architecture [4].

be included in large projects. The key is that using this popular framework for both small and large projects would make future functions easier to add, information sharing between systems easier, and device operation across multiple systems possible.

The main goal of ITS is to assess, analyze, design, and incorporate new sensors and communication technologies and ideas in order to improve traffic efficiency and improve safety and comfort for commuters and pedestrians. Socioeconomic and environmental demands have a strong influence on ITS growth. In the government side, ITS will be tailored to the region's needs based on the standards set by the national as well as regional authorities. The consumer market will fuel new ideas on the private side technology advancements, simulation, connectedness of various areas of engineering, such as information technology, communication systems, transportation, electronics as well as capacity building, should all be considered when designing an intensive ITS program. This, in turn, is contingent on collaboration between the government, academic research institutions, and business. The successful service delivery and implementation of ITS in the smart cities requires a thorough understanding of the transportation system and traffic management. When such an automated data capturing system for the traffic data has been implemented, the generated data can be stored and indexed to create models that will aid in the deployment of many solutions for ITS.

To provide reliable, productive, and efficient mobility of services and products while enhancing natural resource management and reducing ecological effects such as sulfur, nitrous, and carbon dioxide emissions, streamlined interconnection of the numerous nodes of transportation sector is needed. To ensure such seamless interconnectivity, ITS technology can play a critical role by collecting and exchanging data. For ITS design and widespread implementation, the ability of the workforce to devise, operate, and safely execute existing, and emerging technologies is critical. The major systemic and structural challenges that face the implementation of ITS are an underdeveloped road infrastructure, extreme budget constraints, accelerated industrialization, and growth, a shortage of technology, and capital resources for maintenance and service, a lack of need for innovation in automation, a lack of involvement and participation among policy stakeholders, and an inadequacy of administration awareness. The following are some of the relevant courses of actions that must be taken to address the ITS challenges:

- Creating a nationwide ITS standards for various ITS technologies and elements.
- Establishing a national ITS warehouse that records all ITS developments and deployments, including design, implementation, best practices, and economic evaluation.
- Formulating automated traffic data gathering strategies.
- Creating a national data depository for ITS.
- Creating designs and algorithms for deployments.
- Increasing collaboration between academia, industry, and government entities to build up more inclination and, as a result, projects in the ITS field.

Only by implementing ITS at the network level, rather than in small corridors, will the full potential of the technology be realized. Across the board, current deployments show potential and opportunity for ITS deployment in any country, offering an initial scientific foundation and evidence on ITS deployment while emphasizing data, technological, operational, and research concerns. The ITS system is represented as four sub systems for ease of implementation and operation. They are:

- Infrastructure subsystem: Components which aid in efficient management of traffic and enhances life of commuters.
- Vehicular subsystem: Components which aid in ease of vehicular transports.

- User subsystem: Components which enhances the quality of commutation in perspective of the pedestrians.
- Central subsystem: Components which helps in monitoring and management of the transportation among entire area.

6.3 History of Intelligent Transport System Concept

The origin of the ITS formal program dates from the 1960s when the Electronic Route Advice System (ERGS) was developed in the US to offer guidance to users for traffic on a real-time basis [5]. The system makes use of a variety of facilities located along the path, bidirectional data on- board equipment in automobiles that serves as a vital link between the user and the ERGS network, and a centralized software system for processing data obtained from distant systems. The ERGS program progressed to a complex, electronic system of interactive multimedia graphical maps known as Automatic Route Control System or ARCS, in the early 1970s [5]. The Urban Traffic Control System has simultaneously been designed to connect multiple traffic signals and computers to default signal times for a better traffic organization. In the meantime, the US was working on a bill to resolve the concerns of escalating congestion, travel related incidents, fuel wastes, and emissions in the US as successor to the fifties Interstate Post Bill. In 1986, the IVHS was drafted, and contributed to a spate of innovations in the field of ITS. In Europe, the safety European Traffic Systems Program was intended for vehicle producers followed by DRIVE project [6]. The recent production of ITS facilities has been improved in the last decade, in line with the growing demands, which meet a range of requirements. The high range of considerations has led to standardization to determine an overall process by which systems and components communicate. The architecture adopted for ITS in different countries are described among the below sections.

6.3.1 American ITS Architecture

The US Department of Transportation was the backbone in creating the ITS backbone in US. These were planned to encourage urban mobility by means of a cooperative structure. It consists of a collection of interrelated elements, grouped together, as we see in Figure 6.2; this is collectively termed as the ARC-IT (Architecture Reference for Cooperative and Intelligent Transport).

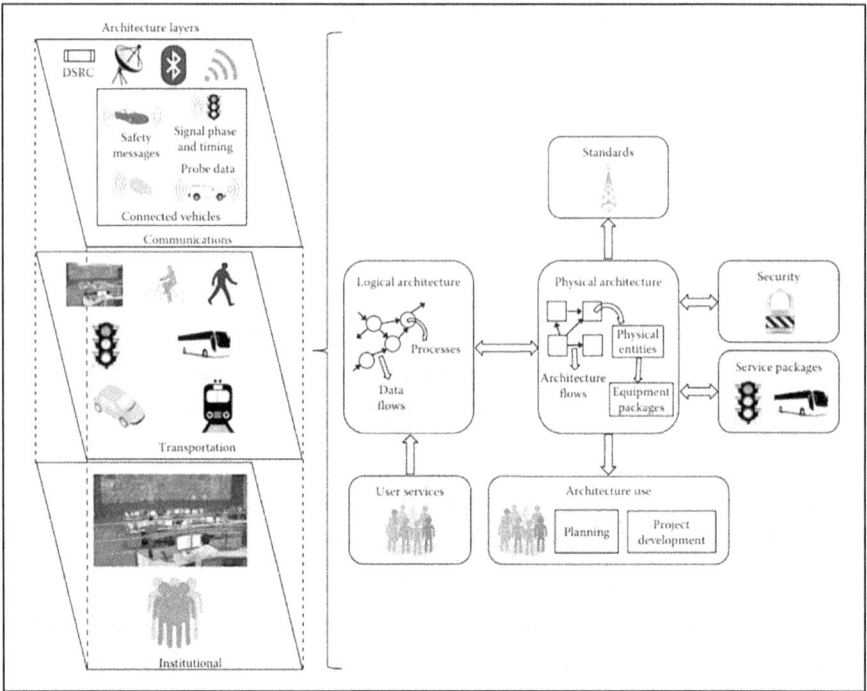

FIGURE 6.2
American ITS architecture [7].

The US architecture offers an ITS definition structure that describes the tasks that Physical Structures must carry out. Although the architecture offers its users with multiple resources, the support for the concurrent utilization of different technologies to satisfy its user demand is not explained.

6.3.2 Canadian ITS Architecture

The ITS architecture of Canada which is represented by Figure 6.3 was adopted by Transport Canada and a structure for the preparation, definition, and integration of ITS was developed. The following elements are in the architecture:

- Functions represent the steps that need to be taken for ITS, such as traffic data collection or path request.
- The physical entities refer to the systems impacted by the functions.
- Data flows and information transmission are connected together in an interconnected structure between these physical and functions subsystems.

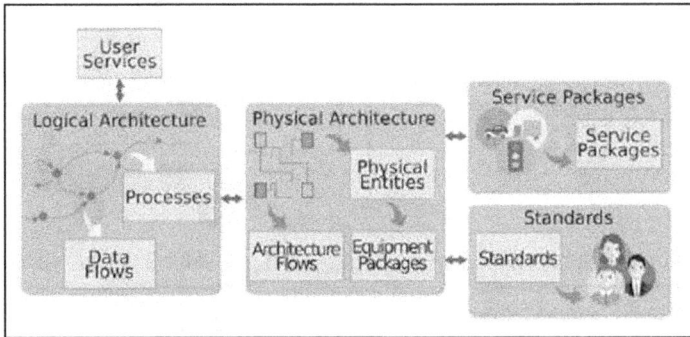

FIGURE 6.3
Canadian ITS architecture [8].

It provides a standard framework to design an ITS architecture which describe the functions that the components and subsystems need to perform. To chart the specifications of each role and the necessity of each service delivered to its customers, the used knowledge flow uses a coordination framework. Although the architecture will provide its users multiple services, it is not clear if the architecture requires the different communications systems to be used concurrently to satisfy consumer requirements. The availability of resources also affects the architecture; all the facilities are in the city centers and communication between the two groups takes place via an interface which restricts the use of new innovative approaches.

6.3.3 European ITS Architecture

The European council on ITS oversees designing the ITS architecture, which is divided into four elements as seen in the below Figure 6.4.

- Personal provides users with access to services through as smartphones.
- Vehicle refers to equipment installed to collect data from, transmit data to, and share data with other drivers.
- Central refers to a system that controls, tracks, and makes ITS resources accessible to customers.
- The Roadside refers to machines mounted along the roadside that power ITS applications.

The European architecture consists of a set of control system that provide ITS services. As the previously mentioned architecture, the use of a centralized control element and a lack of familiarity with new paradigms of communication are hindered.

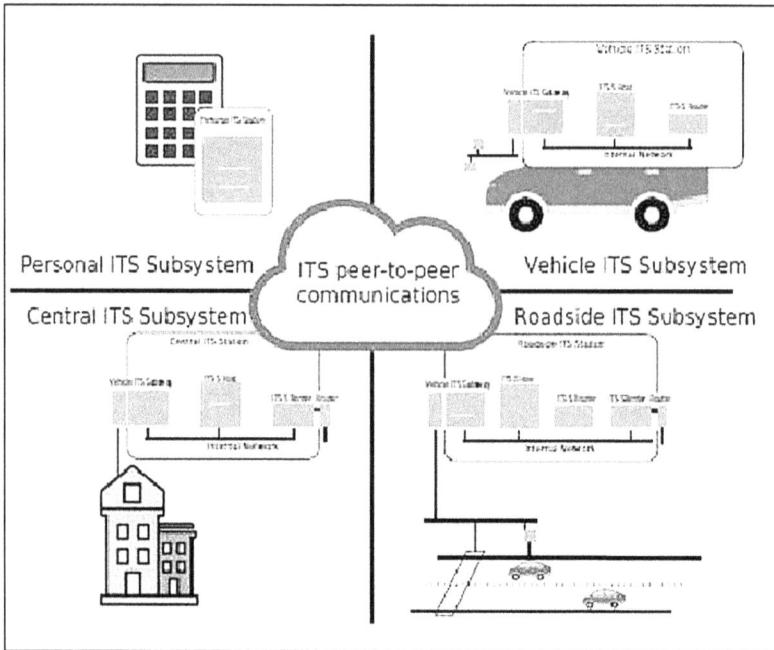

FIGURE 6.4
European ITS architecture [9].

6.4 Components of Intelligent Transport System

A Traffic Control Center is the administrative hub for transportation, in which data is gathered, interpreted, and integrated with other control and operational principles to handle the extensive transport infrastructure. Generally, multiple organizations share transportation infrastructure administration via a matrix of traffic management centers data and information are often transmitted globally, and the centers use a range of standards to meet traffic control objectives. Generally, multiple organizations share transportation infrastructure administration via a matrix of traffic management centers. This interrelated autonomy in processes and policy decisions is important due to the diversity of specifications and performance characteristics of interconnected modules. The following components are crucial to the efficient operation of the traffic management center, and thus the effectiveness of the ITS:

1. Sensors networks and data acquisition systems
2. Communication systems

3. Data management using Integrated command and control centers

4. Applications and Services for citizens and authorities

6.4.1 Sensors Networks and Data Acquisition Systems

A vast number of initiatives have been implemented in recent years with the goal of developing efficient intelligent transport systems that provide the final users with useful and cost-effective services. It is important to build layered architectures in which an acquisition system can gather data in the particular region to establish successful ITS solutions. Nowadays, the economic implementation is achievable by using smart sensors based on less complex microcontrollers and equipping them with advanced networking technology capable of supporting well-known standard protocols. Rapid, comprehensive, and precise data collection, and dissemination is vital for real-time tracking and strategic planning. An effective data acquisition system combines proven hardware and effective software to gather credible data on which to base future ITS operations. Some of the key components of state-of-the-art are discussed below.

6.4.1.1 Visual Sensor Networks

The concept behind the visual sensor network or smart camera networks is to create a network of nodes capable of processing visual information locally with a goal of drawing out a number of functions to be transmitted, thereby preventing the transfer of unprocessed digital images[10]. Each VSN integrates image processing, visual sensing, and network communication in the built-in platform, allowing primitive camera transformation into intelligent devices. The main elements of the sensor networks are just the integrated visual sensors and the deep learning algorithms that extract the appropriate feature. Several applications relating to the ITS have been suggested over recent years, based on embedded sensors. A system has been designed to identify the status of parking areas by two embedded low-complexity computer view applications and VSs are used for counting passing cars and measuring car speed. The use of pervasive VSNs would allow a series of interoperable system applications to resolve some open problems in the area of ITS. Without prejudice to future applications, although the overall use of VSs to extract mobility-related parameters will, on one hand, produce open data for municipalities to consume for transport plan purposes, on the other, it may re-design old road lighting systems in the so-called "smart cities" to create advanced service. A real-world example is the monitoring of pedestrians when they cross a road and tracking of vehicles, as shown in the Figure 6.5, thus enhancing future road safety.

FIGURE 6.5
CCTV based vehicle tracking system.

6.4.1.2 *Internet of Things*

The Internet of Things paradigm has been accelerated with the global spread of the Internet, as well as the growth of new miniaturized and embedded chips with communication abilities. The prime idea behind the Internet of Things concept is globally interconnected systems, each of which is found and discussed as a resource in the connected environments. IoT devices can be remotely accessed, allowing for the gathering, and handling massive scales of data about the everyday world. Furthermore, innovative applications can be created by utilizing the data and the new capabilities provided by solutions. The IoT-based sensor networks are an appropriate aid for development of ITS collection layers which are low-cost, that are ubiquitous, interoperable, and can support economic services to the end users. Under such a vision VS machine viewing algorithms run onboard to extract mobility-related characteristics that can, run with the help of the IoT and the REST protocols, be portrayed as network resources [11]. In IoT-based VSNs, a middleware system that disguises the complexity of abstracting network capabilities can be installed where an internal processing of exposed resources is required. The IoT-based solutions are capable of (1) managing network transactions between nodes via REST and IoT protocols; (2) managing VS functionality, which enables onboard computer vision algorithms to be configured using

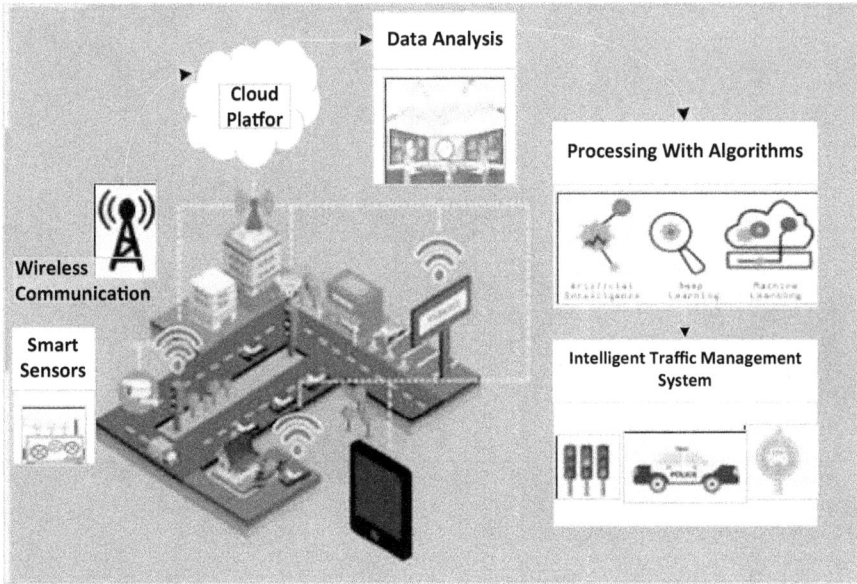

FIGURE 6.6
IoT connected transport ecosystem.

exposed operating systems interfaces; and (3) providing sufficient flexibility to alter the internally processed engine through a virtual machine-based solution. The IoT-based transport system is schematically presented in the Figure 6.6.

6.4.1.3 Global Positioning System

The Global Positioning System (GPS) offers rapid, versatile, and comparatively crisp information to ascertain the position and speed of the vehicle. GPS is a 24/7 Earth- monitoring system which is space based with satellites [12]. At approximately 20,200 km, the 24 satellites are evenly dispersed in six orbital planes, so as to allow at least four satellites to be seen from anywhere in the world at any moment. GPS positioning is loosely based upon trilateration-related strategies, the three-dimensional positioning of man-made satellites. The code-phase or pseudo-ranges and the carrier-phase are the two underlying observables used by GPS for positioning and navigation. It displays basic position data such as longitude, latitude, altitude, and time with respect to the zone. Using these temporal and spatial information, engineers can compute the necessary traffic information, such as trip duration, trip length, trip speed, and lag. It is critical to satisfy the sample size criteria and pursue an adequate field process to generate credible traffic

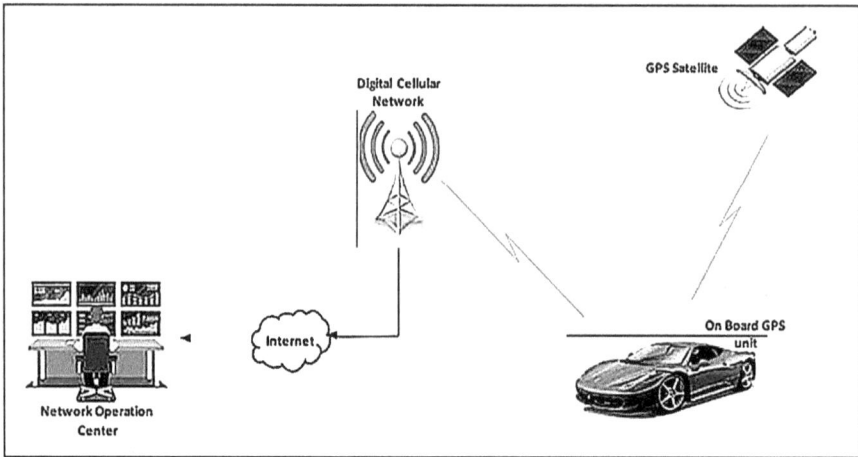

FIGURE 6.7
GPS enables vehicle tracking system.

information from GPS data. The functional depiction of GPS in action is shown in the below Figure 6.7.

6.4.1.4 Vehicle Identification System

The vehicle identification system is made up of readers, identifiers or transponders in automobiles, and a centralized software application. RFID antennas can be found on the side of the road, on overhead buildings, or toll booth which exist separately. The antennas transmit signals in radio frequency band across one or more highway lanes within a capture range. When a test object enters the detection range of the antenna, its sensors respond to the wireless signal and the reader assigns a date and time stamp to its specific ID Following that, the data is routed to a central processing network, where it is analyzed, and archived. RFID systems have the ability to retrieve vast quantities of data in a constant manner with no human intervention. An example of vehicle identification system is depicted below Figure 6.8.

6.4.2 Communication Systems

The productivity of the ITS infrastructure is not only determined by the processing and analysis of traffic-related data, but also by a prompt and accurate dissemination to the management center of data and models from the traffic management center to the general public. This includes communication between data collection centers and the traffic management center, as well as travel, and traffic-related notifications to cars via onboard units

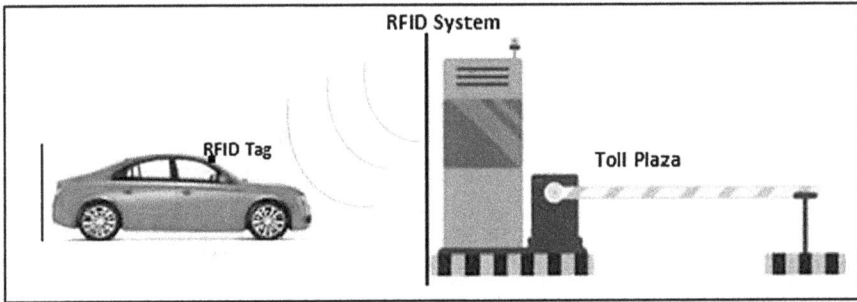

FIGURE 6.8
RFID vehicle detection.

and to visitors via media such as SMS, web pages, and VMS. Limited range communications from the automobile to the roadway corridor in particular areas are provided by DSRC (Dedicated short-range communications) [13]. DSRC comprises Roadside Units (RSU) and Onboard Units (OBU) with transponders and transceivers that operate commercial bands of microwave frequency range [14]. Vehicular network connectivity is provided by wireless networks dedicated to, Traffic Telematics, Road Transport, and Intelligent Transportation Systems. The dedicated and reliable communication involves the use a variety of communication devices, including mobile cellular, and infrared links, to maintain uninterrupted communications between an automobile and the corridors. Some of the key components of cutting-edge technology are outlined below in the Figure 6.9.

6.4.2.1 Vehicular Communications

The modern Internet of Vehicles model (IoV) is created as the conceptualization to introduce the Internet of Things in the automobile industry. IoV believes that all onboard electronic equipment is capable of communicating through mobile Internet technologies with road network infrastructure equipment, the electrical grid, and the mobile equipment for travelers, pedestrians, and cyclists. This converts vehicles into intelligent nodes of a network around the city. The following details the various modes of connectivity between vehicle computers and all other network nodes. The intravehicular communication describes the interplay among multi-various in-vehicle entities (sensors, actuators and ECUs), as well as datum transfer, from one device to another to decide on vehicle behavior.

Vehicle-to-vehicle (V2V) connectivity involves the collaboration with vehicle ECUs and decision support systems between sensors, actuators and other (normally neighboring) vehicle units [15]. This contact seeks to warn the driver, in near-real-time, of the important coordinates, position, speed,

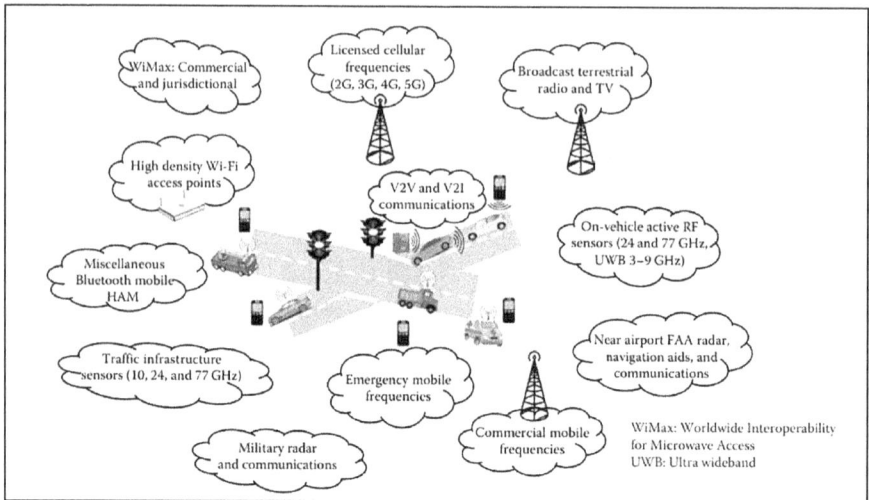

FIGURE 6.9
Different communication systems in ITS.

and other emergency conditions that must be considered before any crisis is handled. Vehicle-to-infrastructure (V2I) refers to communication between active vehicles and the static transport system [16]. The traffic lights, signals, data centers, and antennas are examples of infrastructure institutions. For providing a sufficient overview of its meaning and to adapt it immediately to external requirements, the vehicle shares information with the aforementioned agencies.

6.4.2.2 Vehicle-to-Infrastructure Communication

V2I communication refers to the creation of a Vehicular Ad hoc Network (VANET) between static infrastructure and moving vehicles alongside the freeway. The architecture is centralized, with roadside system serving as a hub for several vehicles. Despite the fact that the word "vehicle-to-infrastructure" appears to be misleading, communication is bidirectional. However, communication in the V2I direction is unicast, while communication in the other direction is both aired and unicast when here is a need for responding to the vehicle's requests. The roadside unit is generally augmented by a constant power supply and a backbone communication network, and thus the use of these resources does not have to be considered. Normally, the downlink channel is much better than the uplink channel, since separate antennas, and/or directive can be mounted on RSUs. In certain V2I systems, the uplink channel is either irrelevant or not, which is almost like broadcasting. V2I topology on the other hand, should not be

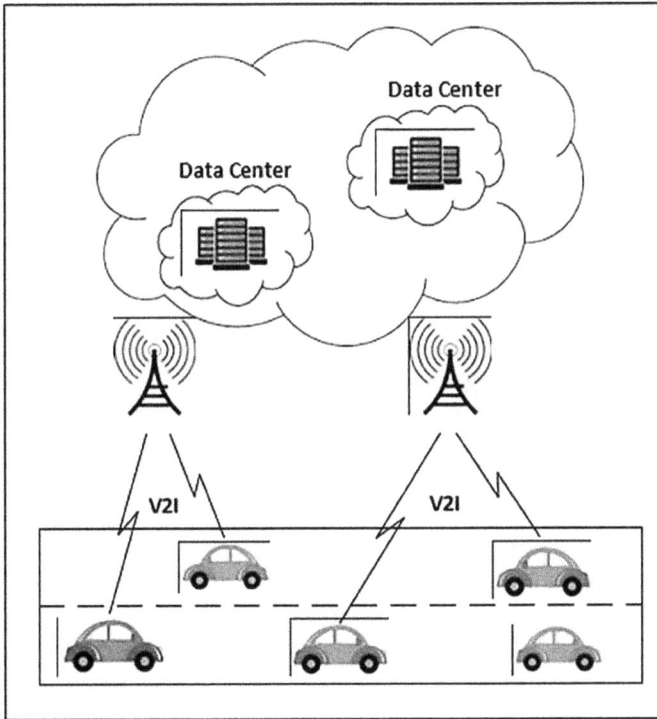

FIGURE 6.10
V2I communication model.

confused with broadcasting systems, with the presence of an uplink, or at the very least the ability to deploy one, being a necessary component when considering V2I. The V2I communication model is illustrated in the below Figure 6.10.

V2I communication is also used to convey information to vehicles from road operators or agencies. A traditional V2I service is an example of road-work warning application in which transceivers which are able to access the vehicular network are deployed as roadside unit, notifying incoming vehicles about the road worthiness. V2I Communication has certain parallels to cellular connections between a telephone node and access point in a traditional wireless network. RSU has superior signal strength and thereby data capacity resources as well as access points because of its stable nature. However, due to the transient existence of the V2I communication, RSU cannot have uninterrupted access to backbone vehicle networks. Instead, RSU may literally act as a hotspot infrastructure by swapping pre-configured high-band information between vehicles and a permanent network around an RSU in the zone.

6.4.2.3 *Vehicle-to-Vehicle Communication*

The vehicle-to-vehicle (V2V) communication method is predominantly suitable for communication with limited ranges by vehicles. The concept is that commuting motor vehicles build an ad hoc, highly opportunistic wireless network for communication between each other. As individual cars communicate equally, the communication architecture is distributed on an ad hoc basis. Data sharing is normally of the unicast type among passing cars, but also multicast, and broadcast transmissions are used. No infrastructure is necessary for a pure V2V network, so it's quick, and comparatively reliable when unexpected events require distribution of information on the highway. It is therefore the leading candidate for communication for car safety applications in real-time. The chance to allow cooperative vehicle safety applications to avoid accidents is one of the main motives of V2V communications. The V2V communication model is depicted in the below Figure 6.11.

The following applications for cooperative collision avoidance are expected to be achieved when it is implemented [17]:

- Identifying all automobiles in the direct vicinity.
- Monitoring the status chart of all vehicles.
- Conducting a constant hazard assessment.
- Detecting potential hazards needing operator intervention.
- Assisting the driver in a timely fashion.

Over the long term, automated intervention for vehicles to prevent or minimize accidents is envisaged with these applications, but much work remains needed to validate the necessary communication reliability. Specifically, when

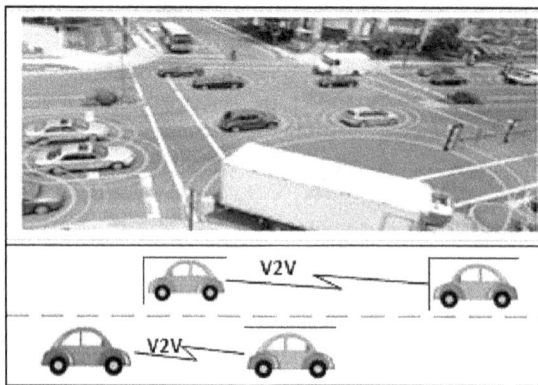

FIGURE 6.11
V2V communication model.

an accident occurs, a warning signal may be sent by vehicles that receive the message within a certain time period, encouraging those within a distance, up to a few kilometers, to take smart driving choices well in advance. This warning message is sent to vehicles that receive a message. There is also a risk of a diffusion storm in thick traffic conditions where several cars are simultaneously attempting to convey the message causing several package crashes and an excessive total communication channel failure. V2V is a very difficult communications entity. In V2V it may not be always easy to communicate between the cars, as the vehicles move at different speeds, as a result of fast network topology changes. Multi-hop transmission must be allowed for the propagation of messages or signals without any roadside infrastructure. Regular transmissions from each vehicle might inform neighbors directly about their addresses, but the address map would inevitably change constantly due to relative movements between vehicles. It is incumbent on the recipient to decide the relevancy and activities of emergency messages. In general, this type of communication is suited for highways and makes it negligible for rural areas.

6.4.2.4 Hybrid Vehicle-to-Vehicle and Vehicle-to-Infrastructure Communication

Combined V2I and V2V can be correlated as V2V with V2I features in addition to the simple V2V [18]. Apart from V2V, with previously specified applications, and integrated V2I, a wider selection of vehicle crash prevention applications will allow for safety with the communication network. Additionally, the other feature V2I allows is the prevention of a collision, so the system can alert the driver to another potentially dangerous driver, since it knows the dynamic status map of all cars and the geometry of the intersection. RSU has normally fixed power on the infrastructure side of V2I and may use antennas specially adapted to RSU. Furthermore, RSU prefers to contact all cars, while vehicle communication uses communication resources to improve their utilization by reducing interference with other vehicles. In addition, since RSU normally has a fixed connection, it can be viewed in a specific type of vehicle wireless network as the wireless network's access point as well. The hybrid communication topology is briefly explained in the Figure 6.12.

6.4.2.5 Cellular Networks

The mobile cellular networks are constantly advancing to meet the growing needs of usage trends. Although these technologies were intended for data transfer through voice, they can also be used in vehicular networks. Cellular networks are currently migrating from GPRS to LTE communication standards, improving bandwidth, and decreasing latency, render these networks appropriate for certain applications including productivity and route

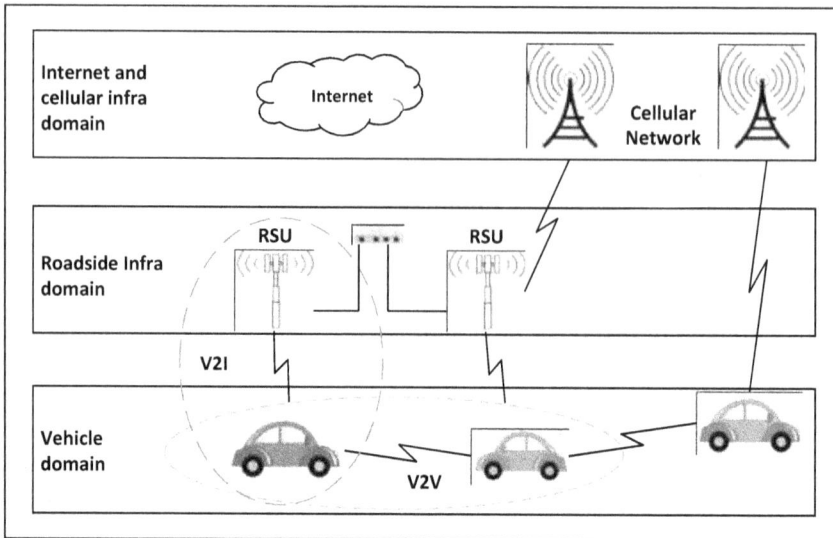

FIGURE 6.12
Hybrid communication network.

scheduling [17]. Cellular networks have certain qualities that make them suitable for cooperative systems, such as significant utilization, and long-distance communication. However, the following are some disadvantages of cellular networks that are pertinent to vehicular connectivity:

- Higher latency.
- Only point-to-point communication is supported.
- Service charges for operations.

Amidst these inconveniences, wireless networks can be used for certain applications requiring considerable latency in communications, long range, and limited data speeds.

6.4.2.6 WiMAX

WiMAX refers to Worldwide Interoperability for Microwave Access is a wireless data transmission standard based on the family of IEEE 802.16 standard that is capable of providing wireless communication over number of types including point-to-point access points to complete cellular access [19]. WiMAX will bridge the gap between cellular and wireless LAN specifications. It provides the high data rates, portable connectivity at low speeds and broad area coverage needed to provide mobile clients with high-speed

Internet access. WiMAX is a wireless last-mile Internet access capacity which can be used for vehicular communication applications.

6.4.2.7 Dedicated Short-Range Communications

Dedicated Short-Range Communications which is denoted as DSRC is a service that enables both secure activities in V2I & V2V communication contexts [13]. It is intended to supplement mobile carrier communications by delivering high data rates in situations by reducing latency and isolating to relatively smaller communication zones or fragments. DSRC is intended to operate in the band around 5,850 to 5,925 GHz radio frequencies, for vehicular wireless communications. With the defined data rates of 6–27 Mb/s, the range of communication with DSRC reaches 1,500 meters, with cars going up to 140 km/h [13]. The DSRC is split into two communication types, V2V, and V2I, as previously stated. In order for cooperation applications to operate correctly, the V2V communication shall also be used when cars have to exchange the information among themselves, whereas the V2I connectivity is employed when roadside units form a part of cooperative application. Some applications in cooperative systems have to regularly (for example every 100 ms) send messages, while other applications send messages while an event is happening.

6.4.2.8 Wireless Access in Vehicular Environments

Designing an effective communication protocol for privacy, safety, multicanal operation, and resource management in the automotive industry is a daunting task, which is being investigated extensively. This operation is entrusted to the IEEE specialized group and the IEEE 1609 protocol suite, often known as WAVE or Wireless Access in Vehicular Environments [15]. The WAVE specifications describe architecture and a supplementary, defined combination of solutions, and interfaces that pave way for secure wireless V2I and V2V models. These standards together contribute the basis for a wide range of applications, including automated tolling, driver safety, improved mobility, traffic management, and many other applications in the transport environment. The WAVE architecture, interfaces, and messages support secure wireless communication between and between vehicles and road infrastructure.

6.4.3 Data Management Using Integrated Command and Control Center (ICCC)

The analysis of data involves data purification, fusion, and interpretation. The information from sensors and other collectors relayed to the traffic management centers must be verified. Failure to remove and keep clean data should be done. In addition, data from various sensors can be merged or fused for further evaluation. To estimate and predict traffic states, cleansed, and fused data will be analyzed. The purpose of estimating traffic status is used to provide users with appropriate and accurate information. The data collected from ITS are becoming ever more complex and feature big data. Data are also available. Data from various sources such as GPS, smart cards, and sensors is available in ITS. Apparently disorganized data can best serve ITS by means of precise and efficient data analyses. The volume of data produced in ITS develops from the level of giga bytes to petabytes with the development of ITS. Figure 6.13 depicts the data management architecture in ITS. It is grouped into two layers: the data collection layer and the data analytics layer.

Increased complexity, variety and the amount of data generated and gathered from the vehicles and people activities have resulted through technological advancement in ITS. The various sources of data from the ITS applications are explained in Sections 6.4.3.1–6.4.3.6.

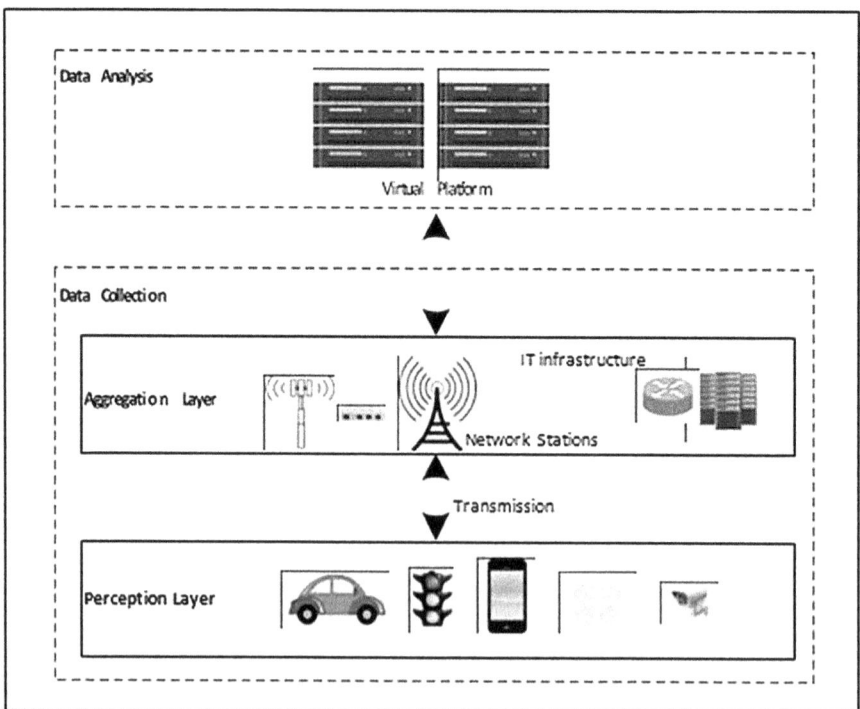

FIGURE 6.13
Data management layers in ITS.

6.4.3.1 Data Source from Smart Card

Automatic Fare Collection (AFC) systems have been widely implemented in urban railway systems, making smart card data the primary sources of data to investigate movement patterns of passengers. Passengers using smart cards when taking a bus or rail are needed for AFC systems. When the smart electronic readers touch their smart cards, they can capture passenger. Every day in major cities, intelligent cards in AFC systems create large volumes of data records.

6.4.3.2 Data Source from GPS

GPS is the best-known monitoring tool. With location tracing via GPS, traffic data can be gathered more effectively and securely. The GPS is a promising data collection tool that combines geographic data system (GIPs) or other map viewing technologies, and the collected data can be used to address certain problems related to traffic, such as delay measurement, travel mode detection, and traffic surveillance.

6.4.3.3 Data Source from Videos

The ITS commonly utilizes video cameras. Video imaging systems provide an excellent alternative to traditional sensors in tasks, such as identification of vehicles and detection of traffic flows, which been part of advanced Traffic Management Systems (ATMS). Freeway image sensors with enormous video data have been used successfully for incident detection and in some circumstances have shown high precision.

6.4.3.4 Data Source from Sensors

Vehicular speeds, traffic density, flow of traffic, and travel time are all collected using sensor equipment installed in ITS. On-road units or sensors like microwave and infrared sensors have evolved to collect, process, and transmit traffic data.

6.4.3.5 Data Source from CAV and VANET

Connected and autonomous vehicles (CAVs) are emerging ITS technologies that blend drastic changes in vehicle design with their relationships with road infrastructure. Connected and autonomous vehicles combine a variety of technologies to allow for the safe and efficient transportation of persons and goods. Vehicle ad hoc Network (VANET) is a type of mobile ad hoc network utilizes nodes comprising of infrastructure components and vehicular elements to expand range and reliability of communication. VANET produces a lot of data, and it is such an integral part of ITS.

6.4.3.6 Data from Other Sources

There are certain sources of data that do not get categorized into any of the above types. Real-time infrastructure status, for example, is regarded as a valuable data source. The smart grid, for example, would allow us to gather daily electricity consumption data for electric cars and train traction in the urban rail transportation system.

In general, IT-based traffic management is accomplished by a centralized facility that connects the transportation network systems (signals, cameras, detectors, etc.) with the operators that control the conditions of traffic as well as the system's performance. These centralized installations are hubs or nerve centers of a transport management system which are capable of:

- Real-time collection of data from field facilities on traffic conditions.
- Analysis of traffic management systems data and control.
- Information monitoring by operators on conditions of traffic.
- Appropriate actions to rectify errors.
- Information dissemination to travelers and other stakes.

ICCCs collect data, process it, and analyze it with other necessary parameters to extract valuable data from facilities installed in transit networks. Based on this information, the ICCC takes action to manage traffic flow, support the management of accidents, to automate commands to connected systems and to provide users with traffic notifications. In order to increase system performance, different agencies often use the information to monitor and control their activities.

6.4.4 Applications and Services for Citizens and Authorities

In many countries today, ITS application is widely accepted and used not only for control and information on the congestion of traffic but also for effective use of infrastructure and road safety. ITS has become a multidisciplinary conjunctive field because of its infinite options, and so many companies worldwide have come up with solutions to satisfy needs for ITS applications. The design of ITS services and apps helps travelers and drivers with an emphasis on accident reduction and traffic management in smart cities. Other types of applications also contribute to and facilitate driver services, making travel peaceful, and enjoyable. Some of the major ITS-focused applications are presented in the below Figure 6.14:

6.4.4.1 Traffic Enforcement

In smart cities, traffic control is usually conducted by the traffic police. The following activities will include typical activities related to traffic enforcement.

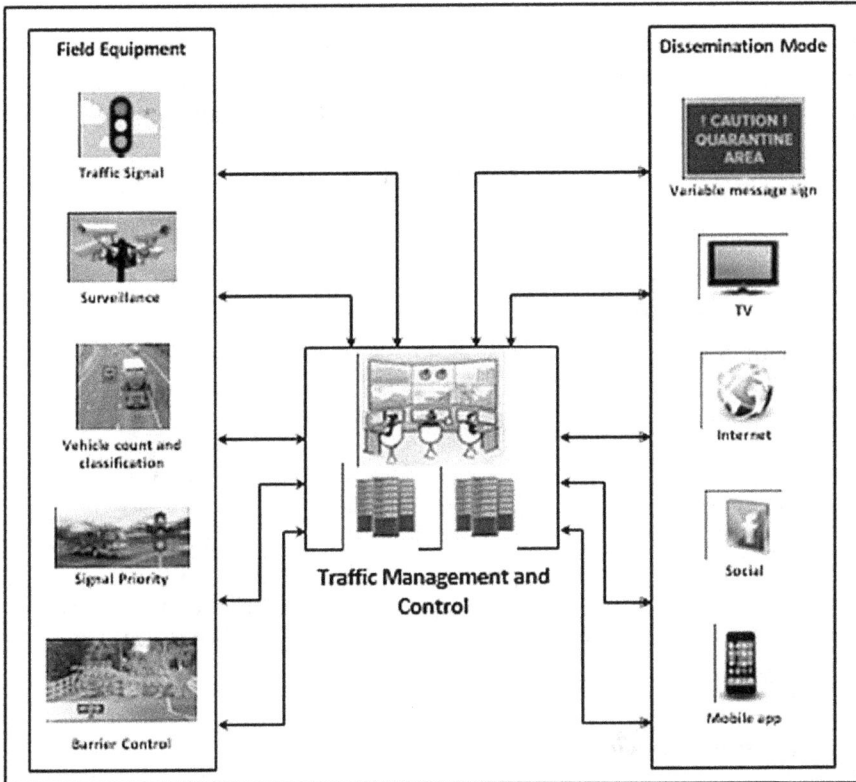

FIGURE 6.14
Overview of traffic management services.

- Checks on the road.
- E-Challan system.
- Parking violation.
- Red light violation.
- Speed violation.
- Restricted entry.

ICCC is made up to facilitate systems to retrieve or share pertinent information on this subject for the above activities with the Traffic Enforcement Center.

6.4.4.2 Roadways and Junctions Traffic Management and Monitoring

The road systems include all sensors and field devices deployed in the roadways to support road network management. These include key inputs covering areas like volume of traffic, signal status, and road images, that assist in taking adequate measures to handle traffic. Data and information are obtained by ICCCs from a number of sensors deployed in field and on the basis of data analytics, the ICCC carries out various operations, including signal management, monitoring, fostering incidence, enforcing traffic rules, dissipating data, etc.

The following components covered by this system are:

- Signal timing and operations.
- Road traffic monitoring.
- Management of active traffic and demand.
- Management of emergencies and incidents.

6.4.4.3 Integration with Transit Operations

Transit operations use road and infrastructure and are also influenced by signal plans at signal crossings, traffic incidents/events, road closures/diversion, road buildings/keeping work, road congestion levels, existing, and forecast traffic. In order to ensure compliance with the schedule or preemption during crises, you may need signals of priority in some predetermined corridors. The ICCC is usually responsible for all of the above traffic control issues.

6.4.4.4 Integration with System Applications

To determine the effect of the traffic management processes as well as the actions to be taken, ICCCs collect data from certain transit management centers and subsystems processed by the ICCC system. The subsystems which are managed are as follows.

- Parking management.
- Integrated corridor management.
- Police operations and enforcements.
- Weather information system.
- Congestion pricing.
- Air quality monitoring.
- Toll operations.
- Lighting control.

6.4.4.5 Information Dissemination

The goal of information dissemination is to give travelers detailed, appropriate, reliable, and updated traffic information to plan their journeys. The idea

FIGURE 6.15
Example of information dissemination through ICCC.

is not to warn flooders but only to provide information which might affect the choice of travel. The information comprises the factors affecting congestion, road network capacity, road closure, travel times, or distractions. This contains information. Information about these factors must, therefore, quickly be collected and disseminated through appropriate channels, to allow travelers to make their plans for travel accordingly. The information can be distributed over many channels: email, mobile apps, social networking, variable signage, and radio and TV channels. Dissemination system contains components used by ICCC for traffic dissemination and associated information, including, fixed format displays, variable message signs, blinking devices, helpline etc. Dissemination system ICCC sends on field makers, traffic commands, and setup particulars and reports the status of system back to ICCC. Figure 6.15 explains the information dissemination through ICCC.

6.5 Advanced Traffic Control and Management Systems

The aim of ITS in-road transport is to increase transportation system mobility, safety, and efficiency by integrating advanced vehicle and road traffic tracking, communication, computers, and display and control technologies. ATMS has proved to be among the most successfully achieving these goals under the umbrella of ITS.

Strategic traffic control and management includes knowledge on operating status and traffic flow characteristics. The most important parameters for ATMS are explained in the Figure 6.16.

- Traffic flows/volumes—the number of vehicles that pass one point per time unit.
- Speed vehicles—the distance that traverses a vehicle per unit of time.

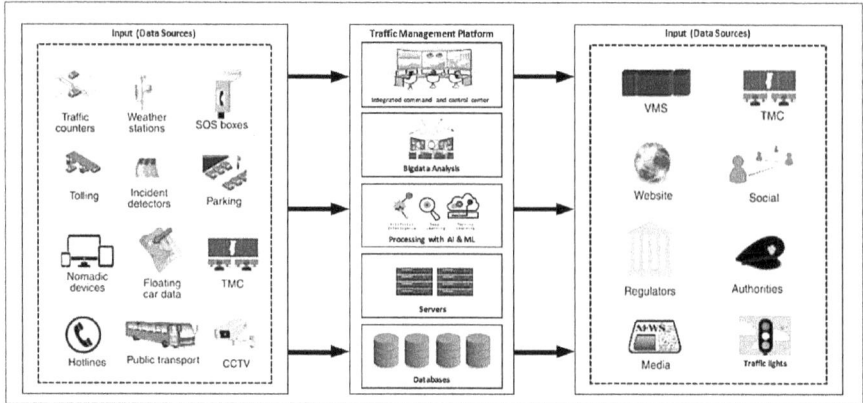

FIGURE 6.16
Advanced traffic control and management system.

- Density of traffic—the number of vehicles occupying a road lane at the instant of time.
- Occupancy—percentage of the time a sensing zone is occupied on the road.
- Incident—an unplanned occurrence on a path.
- Current weather—related information such as weather conditions.

6.5.1 Establishment of ATMS Goals and Objectives

Operational priorities and objectives would emerge from the needs described in the last part. These objectives would probably focus on the provision of transport goods and services that will have a positive effect on society by ensuring safety, promoting a better quality of life and increasing the attractiveness of businesses in a city. The aim is to increase the level of efficiency of current transport goods and services, to satisfy social and economic needs, in the following priorities, and objectives for ATMS implementation.

- Increased efficiency of capacity building and operations management.
- Increased Productivity.
- Improved safety and security.
- Enhanced Traveler Comfort and ease.
- Enhanced Public Transportation Services and ease of Operations.
- Reduced ecological impacts and maintaining balance.
- National Context.
- Performance Monitoring.

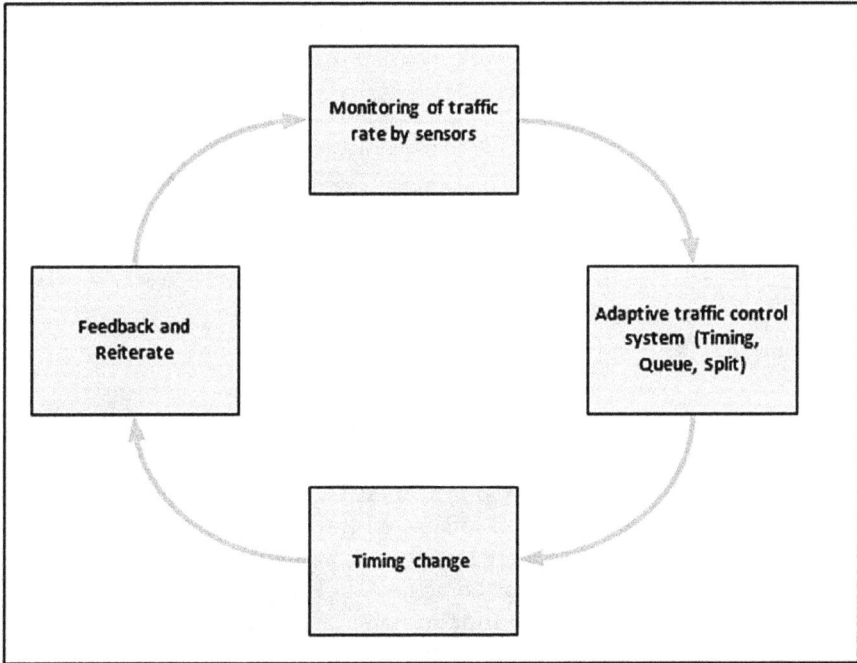

FIGURE 6.17
Flow of ATCS.

6.5.2 Adaptive Traffic Control Systems (ATCSs)

ATCSs are used to improve timing and duration of the traffic light signals by using real-time traffic data. This reduces traffic congestion in big urban zones by minimizing stopping times and delays in successful ATCSs. The traffic system naturally evaluates timing structure, break, queue estimation, and phase optimization based on the trajectory. A significant number of ATCSs were built and investigated using various approaches and structures to minimize travel times and congestion. The flow of traffic control is explained in Figure 6.16.

The benefits of adaptive traffic control systems are:

- Quick response to abrupt changes in traffic conditions.
- Reduction road congestion and fuel consumption.
- Extend the feasibility of stoplight scheduling action arrangements.
- Render stoplight operations more proactive by keeping an eye out for and responding to crevices (Figure 6.17).

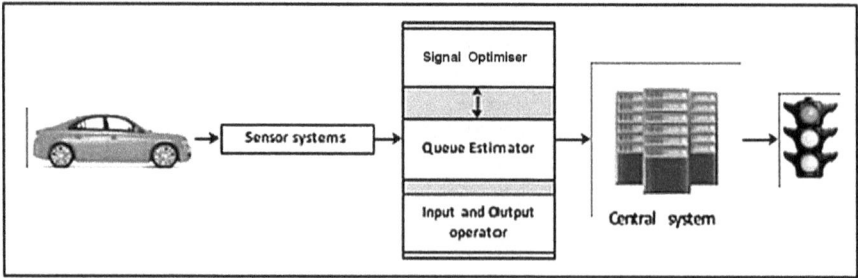

FIGURE 6.18
Architecture of SCOOT.

6.5.2.1 SCOOT

SCOOT was Split Cycle offset Optimization Technique [20]. SCOOT is perhaps the world's most deployed integrated traffic management technology with over 200 applications. There have been long links to provincial boundaries where cooperation cannot be achieved. The performance of SCOOT depended in general on the information obtained from traffic streams from the sensors. The architecture requires an extended number of identifiers found on each relation in pre-ordained areas. The sphere of the discoverers was discriminatory, mostly at the upstream end of the approach. There have been long links to provincial boundaries where cooperation cannot be achieved. The performance of SCOOT depended in general on the information obtained from traffic streams from the sensors. The architecture requires an extended number of identifiers found on each relation in pre-ordained areas. Three enhancements are available for SCOOT: Cycle Time Optimizer, Offset Optimizer, and Split Optimizer. Criticism was made about the computer vehicle and the execution of the action frame was stopped at any relation and measure. SCOOT has modified the predefined signal timing schemes after a general failure of the device. The time schedule settings will actually alter SCOOTS. It will take drifts of the traffic day by day to keep up with synchronization of the sign system. The general architecture of SCOOT is explained below (Figure 6.18).

6.5.2.2 SCATS

The Sydney Coordinate Adaptive Traffic Signals (SCATS) is perhaps the most outstanding and adaptive method of traffic management [21]. SCATS was introduced by the Roads and Traffic Authority of New South Wales, Australia. SCATS can modify the signal transit request because of differences in the time order of the signals and the structure cap as an integrated traffic control scheme. The Architecture of SCATS is described in Figure 6.19.

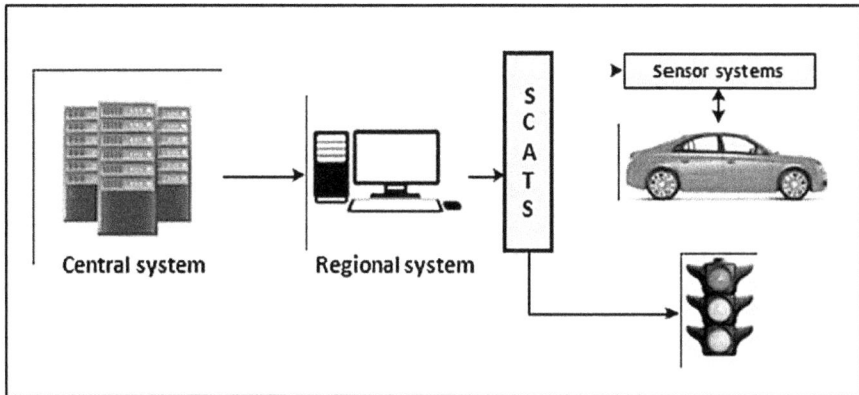

FIGURE 6.19
Architecture of SCATS.

Three control stages were designed for the SCATS: central, regional, and local. SCATS transmitted processes from a regional device at the focus of operation to the field controller for each convergence. The core level was worked by the target system referring to the level of options within the progression, mostly for recognition capability. To provide consumers with a frame that can satisfy diverse operating requirements, SCATS consolidates adaptive traffic management with routine control procedures. Methods of management include multi-faceted activity, day to day coordination, and retreat signing. The platform allows traffic engineers to track framework activities with continuous reporting devices. Persistent convergence control quickly warns managers of any unusual circumstances or deceptions.

6.5.2.3 RHODES

RHODES was a multilevel adaptive traffic control system in real-time refers to Real-Time Hierarchical Optimized Distributed Effective System [22]. Furthermore, to receiving input from a variety of localizers or detectors, RHODES will develop improved signal management schemes, strengthening potential traffic conditions. The RHODES has three remarkable characteristics which the advancement group has noted, making RHODES an adaptive signal control device which is rational and effective. In order to search for the independent device for the sharing, planning, forecasting of traffic information, and sign control, recent inventions, and programs have been decently obtained at RHODES. Secondly, RHODES has considered the irregular way of streaming operation varieties. Thirdly, RHODES had been involved in voicing predictions of individual vehicle departures, platoon arrivals, and rates of traffic sources. The architecture of RHODES is depicted below (Figure 6.20).

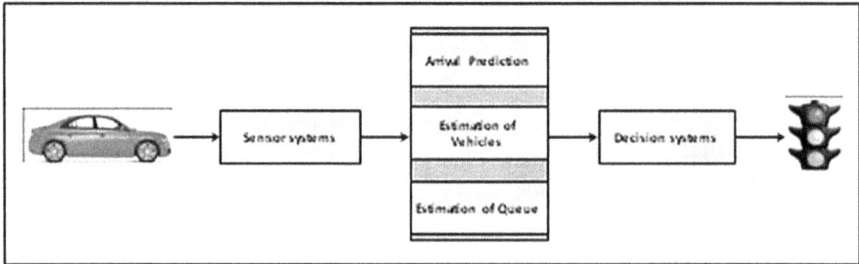

FIGURE 6.20
Architecture of RHODES.

6.5.2.4 ACS LITE

ACS Light offers a low energy traffic management framework that deals for small to medium-sized communities constantly, modify and facilitate signal time for failure to adjust the model of traffic [23]. ACS Light uses new signals for traffic or strengthens existing signals for operation. In fact, at the point at the point where the flux calculation is at the middle, the acquisition or optical movement of vehicles is in the center of the cycle reproducing green around the traffic phase. The algorithm scheme represents the control flow over either process, which assesses entirely different talents by reducing the aggregate acquisition volume and selecting the equilibrium. Benchmarks for exchanging data between ACS Lite structure and traffic control systems are used under the National Transport Communication and ITS Convention conventions (NTCIP). The architecture of the system is presented in the Figure 6.21.

6.5.2.5 OPAC

OPAC is one of the technologies for adaptive traffic control. In the 1970s it was discovered. OPAC refers to Optimized Policies for Adaptive Control [24]. It was a technique for traffic management delivered in real-time. Modified signal periods are used more than once to reduce the executive potential of vehicle deferment to stop more than the predefined horizon. Optimizing any component or sections built to undermine and/or avoid straightforward convergence. Setup parameters for changing afterwards time arrangements, which are designed to alert developments beyond any doubt about unusual or time-of-day circumstances. In OPAC, OPAC— I, OPAC—II, OPAC—III, OPAC—RT, OPAC—IV have been exposed to many ways.

6.5.2.6 InSync

The Rhythm Engineering, InSync Adaptive Traffic Control System is a smart transportation company, which enables traffic signals to adapt to real traffic requests [25]. InSync can be an interface that interacts with existing controls

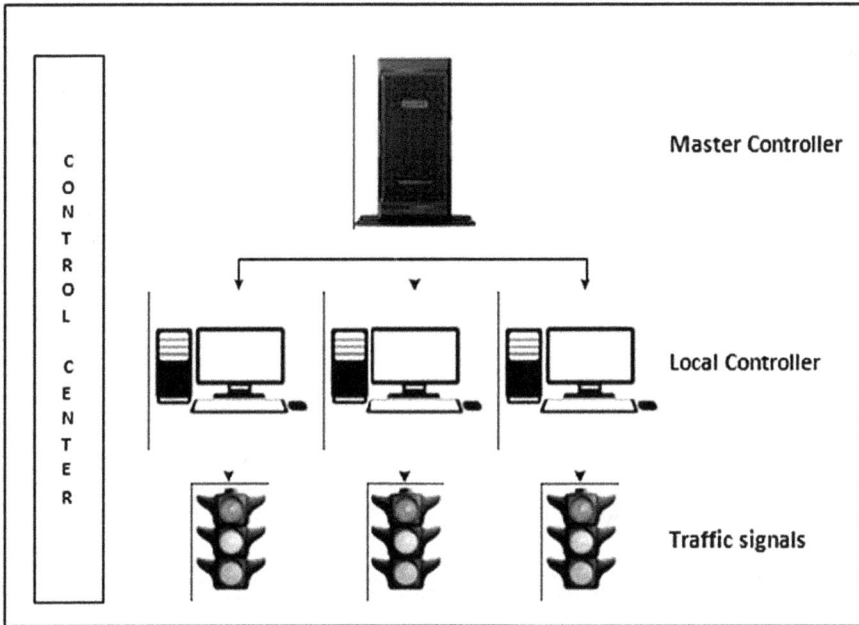

FIGURE 6.21
Architecture of ACS lite.

and controllers. The two major components of its facilities used one IP cam-corder and a CPU. Installed camcorders confirm the number of vehicles arrived and how much the vehicles keep. The server, a state-owned computer, resides within the convergence office controller. The structure circles the source, which is better served by sincere curiosity, and provides an optional overlap. The InSync system has been split into two simplification methods, which are improving, and improving globally. Optimizing the neighborhood InSync uses synchronized computer-based sensors to grasp the precise mix-ture of car management at a partner crossing point and the time it lasts. This line and postponement learning have to be upheld for square measure meth-odologies considering the scale. The complex staging and aspect unpracticed pieces of InSync change the traffic signals to make the most of unpracticed time. Optimization on a multinational scale InSync generates traffic in a sys-tematic manner by using "green passages." Platoons of vehicles form up and then travel freely along the road (Table 6.1).

6.5.3 Traffic Management System as an Application of Intelligent Transport System

Over the past 50 years, detectors, and sensors have been used for traffic counting applications, monitoring, and governance. On the ground sur-face, early sensors focused on sights like optical sensors acoustic sensors

TABLE 6.1

Comparison of Different ATCS Models [25]

Traffic Control System	SCATS	SCOOT	RHODES	ACS Lite	OPAC	InSync
Purpose	Minimized delay in stops	Minimize performance index	Minimize cumulative delay	Minimize flow of traffic	Optimization of splits and reduce total junction delays	Traffic control based on delay of individual movements
Arrival prediction	No	Yes	Yes	Yes	Yes	Yes
Estimation of queue	No	Yes	Yes	Yes	Yes	Yes
Optimization of split	Yes	Yes	Yes	Yes	Yes	Yes
Optimization of offset	Yes	Yes	Yes	Yes	Yes	Yes
Cycle time reduction	Yes	Yes	Yes	Yes	Yes	Yes
Phase sequence optimization	No	No	No	No	Yes	Yes
Saturation	Good	Good	Good	Good	Good	Good

for sound and pressure sensors like seismic and piezo sensors for weight estimation of vehicles. These detectors measure the change in respective fields or quantities caused by moving vehicles and use the knowledge to quantify traffic data and parameters. Most of these devices are invasive, since they are integrated on the surface of the roads and offer traffic alerts in real-time. The most frequently collected traffic parameters are the vehicle's length, occupancy, and pace. Inductive loop detectors, magnetometers, and magnetic detectors are the three major categories of vehicle detectors currently in operation. Traffic control for road monitoring was adopted by video cameras based on their ability to relay closed circuit TV imaging for analysis to a human operator. Today's traffic control systems use video image analysis to evaluate the concentrating and traffic tracking details scene automatically. Usually a VIP (video image processor) device encompasses cameras, a computer built on a microprocessor for digitization and image processing, and image interpretation tools for converting to traffic stream data. Some of the applications of ITS in the field of traffic management are highlighted below.

6.5.3.1 Real-Time Traffic Monitoring and Control

Instant traffic data (text, photography, video) obtained by means of the camera installed on the ground, meteorological sensors and vehicle sensors are transmitted via established communication networks to the center and tracking traffic and monitoring in real-time on some parts of state highways, motorways, and tunnels. The data was viewed and analyzed by the program in the center on the imaging systems. In the area of real-time traffic monitoring and control, traffic data gathered and analyzed include:

- In a certain timeframe, total number of vehicles.
- Classification of vehicles dependent on vehicle length.
- Average speed and occupancy information.
- Vehicle steering and speed identification.
- Mobile and stationary vehicle identification.
- Position of public works, atmospheric patterns, and road conditions, etc.

Once data is analyzed in the center, drivers are alerted by using traffic density maps and variable message signs on the traffic volume via Online and mobile platform. In the event of a collision, and road work to lead them to safe roads they are warned of the weather situation and informed of the conditions on the road. The figure below explains about the real-time monitoring of traffic from ICCC (Figure 6.22).

6.5.3.2 Traffic Signal Monitoring and Control

Fixed time signaling systems are equipped at the intersections on the road network of the Highways. This scheme allows many signaling programmed to operate on a cross-section control unit depending on daytime, seasons,

FIGURE 6.22
Real-time traffic monitoring from ICCC.

Signal phase diagram

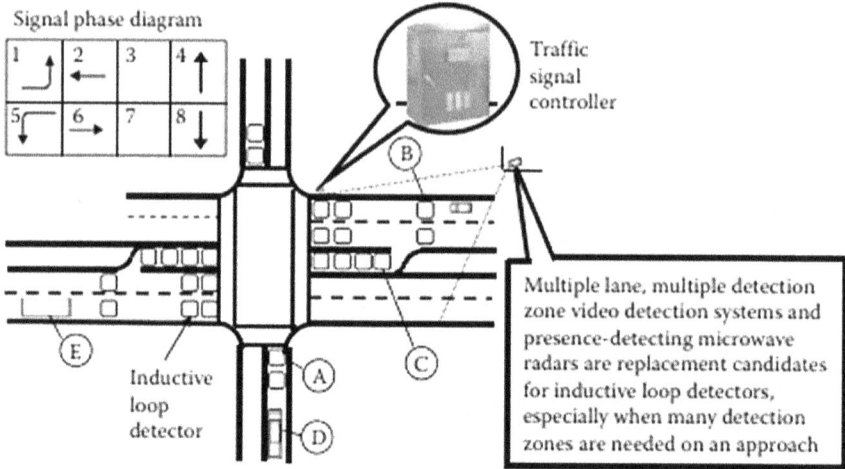

Traffic signal controller

Multiple lane, multiple detection zone video detection systems and presence-detecting microwave radars are replacement candidates for inductive loop detectors, especially when many detection zones are needed on an approach

Inductive loop detector

FIGURE 6.23
Real-time traffic planning and control.

and some days. The remote access to an intersection control unit is made possible through web-based applications in these centers to modify signal times, transmit new signals programs, and track breakdowns. In addition to the fixed signal systems, experiments are also in progress of implementing semi-actuated, fully managed, and accommodating signaling systems and controlling the center's remote and real-time traffic. The schematic of traffic signal monitoring and control is depicted in the Figure 6.23.

6.5.3.3 Speed Violation Vehicle Enforcement

Moreover, corridor speed limit control systems that are mounted at some points in the road parts monitor the average vehicle speeds. Average speed detection may also be performed using electronic toll collection systems input output data. The driver's speed is tracked by radars in tunnels. Furthermore, speed limit alert systems incorporated with radars and variable signals on black spots and the road stretches, where road traffic operators are needed to be notified throughout the road network. Speed detection instruments for the radar and laser based are not only used for inspection of speed limits but also for spot-speed tests.

6.5.3.4 Variable Message Signs (VMS)

Intelligent transport systems include a driver and passenger information system to increase traffic security and driving comfort. VMS are the most

FIGURE 6.24
Variable message displays/signages.

interactive and active part of the passenger information system. This is the graphical device that uses LED to represent text, picture, and picture and consists of a normal matrix sphere. In GDH's integration with other ITS components (camera, sensor etc.) and meteorological information stations located at the points on which the road is closed periodically with adverse weather conditions, variable messages signs are controlled. Data obtained from these systems shall be forwarded into the control center, after analyzing the data collections, the data controllers shall be alerted, automatically on the basis of predefined scenarios or manually, of traffic density, traffic collisions, road and weather conditions, and, if applicable, drivers shall be guided to the other alternate routes accessible. This enables drivers to comply with road and meteorological conditions and effectively monitor traffic flow. This would increase the efficiency of the road network capability and enhance traffic safety. Speed and license plate information is reflected on the VMS, and thus warnings are disseminated for vehicles exceeding the speed limit. Besides VMS variable traffic signals, speed limits, and lane assignments are also reflected in road networks. The depiction of various signages is shown in the below Figure 6.24.

6.5.3.5 Advanced Traveler Information System (ATIS)

The system provides passengers with updates on variable signals, road conditions and path analytics through their website, and smartphone apps to allow road users to figure out the right route, alternate directions, roads closed to traffic and road work, road status, and map conditions. The goal

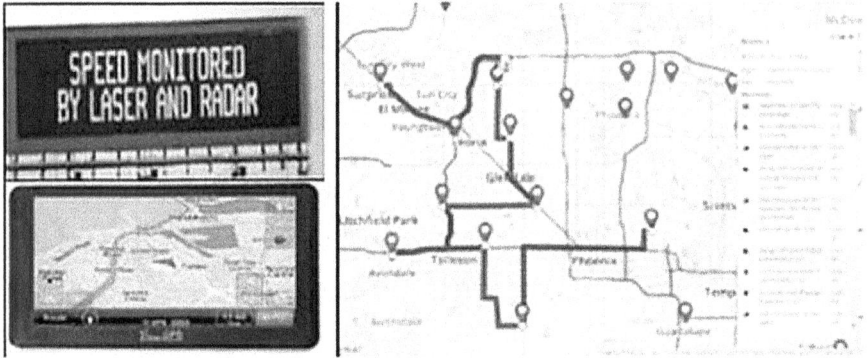

FIGURE 6.25
Advanced traveler information system.

of this system is to provide road users traveling in the radio coverage area with real-time updates on road collisions, road maintenance operations, and weather. The example of such real-time updates is shown in the Figure 6.25.

6.5.3.6 *Travel Demand Management*

This consumer service is developing and implementing policies aimed at reducing the number of single vehicles while promoting the use of high-capacity vehicles and the reliable mode of travel. The adopted techniques are:

- Price of congestion.
- Managing and controlling parking.
- Help for Mode Shift.
- Telecommuting and work routine alternatively.

6.6 Technology Advancements in Modern Intelligent Transport System

The various technology advancements evolved as an aiding factor to enhance the level of intelligent transport systems efficiency are presented below.

6.6.1 Big Data Analytics for Intelligent Transportation Systems

Ubiquitous sensors can be used in a variety of intelligent computing and decision-making processes, ranging from monitoring to environmental and human sensing. At the same time, crowd data from other sources can be used to track traffic-related incidents and to modify public transit schedules such as mobile network details, RF, IR-based data, satellite, and video data. Crowd analytics can also benefit from modern traffic control systems as they are able to process heterogeneous large volume data and maintain valuable knowledge on time and predictive information for future analyses for longer periods of time. Although the volume of mass data obtained from a sensor system can be enormous, processing it in its discomfort and maintaining it for long periods is almost impossible. In order to improve the reliability of this operation, sensor data must be filtered concurrently to hold valuable information at that particular instant.

6.6.2 Personalized Mobility Services and AI

The application of machine learning and AI algorithms to data obtained from many sensors linked to drivers, cars, and the road network makes it possible to create new intelligent solutions. It also draws large software developers, sensors and computer hardware firms who encourage solutions as a service and add a wide range of assistance accessories to drivers in cooperation with car manufactures. In the other hand, with the emergence of modern intelligent mobility services and AI apps as drivers or co-pilots, new problems occur.

6.6.3 Integrated Mobility Solutions for ITS

Through aiding drivers in avoiding automotive crashes, integrated mobility solutions add greatly to the reassessment of vehicle safety. They also use smart routing to optimize the ability of the grid, eliminating the need for new highways. Real-time data, for example, enables smart traffic signals to optimize traffic flow and smart bus schedules to automatically reallocate buses in low-traffic areas. They also save consumers time and money by reducing travel time, auto fuel usage, and carbon dioxide emissions, as well as improving a country's overall economic condition.

6.6.4 Intelligent Transportation Systems and Blockchain Technology

Emerging innovations, such as Blockchain, and the Internet of Things, seem to have the potential to make the Internet more decentralized. Blockchain technology believes that a group of individuals can collaborate to generate and exchange information without the need for a centralized body, relying solely on a network of peers. The network is decentralized and distributed uniformly, which ensures that no node/person has a higher priority or is superior to others. The goal of all nodes in the blockchain network is to build and exchange a file, and their goal is to do so using a series of rules known as the consensus protocol, which is used to maintain trust between the collaborating parties. In this way, blockchain has the potential to transform centralized ITS by creating a decentralized network that provides security and trust while enabling autonomous ITS to thrive, leverage legacy infrastructure and resources, and efficiently use crowdsourcing.

6.6.5 Autonomous Vehicle Principles and Enablers

Researchers and transport experts have also recognized the benefits of using mechanisms that remotely control the vehicle's main driving functions, such as steering, braking, and acceleration. The power of individual cars to avoid traffic makes it more effective to save petrol. Efficient collision reduction technologies help minimize road pollution and cruise control devices reduce emissions of greenhouse gases. Finally, interconnected vehicles exchanging information and data and communicating with the road networks are designed to enhance traffic safety and to minimize travel and congestion times. Automotive companies are equipping cars with multiple automation systems, intelligent decisions and state-of-the-art advanced driver assistance systems (ADAS), which encourage driving, handle the car under certain circumstances or inform the driver when required to take over, to achieve the above-mentioned degree of autonomy [26]. Sensors become their brain and nervous system in the eyes and ears of the driver and integrated computer platforms and this technology remains so reliable, defective, and stable. Another technology enabler of autonomous vehicle is artificial intelligence. The proliferation of sensor data and the growth in potential for complicated calculations by car-embedded systems are driving researchers from basic machine learning algorithms to profound artificial neural network architectures to make their decisions. Input from a car fleet offers profound and enhancement paradigm learning advantages against other solutions instead of individual instances. Using deep learning algorithms to gather real-life data from various transport scenarios constantly encourages autonomous vehicles to improve.

6.6.6 Connected Vehicles

The benefits of the connected vehicle ecosystem are the ability of wireless communications between cars, networks, and mobile devices to transform road safety, accessibility, and transport system environmental impacts. The primary objective of the Connected Vehicle Program is to increase protection for drivers and travelers. In vehicles, infrastructure, and back-office structures, the idea of connected vehicles requires different types of sensors and technology. This includes onboard equipment, roadside equipment, communications networks, and the maintenance of security credentials. The infrastructure's communication aspect comprises traffic and transport control, interactive signage boards, roadside radio warning transmissions for drivers, road signals, Bluetooth, and DSRC receptors, mobile phone equipment, and roadway sensors [27]. These components together form the interconnected ecosystem for the car. The transfer of safety-critical data using the DSRC protocol is now envisaged. Non-critical data and information can be shared via wireless networks or Bluetooth within the range. The Connected Vehicle System is designed to strengthen the safety and accessibility of public transport system management and service, to decrease the environmental impact and to improve the performance. Figure 6.26 explains the applications of connected vehicles concept in brief.

6.6.7 Cooperative ITS (C-ITS)

In Cooperative ITS (C-ITS), vehicles connect with each other and/or roadside facilities, significantly enhancing the accuracy, and efficiency of vehicle and road conditions knowledge. This brings significant social and economic advantages and increases travel quality and protection. While ITS focuses on the provision of information through emerging technology on the roadside or cars, the Cooperative-ITS focuses on cooperation between these networks. It makes it possible for vehicles to be actively communicating with

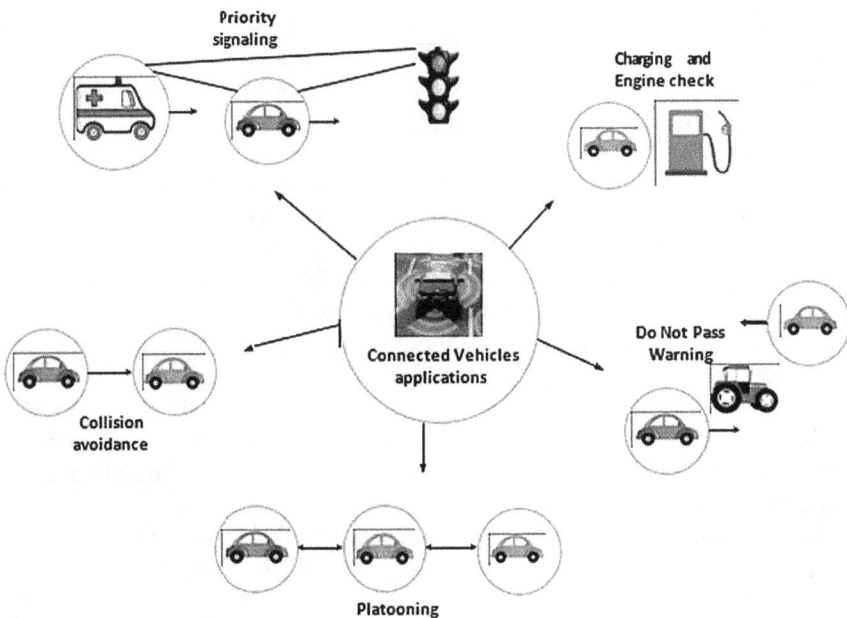

FIGURE 6.26
Applications of connected vehicles concept.

each other and the road systems in the vicinity so that road users and traffic planners can collaborate and use it to organize activities. This cooperation included communication between vehicle-to-infrastructure (V2I), vehicle-to-vehicle (V2V), and communication between vehicles and pedestrians and commuters (V2X).

6.7 Intelligent Transport System Integration with Smart Cities

A range of sensors, cameras, processor units, and networking resources are available to modern vehicles. Both of these built-in capacities enable automobiles to gather, communicate, and process data to help drivers with equipment to acquire data and function. These functionalities transform vehicles into a powerful platform that can be used to collect relevant information to be supported for traffic or resource management. The smart city can be well-explained as a smart community that integrates ICT and produces immersive experiences that connect in the real world, to solve the everyday challenges of people in urban communities and to ensure better public management. In order for such solutions to be supported, urban data must be collected and distributed through communication infrastructure. This data flow essentially involves interconnected, heterogeneous, and smart wireless networking methods. We can include synchronization of traffic signals, emergency service, parking, weather services, location services, and visitor services among the applications in a smart city. To increase the quality of the data and information supplied to citizens of the smart city or the travelers, all resources should be interconnected. The use of sensors in the streets for tracking and alerting hazards is another valuable service. The integration of ITS into the smart city segment is achieved by interconnecting each layer of ITS as a technical application. The layers of ITS for integration with smart city is depicted below in the following Figure 6.27.

The analytics and understanding of data gathered by different field devices with coordinates, such as LTE, and GPS are a major challenge in smart cities. Human mobility in such areas should be taken into consideration to map vehicle and pedestrian activity to maximize interconnection and provision. This urban dynamic thus involves understanding patterns of mobility, social activities, habits, and human experiences with the community and others. The microscopic variables linked to human activity are equipped with the urban spatial characteristics and allow individuals' macroscopic patterns and flow aspects to be determined in a given region over a time span to demonstrate sensitive areas in a road network. A stable and efficient approach to deal with the huge volume of data and diversity of the same is one of the obstacles of ITS planning in smart cities. The precise approach about the

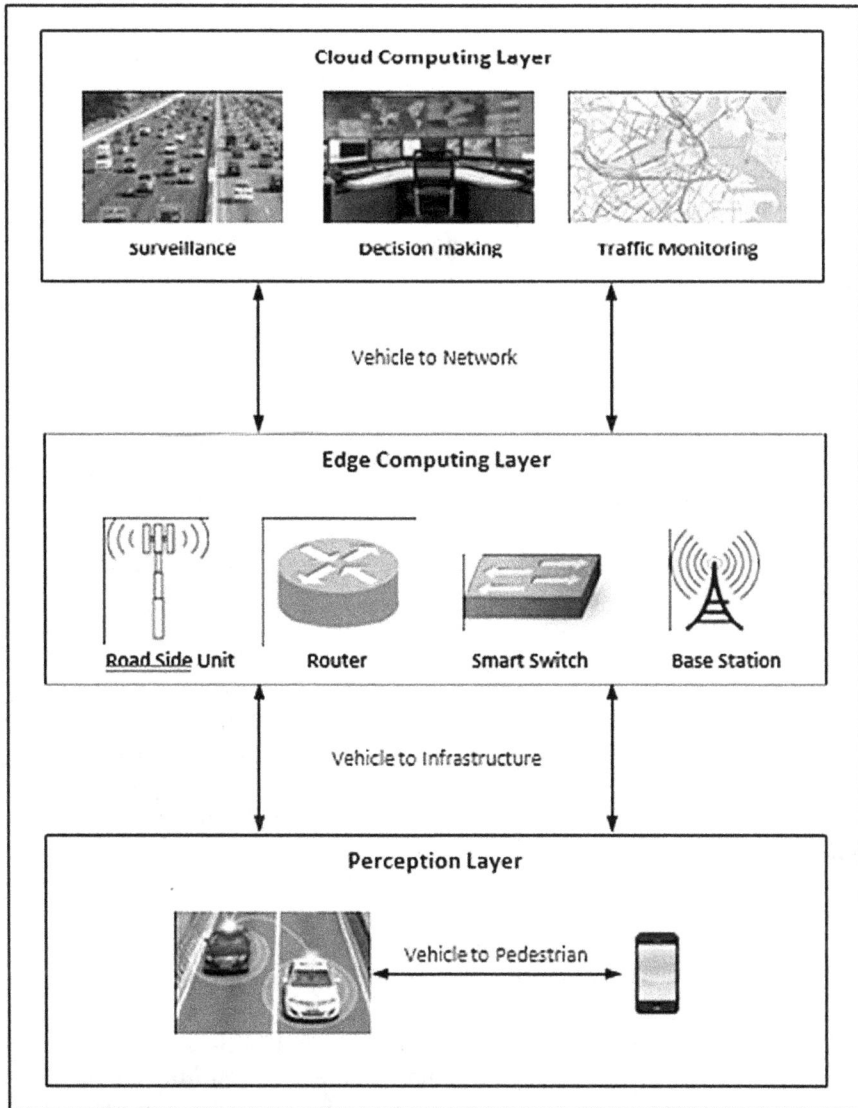

FIGURE 6.27
Layers of ITS with respect to smart city integration.

analysis and evaluation of data which can be used to take decisions is another aspect of the problem. The research can not only be used by residents and local agencies such as fire stations, ambulances, and police in a wide range of applications, such as the distribution of human resources or the movement of physical resources.

6.8 Challenges in Intelligent Transport System Design and Implementation

Currently, the number of vehicles on smart cities are showing considerable and significant dynamism in mobility of urban lifestyle and specific features of transportation networks. The rise in traffic volumes has resulted in an increase in transportation-related problems, including pollution, and significantly greater traffic injuries. Advances in technologies, vehicle sensing, and in-vehicle control have resulted in a slew of innovative solutions that focus on imperative and complex communication, enabling mobility to be abstracted in the smart city. The core ITSs can connect using diverse communication channels that are not limited to vehicle networks, like the mobile, and portable devices. This wider network connectivity makes it more scalable and less time consuming to send and receive transport system information. However, with those structures the architecture is faced with a range of problems in terms of efficiencies and efficiency and protection, with increased costs, which means it is not always practical for such programs to be introduced.

Communication between the instruments and even subsystems of this design is crucial and defines in large part the viability and efficiency of ITS solutions. The connectivity heterogeneity calls for various technical techniques, including 4G, 3G, GSM, WiMax, Wifi, and Bluetooth, to enhance design functionality and complexity. The abstraction of data from various network technologies is one of the major challenges in developing an architecture of transport system. As stated earlier, this difficulty stems from the complexities of forming a heterogeneous relation. To operate efficiently, a framework needs to establish specifications that promote component integration. Furthermore, the separate features of the urban landscape and high vehicle mobility need additional attention for proper infrastructures deployment, taking into account the whole transport system and the tolerance for delays and faults.

The ITS applications will provide a range of wireless networking technologies that enable more than single data channel of communication. In the work discussed, the relevance, and role played by the Internet infrastructure in vehicle networks is studied. Future Internet trends include digital peer-to-peer networking and support networks to deliver software and resources in a correct manner. Current works demonstrate an increasing trend toward the combination of vehicle and cloud computing networks. The aim of this combination of the two technologies is to make transport networks more scalable. Via mobile cloud connectivity, the computing ability of ITS architectures can also be improved dynamically between computers, vehicles, and a static cloud, which is the same as control centers or data centers.

6.9 Conclusion

ITS converges to information and communication technology in transportation networks. Travel planners, transport information systems, intelligent message boards and traffic signs, surveillance software, and traffic management are all examples of ITS applications and services. Vehicles today are now associated in a variety of ways. However, in the not-too-distant future, they will communicate with one another, with road networks, and potentially with other vehicles, as part of a shared, wired, and autonomous mobility environment. Collaborative intelligent transportation networks are the domain of this relationship (C-ITS). Road users and traffic management will be able to share and use information, as well as organize their activities, thanks to collaborative ITS. This factor of synergy, activated by digital communication, is expected to improve road safety, traffic quality, and driving comfort by assisting drivers (and future autonomous vehicles) in making the best decisions and adjusting to changing traffic conditions. Their production aims to achieve success in terms of reducing energy use and environmental effects. They will contribute to more reliable, cleaner, and cost-effective traffic for people, automobiles, and goods by integrating information and communication technologies.

This integration of telematics, information technology, and telecommunications allows for real-time data sharing and is common to all forms of transportation. Multimodality of travel refers to the use of several modes of transportation with the aim of maximizing capacity use, reducing congestion, and promoting more environmentally efficient modes of transportation. These systems allow data sharing between vehicles and road infrastructure using wireless technology, with the goal of improving road safety. The proper installation of equipment at key stages, as well as the creation of "intelligent" smartphone apps, are also necessary criteria. ITS can help minimize transportation-related pollution, such as CO_2 and other polluting pollutants, by putting a premium on road congestion, decreasing travel time, and improving traffic safety. Year after year, ITS is becoming one of the most relevant segments of the global market for goods and services, and their widespread acceptance will contribute significantly to the growth and optimal use of competitive advantages. Furthermore, the following patterns (among others, of course) are predicted to dominate in the (not-too-distant) future:

- Countries will gradually turn to ITS applications as the most cost-effective way to maximize their infrastructure investments.
- Many countries and manufacturers would be interested in vehicle-to-vehicle safety systems, but their market production will take time.

- Cities will develop in lockstep with the integration of ITS applications into transportation plans and programs, resulting in innovative, and streamlined transportation management systems.
- The businesses will push for universal and comprehensive transportation solutions, but implementation will be sluggish and incremental.
- Smart technologies that are constantly evolving will start to appear, but conventional implementations will take longer to implement them.
- Ambitious applications, such as autonomous cars, would be difficult to achieve, "primarily due to political and personal objections, not a lack of know-how," according to the study.
- Third-party apps for mobile devices will continue to grow and become more popular.
- Cities will struggle with traditional (legacy) technology and compatibility problems.
- For the ambitious administration of the new era of coming applications related to transportation, strong political leadership would be expected.

It is our duty to include the social component in the plans and strategies for the future of urban transportation and travel, to ensure that the most disadvantaged social classes and the weakest economic people are not separated from their everyday movements. Both residents should be able to afford transportation and travel, and people with disabilities should be able to get around more easily. ITS is undeniably a necessary step in this direction, as well as in enhancing the overall quality of human life.

References

1. Intelligent Transportation Systems, (RITA), U.S. Department of Transportation, http://www.itsoverview.its.dot.gov/CVO.asp.
2. Department of Transportation, US (May 9, 2013). National ITS Architecture Documents. Retrieved from National ITS Architecture.
3. Ghosh, S. and Lee, T. (January, 2000). *Intelligent Transportation Systems: New Principles and Architectures*. Boca Raton, FL: CRC Press. Reprinted June 2002.
4. A Report to the ITS Standards Community ITS Standards Testing Program by Battelle Memorial Institute for US Department of Transportation (USDOT), Chapter 2.
5. Intelligent Transportation Primer. U.S. Department of Transportation and Institute of Transportation Engineers, Washington, DC, 2000.

6. European Telecommunications Standards Institute (ETSI). (2011). Intelligent Transport Systems ITS.

7. Architecture Development Team, National ITS Architecture Service Packages, Prepared for Research and Innovation Technology Administration (RITA) U.S. Department of Transportation, Washington, DC, January 2012.

8. ITS Architecture for Canada. https://www.itscanada.ca/about/architecture.

9. The FRAME Architecture and the ITS Action Plan, Booklet of the E FRAME Project, June 2011. http://frame-online.eu/wpcontent/uploads/2014/10/FRAME-ITS-Action-Plan.pdf

10. Magrini, M., Moroni, D., Nastasi, C., Pagano, P., Petracca, M., Pieri, G., Salvadori, C. and Salvetti, O. (2011). Visual sensor networks for infomobility. *Pattern Recognition and Image Analysis* 21: 20–9.

11. Muthuramalingam, S., Bharathi, A., Kumar, S., Gayathri, N., Sathiyaraj, R. and Balamurugan, B. (2019). IoT-based intelligent transportation system (IoT-ITS) for global perspective: A case study. *Internet of Things and Big Data Analytics for Smart Generation*, pp. 279–300. Cham: Springer, doi: 10.1007/978-3-030-04203-5_13.

12. Binjammaz, T., Al-Bayatti, A. and Al-Hargan, A. (2013). GPS integrity monitoring for an intelligent transport system. *2013 10th Workshop on Positioning, Navigation and Communication (WPNC)*, Dresden, Germany, pp. 1–6, doi: 10.1109/WPNC.2013.6533268.

13. Dedicated Short-Range Communications (DSRC) Message Set Dictionary, SAE Std. J2735, SAE Int., DSRC Committee, November 2009.

14. Kenney, J. B. (2011). Dedicated short-range communications (DSRC) standards in the United States. *Proceedings of the IEEE* 99(7): 1162–82.

15. Dar, K., Bakhouya, M., J. Gaber, Wack, M. and Lorenz, P. (2010). Wireless communication technologies for ITS applications. *IEEE Communications Magazine* 48(5): 156–62.

16. Gozalvez, J., Sepulcre, M. and Bauza, R. (2012). IEEE 802.11p vehicle to infrastructure communications in urban environments. *IEEE Communications Magazine* 50(5): 176–83. IEEE Std 1609.4 2010, IEEE Standard for Wireless Access in Vehicular Environments (WAVE)—Multi-channel Operation, February 2011.

17. Emmelmann, M., Bochow, B. and Kellum, C. C. (2010). *Vehicular Networking: Automotive Applications and Beyond*. Chichester, UK: John Wiley & Sons Ltd.

18. Sukuvaara, T. and Nurmi, P. (2009). Wireless traffic service platform for combined vehicle-to-vehicle and vehicle-to-infrastructure communications. *IEEE Wireless Communications* 16(6): 54–61.

19. Garcia Zuazola, J., Elmirghani, J. M. H. and Batchelor, J. C. (2008). WiMAX antennas for Intelligent Transport System communications. *2008 Loughborough Antennas and Propagation Conference*, Loughborough, UK, pp. 133–136, doi: 10.1109/LAPC.2008.4516884.

20. Stevanovic, A., Kergaye, C. and Martin, P. (2009). SCOOT and SCATS: A closer look into their operations. *88th Annual Meeting of the Transportation Research Board*, Washington, DC.

21. Lowrie, P. R. (1982). SCATS: The Sydney co-ordinated adaptive traffic system. *IEEE International Conference on Road Traffic Signaling*, London, UK.

22. Mirchandani, P. and Head, L. (2001). RHODES: A real-time traffic signal control system: Architecture, algorithms, and analysis. *Transportation Research Part C* 9(6): 415–32.

23. Shelby, S. G., Bullock, D. M., Gettman, D., Ghaman, R. S., Sabra, Z. A. and Soyke, N. (2008). An overview and performance evaluation of ACS lite—A low cost adaptive signal control system. *87th Transportation Research Board Annual Meeting*, Washington, DC.
24. Ghaman, R., Gettman, D., Head, L. and Mirchandani, P. B. (2002). Adaptive control software for distributed systems. *IEEE 2002 28th Annual Conference of the Industrial Electronics Society. IECON 02*, Seville, Spain, vol. 4, pp. 3103–3106, doi: 10.1109/IECON.2002.1182892.
25. Stevanovic, A. and Zlatkovic, M. (2013). Evaluation of InSync adaptive traffic signal control in microsimulation environment. *92nd TRB Annual Meeting*, Transportation Research Board, Washington D.C.
26. Klein, Lawrence A. *ITS Sensors and Architectures for Traffic Management and Connected Vehicles*. Boca Raton, FL: CRC Press.
27. Lu, N., Cheng, N. and Zhang, N. (2014). Connected vehicles: Solutions and challenges. *IEEE Internet Things Journal* 1(4): 289–99.

7

Application of IoT and AI in the Development of Smart Cities

Rahul and Kritesh Rauniyar
Delhi Technological University, Delhi, India

Monika
University of Delhi, Delhi, India

Javed Ahmad Khan
Government Girls Polytechnic College, Ballia U.P., Tikhampur, India

CONTENTS

DOI: 10.1201/9781003287186-7

7.1 Introduction

A smart city is an urban growth that makes use of information and communication technology (ICT) and the Internet of Things (IoT) aimed at providing necessary data for resource and financial services. Information collected from people and devices operates and analyzed to maintain traffic systems, power plants, water supply systems, garbage dumping, and other structures [1]. The IoT will be the next logical step for the growth of the Internet and its worldwide extension. It is a wide network of gadgets that communicate with one another and exchange information with a wider network, from which the collective information could be used to reap benefits. Every gadget should have a distinctive label and rely on installed systems to collect details of their own and the surroundings, as well as to transmit that electronic devices or hosts. The information must also be correlated and evaluated to make more informed choices. IoT systems have four vital parts: Gadgets, Information, Person, Procedure. Gadgets: The physical devices that make up the IoT are referred to as gadgets. Each machine must be able to communicate with other machines and the network as a whole. Information: Information is the collection of data where the original details are essential to be inspected, i.e. error-verified and arranged, before being saved for later study. Person: As a specialist in their field, people work in their domain. However, with IoT, there would be a wider sense of interconnectedness across functions, and the public is increasingly communicating with people from other industries. Procedure: The final stage of IoT is a procedure where the most vital work like decision-making, bug detection, automation of IoT is done. AI is a wide area of computer science that specializes in developing tools that can handle functions that would normally involve human knowledge. Despite the fact that AI is an interdisciplinary study with multiple methods, developments in ML and deep learning are influencing nearly every sector of the software industry [2]. The main purpose of IoT is to interconnect Healthcare, Power management, Traffic Management, Security of devices, Urban mobility, Sanitation, E-governance, and so on. When a company adopts the IoT, it could hope for increased security, familiarity, and productivity. Dangerous settings and workplaces could be better assessed, and risks can be properly organized.

7.2 Some Important IoT Platform

7.2.1 Microsoft Azure

IoT describes a set of edge and cloud-based managed and platform services that communicate, track, and control billions of IoT assets. Azure allows strong reliable network communication between IoT and the machines [3]. Different services are available in Azure:

- Azure Internet of Things (IoT) Hub.
- Azure IoT Edge.
- Azure Machine Learning.
- Azure Stream Analytics.
- Azure Logic Apps.

7.2.1.1 Azure IoT Hub

It is a cloud-based program that interlinks almost any hardware. With per-device encryption and built-in network monitoring, you can extend your problem from the cloud to the edge. IoT Hub enables the communication between devices to the cloud and vice versa [4].

There are two software development kits (SDKs) for IoT Hub:

1. IoT Hub Device SDKs: build IoT application using IoT devices which allows to transmit telemetry, and also collects messages, job, method from IoT hub. It uses the coding language as well as the development language for the completion of the task.
2. IoT Hub Service SDKs: develops backend application that controls the IoT hub and transfers text, plans work, and forward requested asset improvement to the IoT machine.

7.2.1.2 Azure IoT Edge

It brings cloud analytics and personalized trade logic to gadgets, allowing the company to concentrate on advanced analytics rather than data management. Azure IoT Edge is a highly scalable Azure IoT Hub product. Devices will spend less time communicating with the cloud, respond faster to local changes, and operate reliably even during the long times of downtime if certain tasks are moved to the network's edge [5]. Figure 7.1 represents the framework of IoT Edge where it shows the execution steps of the network.

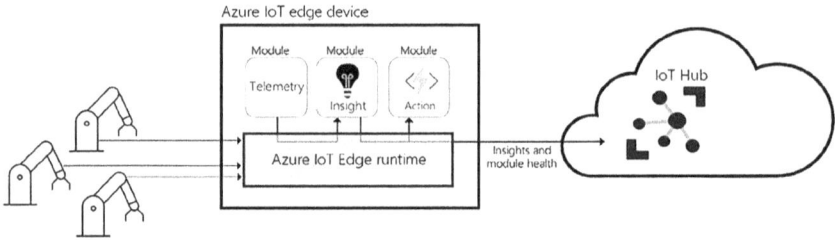

FIGURE 7.1
Framework of IoT edge.

The Azure IoT Edge includes three parts:

1. IoT Edge modules: a container that runs Azure, third-party, or custom code. Modules are downloaded and installed on IoT Edge devices, where they are run locally.
2. IoT Edge runtime: a program that is executed on each IoT Edge machine and controls the systems installed on it.
   - Workloads are installed and updated on the module.
   - Sustain IoT Edge protection policies.
   - Guarantees that IoT Edge components are fully operational all the time.
   - Module safety is reported to the cloud for remote access.
3. Cloud-based interface: used to automatically analyze and operate IoT Edge devices, and also performs several operations:
   - Create and customize a workload for a particular gadget type.
   - A workload may be sent to a group of gadgets.
   - Workloads operating on mobile devices are monitored.

7.2.1.3 Azure Machine Learning

Azure Machine Learning supports all types of machine learning, including classical, advance, supervised, and unsupervised learning. You can configure, train, and track ML and deep-learning systems in an Azure ML workspace. If you want to use the SDK to write Python or R code or the studio to work with no-code alternatives [6].

7.2.1.4 Azure Logic Apps

When you combine apps, data, modules, and products across businesses or institutions, Azure Logic Apps is used which is a cloud facility that supports planning, automatize, and arrange activities, business processes, and work-flows [7]. Logic Apps makes it easier to develop and implement scalable app integration, data and system integration, enterprise application integration (EAI), and business-to-business (B2B) communication solutions.

7.2.2 AWS

Amazon Web Services (AWS) IoT is a cloud-based service that permits linked gadgets to communicate with cloud applications and other machines simply and securely. AWS IoT handles millions of gadgets and billions of messages, and it can accurately and reliably handle and track those content to AWS datasets and different systems [8]. The program will use AWS IoT to take care of and connect with all the hardware at a time, even when it isn't linked. AWS IoT provides services across all layers of security, involving preventive security such as encryption and authentication for data sources, as well as a component that monitors. AWS provides the most comprehensive cloud platform with the largest and widest set of machine learning services, allowing them to build more effective solutions. Developers can easily add intelligence to their applications without having to be experts in machine learning. Using AI services that have been pre-trained to do things like creating more intelligent contact centers, increase demand forecasts, detect fraud, and personalize the customer experience [9].

7.3 Application of IoT and AI

7.3.1 IoT In-Home Application

A smart home is a simple home configuration in which appliances and gadgets could be operated remotely using a smartphone or other interconnected gadget from anywhere with the help of the internet. The IoT-based system can use a 433 MHz wireless detector and actuator networking as its lower level, making reconfiguration and reorganization very simple [10]. Figure 7.2 shows the connectivity of houses using IoT devices where fundamental devices are interconnected such as smoke detectors, WiFi routers, weather sensors, and so on. There are many sectors where IoT can help us in with our daily work at home:

- Smart Light: smart lighting implies that you are connected to the internet, potentially through a sensor. IoT Lighting system is the best solution for those who decrease power usage and expenses, or who wish automatic gadgets to be more comfortable.

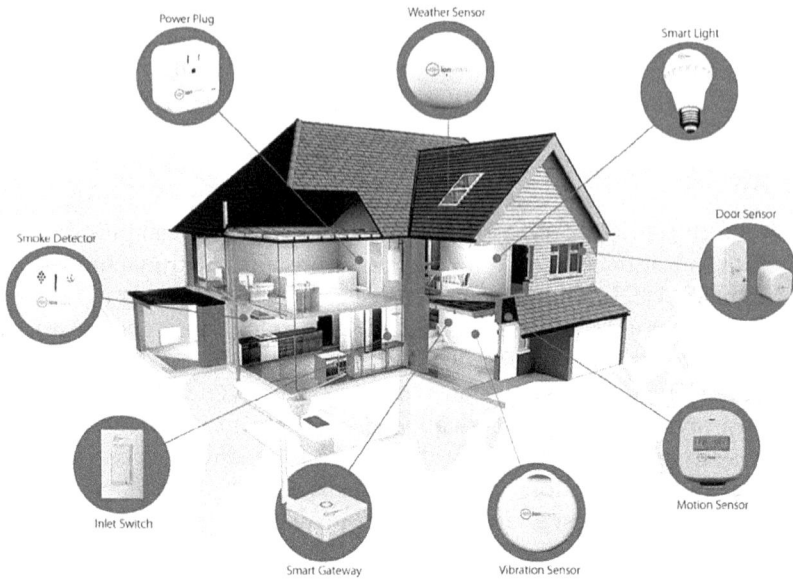

FIGURE 7.2
Smart home. (https://iot5.net/iot-applications/smart-home-iot-applications/.)

- Gas detection: this application is adjusted to identify the condition of the air. In the event of burning or smoke, it sends out notifications immediately, and users can be notified of health risks through email or SMS using the ubidots IoT platform [14].
- Voice assistant: voice command can be used to operate the devices and execute the task. Popular voice assistant applications are Alexa, Siri, Google assistant.
- Energy-saving: energy-IoT service gathers processed information via E-IoT modules, delivers it to the analysis model via the E-IoT platform, and forecasts energy usage in the E-IoT system [11].

7.3.2 IoT Wearable Device

Employees' safety and health are critical in the workplace; thus, an IoT networking that could track both environmental and physical data will significantly increase workplace safety. When hazardous environments are identified, the detector would send out a timely notification and alert to the users. Data processing, web server, and cloud connectivity are all provided by a smart IoT gateway [12]. The wearable is also used in the field of entertainment. Some commonly used wearable devices are smartwatches, fit bands, VR glasses, and many other devices. Nowadays, using wearable devices, we can make the payment process very easy.

7.3.3 IoT in Industry

In industry, coordination processes are extremely intricate, and they require detailed information about all contributing components to function properly. As a result, the engineers want to connect them using appropriate technologies [13]. Using IoT will be energy-efficient and capable of providing long-term operations. Using IoT in the industry can provide features such as [15]:

- Interconnected factories.
- Inspection of production.
- Controlling inventory.
- Security.
- Optimizing the packing.

7.3.4 Agriculture Using IoT

Agriculture plays a very important role in the human lifestyle, so we look forward to the benefit for agriculture using IoT and AI. Our main focus is to increase the productivity of the crops and cattle by using IoT devices. Important factors such as pH level of soil and nutrient content could be calculated and analyzed using IoT devices in agriculture to optimize yield [16]. Figure 7.3 highlights the different areas for development in farming using IoT like Precision Farming, Cultivation using drones, Smart Greenhouse [15].

7.3.4.1 Precision Farming

Precision farming is a procedure that improves the accuracy and management of the agriculture process. The use of different types of advanced sensors, automatic machines, and control systems are vital parts of the system. The system can control temperature and humidity with precision using the timer, hysteresis control, and condition control [17].

7.3.4.2 Cultivation with Drones

Drones have many benefits, including ease of operation, time saving, crop monitoring, and advanced visualization. Drones can be used in the field of cultivation where they can monitor the growth of crops, spraying pesticides with high accuracy, GPS and laser can be used to evaluate the quality of land, and identify the weed in the field [18]. They are mounted with various camera sensing devices (multispectral, hyper, and thermal).

FIGURE 7.3
Structure of IoT in agriculture. (https://iotdesignpro.com/articles/smart-farming-iot-applications-in-agriculture.)

7.3.4.3 Inspection of Livestock

In general, the monitoring system for animal agriculture is a very complicated system that is heavily affected by a variety of factors such as climate, temperature, and cattle situations. The Advanced livestock system uses IoT sensor data based on a wireless network where the data like temperature, humidity, NH_3, CO_2, and H_2S are compiled, and saved in a database. It offers environmental surveillance data and alerts in the event of an unusual situation [19].

7.3.5 IoT in Greenhouse

Greenhouse farming is a special agricultural technique that involves growing crops in insulated systems that are partly or fully translucent. The key aim of greenhouses is to provide optimal growth conditions for crops while also protecting them from inclement weather and pests [20]. Real-time IoT devices and cloud-based information collection tools for understanding crop growth environments are attached with a simulation model. The value of the sensor value can be transferred using Wifi or 5G technology [21]. Smart hydroponics, air conditioners, sensors, generators can be interconnected to run the greenhouse efficiently which is shown in Figure 7.4.

IoT Smart Greenhouse

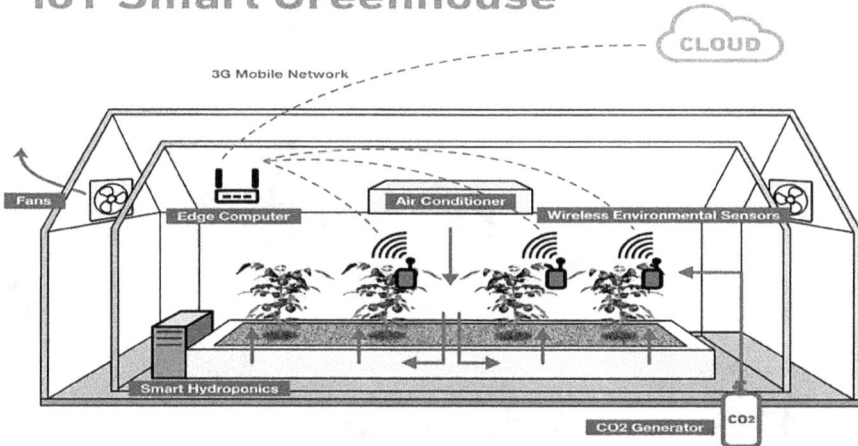

FIGURE 7.4
Smart Greenhouse. (https://aws.amazon.com/blogs/iot/aws-iot-driven-precision-agriculture/.)

7.3.6 IoT in Trade

When you purchase an item from the store, you check out from the store immediately and the product you have purchased will be scanned by an AI automated device. The payment will be deducted by your smartphone or any wearable device which is connected to your bank account. IoT will play an important role in the business market where the customers can easily purchase their items and can make easy checkout without other person help. IoT ecosystem combines five important parts: (1) Hardware; (2) Software; (3) Application developer; (4) Internet; and (5) Users [22]. There are different sectors where IoT and AI help to improve trade and business:

- Automatic Checkout.
- Smart wallet.
- Smart Employee/Robot.
- Electronic discount.

7.3.7 IoT and AI in Smart City

7.3.7.1 Smart Light

Streetlights are an example of a blazing supply of energy that is an important part of everyday life. Traditional road lamps required physical maintenance and used a significant level of energy when they were left on from sunset to

FIGURE 7.5
Smart light [31].

sunrise [23]. Figure 7.5 shows that automatic light could be guided to unlock the lights only when required by using sensors to detect use and operation at the lights. When visibility is reduced due to ambient factors such as fog or rain, they should adjust the lighting settings. In towns, a smart light will help minimize power usage. In criminal areas or during a crash, lamps may also be used to support emergency crews or police departments by adding additional illumination.

7.3.7.2 Smart Parking

In this growing population, almost every house owns more than two vehicles, but the parking facility for the vehicle is in the worse condition. Here IoT can solve this problem very easily by deploying the active real-time sensor which will detect the empty slot in the parking area and send the message to the user about the location of the empty slot. In an IoT-enabled automatic parking system, the user pays the bills directly without any other participation. A two-fold sensing technique can be used to achieve reliable car identification and parking occupancy control which is a combination of motion detection and GNSS (global navigation satellite system) [24].

7.3.7.3 Smart Traffic

A standard traffic grid comprises four paths, including one with a signal that works successively and has a set timer across each end. The issue with this traditional method is that it is unable to sense the volume of vehicles on each lane, resulting in time being lost even though a lane is clear [25]. Real-time traffic information could be used to help reduce traffic congestion. The objectives are to minimize travel time, increase traffic flow, and decrease congestion and gasoline consumption in the region [1]. Using an advanced parking system, vehicle collisions can be identified more easily, and relief can be deployed to the scene more effectively.

7.3.7.4 Waste Management

The present scenario of waste management comprises the manual collection of waste in a particular time slot. The recycling process is not that practically implemented. People don't dispose of their waste in the proper place, and this increases pollution. IoT helps to track the waste using different sensors like infrared, ultraviolet, and other important sensors. Using IoT devices identify not only the waste, but also the smell. With the help of sensing data, companies can effectively collect the waste from the proper places [26]. Working on this type of technology can decrease pollution.

7.3.7.5 Security

The objective of smart city systems is to provide real-time monitoring, adaptive control, and monitoring in the city. The integration of Software Defined Network (SDN) and IoT provides ease in IoT network management while still posing a threat to SDN due to the wide range of access devices [27]. Both the plane and information planes could create a specific limit for a system in a real deployment. The controller can identify and minimize unusual activity based on this threshold.

7.4 Challenges

For implementing a new technology many problems arise which can obstruct the development of Smart City.

Networking: Communicating millions of active linked IoT gadgets is a major obstacle, and existing connectivity models and systems need to be tweaked to address scalability concerns [1]. Present end-to-end distribution congestion management rules are based on TCP, which is not well suited to the IoT system [3]. Such problems can be solved by applying fog computing, peer-to-peer network, and blockchain.

Heterogeneity: IoT information is obtained from a wide range of sensors using a variety of protocols. Consequently, efficiently interpreting, and evaluating the data is complicated. Heterogeneity concerns emerge due to the absence of a standard data collection protocol.

Security: the fact that citizens have no idea how their information has been used creates a sense of mistrust, particularly when the data exposed on the networks is extremely sensitive [28]. In Smart City settings, private data is handled by third-party agencies' "trusted" Centralized Databases. Smart City blockchain-based technologies significantly minimize this data leakage while also helping us to benefit from trustworthy transactions and improved data control.

Big data: IoT must be adequate to discuss time, energy, and computing capacities because connected devices constantly and concurrently produce vast volumes and various information.

Cybersecurity [29]: IoT gadgets are heterogeneous, numerous, have small computational power, and run on the boundary of computer networks, these data are prone to cyber-attack. Would need to be shielded from. These attacks can be solved by using ML where Logistic Regression (LR) algorithm can be used to detect such attacks.

Sensors: sensors are the most important device for the IoT system. The sensor collects the input data and sends it to the processor. Using inexpensive sensors leads to a decrease in accuracy and hackers can easily penetrate through those sensors.

7.5 Role of 5G

A 5G network is evolving to connect static and wireless gadgets for a range of applications with widely varying speed, latency, protection, and network availability specifications. 5G promises to accommodate vast amounts of data, link many devices, decrease service latency, and provide the latest levels of reliability for providing personalized services supported on individual QoS demands. Figure 7.6 shows the major operation of 5G where devices, access networks, core networks, and IoT services are the four levels that make up the 5G framework. Via the core network and access network, many consumer computers are linked to the data network. In the 5G network, 35 forms of the hack have been reported as serious threats to anonymity, authentication, credibility, and accessibility [30]. Although 5G expands the scope and scalability of IoT offerings, it also exposes users to a range of security and privacy risks on different sources.

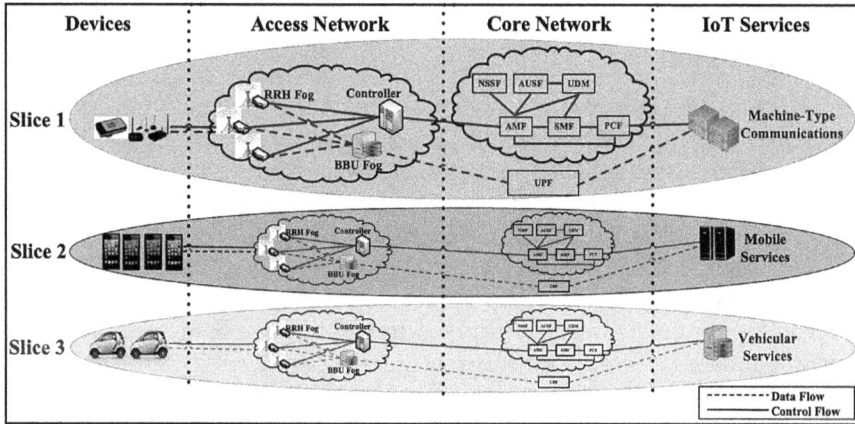

FIGURE 7.6
Architecture of a 5G reference system [32].

7.6 Implemented Smart City Projects

- Spain saves money on taxation by making better use of energy with the assistance of sensors. Garbage cans are only emptied when the bins are fully loaded, and public areas are only irrigated when they're too dry.

- Garbage trucks are no longer used in Helsinki and trash is delivered to underground recycling plants without creating noise or emissions.

- Residents in Rio de Janeiro will mold their surroundings with the aid of a smartphone app. A smart favela is an interface that builds a three-dimensional avatar of a city. As city planners come up with new plans, they can be seen on the app and voted on by users.

- Parking lots in Palo Alto, California, now have sensors that warn people when a parking spot becomes accessible, and the city's traffic is continuously monitored.

- Dubai is expected to be better due to smart robocops. The police robots have cameras and can recognize people using facial recognition. A touchscreen can be used to file technical reports.

- People of Rehoboth, Ukraine, will view civil servant wages and see how lawmakers voted in the City Council. In the battle against corruption, city-owned companies open up their books using open data.

- Interlinked sensors in Darmstadt, Germany, measure air quality, and transfer the data to a data center, which evaluates emissions and, if required, records low air quality.

7.7 Conclusion

This chapter discusses the application of IoT and AI where it describes the problems that IoT systems face and proposes a lively approach by enhancing the current architecture. We presented the role of 5G for the fast and effective way of communication between the different devices. We also gave the technique for solving the challenges faced by IoT in smart cities. IoT connects different devices to one network where the information is collected using sensors, input from the sensor is processed and the output is given to the operator. The Internet of Things should be a transformative breakthrough in the industry and elsewhere to transform business operations. It denotes a reality in which everything is interconnected and intelligently communicated.

References

1. Firouzi, F., Farahani, B., Weinberger, M., DePace, G., & Aliee, F. S. (2020). IoT Fundamentals: Definitions, Architectures, Challenges, and Promises. In *Intelligent Internet of Things*. https://doi.org/10.1007/978-3-030-30367-9_1
2. [Online]: https://solarimpulse.com
3. [Online]: https://azure.microsoft.com/en-us/overview/iot/
4. [Online]: https://docs.microsoft.com/en-us/azure/iot-hub/about-iot-hub
5. [Online]: https://docs.microsoft.com/en-us/azure/iot-edge/about-iot-edge?view=iotedge-2018-06
6. [Online]: https://docs.microsoft.com/en-us/azure/machine-learning/overview-what-is-azure-ml
7. [Online]: https://docs.microsoft.com/en-us/azure/logic-apps/logic-apps-overview
8. [Online]: https://docs.aws.amazon.com/whitepapers/latest/aws-overview/internet-of-things-services.html
9. [Online]: https://aws.amazon.com/machine-learning/
10. Wang, M., Zhang, G., Zhang, C., Zhang, J., & Li, C. (2013). An IoT-Based Appliance Control System for Smart Homes. *Proceedings of the 2013 International Conference on Intelligent Control and Information Processing, ICICIP 2013*, 744–747. https://doi.org/10.1109/ICICIP.2013.6568171
11. Choi, H. S., & Yeom, K. R. (2018). Study on an Energy-IoT Service Platform for Energy Saving in Legacy Manufacturing Site. *International Conference on Ubiquitous and Future Networks, ICUFN, 2018-July*, 811–813. https://doi.org/10.1109/ICUFN.2018.8436969
12. Wu, F., Wu, T., & Yuce, M. R. (2019). Design and Implementation of a Wearable Sensor Network System for IoT-Connected Safety and Health Applications. *IEEE 5th World Forum on Internet of Things, WF-IoT 2019 – Conference Proceedings*, 87–90. https://doi.org/10.1109/WF-IoT.2019.8767280

13. Routray, S. K., Sharmila, K. P., Javali, A., Ghosh, A. D., & Sarangi, S. (2020). An Outlook of Narrowband IoT for Industry 4.0. *Proceedings of the 2nd International Conference on Inventive Research in Computing Applications, ICIRCA 2020*, 923–926. https://doi.org/10.1109/ICIRCA48905.2020.9182803

14. Kodali, R. K., Tirumala Devi, B., & Rajanarayanan, S. C. (2019). IoT-Based Automatic LPG Gas Booking and Leakage Detection System. *Proceedings of the 11th International Conference on Advanced Computing, ICoAC 2019*, 338–341. https://doi.org/10.1109/ICoAC48765.2019.246863

15. Jabraeil Jamali, M. A., Bahrami, B., Heidari, A., Allahverdizadeh, P., & Norouzi, F. (2020). *Some Cases of Smart Use of the IoT*. https://doi.org/10.1007/978-3-030-18468-1_4

16. Kjellby, R. A., Cenkeramaddi, L. R., Froytlog, A., Lozano, B. B., Soumya, J., & Bhange, M. (2019). Long-Range Self-Powered IoT Devices for Agriculture Aquaponics Based on Multi-Hop Topology. *IEEE 5th World Forum on Internet of Things, WF-IoT 2019 – Conference Proceedings*, 545–549. https://doi.org/10.1109/WF-IoT.2019.8767196

17. Wiangtong, T., & Sirisuk, P. (2018). IoT-Based Versatile Platform for Precision Farming. *ISCIT 2018 – 18th International Symposium on Communication and Information Technology, ISCIT 2018*, 438–441. https://doi.org/10.1109/ISCIT.2018.8587989

18. Mogili, U. R., & Deepak, B. (2020). An Intelligent Drone for Agriculture Applications with the Aid of the MAVlink Protocol. In *Lecture Notes in Mechanical Engineering*, 195–205. https://doi.org/10.1007/978-981-15-2696-1_19

19. Lee, M., Kim, H., Hwang, H. J., & Yoe, H. (2020). IoT-Based Management System for Livestock Farming. In *Lecture Notes in Electrical Engineering*, 536 LNEE, 195–201. https://doi.org/10.1007/978-981-13-9341-9_33

20. [Online]: https://www.agrivi.com/en

21. Shamshiri, R. R., Bojic, I., van Henten, E., Balasundram, S. K., Dworak, V., Sultan, M., & Weltzien, C. (2020). Model-Based Evaluation of Greenhouse Microclimate Using IoT-Sensor Data Fusion for Energy Efficient Crop Production. *Journal of Cleaner Production*, 263, 121303. https://doi.org/10.1016/j.jclepro.2020.121303

22. Lee, I. (2019). The Internet of Things for Enterprises: An Ecosystem, Architecture, and IoT Service Business Model. *Internet of Things*, 7(2019), 100078. https://doi.org/10.1016/j.iot.2019.100078

23. Mary, M. C. V. S., Devaraj, G. P., Theepak, T. A., Pushparaj, D. J., & Esther, J. M. (2018). Intelligent Energy Efficient Street Light Controlling System Based on IoT for Smart City. *Proceedings of the International Conference on Smart Systems and Inventive Technology, ICSSIT 2018, ICSSIT*, 551–554. https://doi.org/10.1109/ICSSIT.2018.8748324

24. Kanan, R., & Arbess, H. (2020). An IoT-Based Intelligent System for Real-Time Parking Monitoring and Automatic Billing. *2020 IEEE International Conference on Informatics, IoT, and Enabling Technologies, ICIoT 2020*, 622–626. https://doi.org/10.1109/ICIoT48696.2020.9089589

25. Talukder, M. Z., Towqir, S. S., Remon, A. R., & Zaman, H. U. (2017). An IoT-Based Automated Traffic Control System with Real-Time Update Capability. *8th International Conference on Computing, Communications and Networking Technologies, ICCCNT 2017*, 1–6. https://doi.org/10.1109/ICCCNT.2017.8204095

26. Chen, W. E., Wang, Y. H., Huang, P. C., Huang, Y. Y., & Tsai, M. Y. (2018). A Smart IoT System for Waste Management. *Proceedings – 2018 1st International Cognitive Cities Conference, IC3 2018*, 202–203. https://doi.org/10.1109/IC3.2018.00-24

27. Wang, S., Gomez, K. M., Sithamparanathan, K., & Zanna, P. (2019). Software Defined Network Security Framework for IoT-Based Smart Home and City Applications. *2019, 13th International Conference on Signal Processing and Communication Systems, ICSPCS 2019 – Proceedings*. https://doi.org/10.1109/ICSPCS47537.2019.9008703

28. Mora, O. B., Rivera, R., Larios, V. M., Beltran-Ramirez, J. R., Maciel, R., & Ochoa, A. (2019). A Use Case in Cybersecurity Based in Blockchain to Deal with the Security and Privacy of Citizens and Smart Cities Cyberinfrastructures. *2018 IEEE International Smart Cities Conference, ISC2 2018*, 1–3. https://doi.org/10.1109/ISC2.2018.8656694

29. Chesney, S., Roy, K., & Khorsandroo, S. (2021). *IoT Cybersecurity Attacks*. Springer International Publishing. https://doi.org/10.1007/978-3-030-55190-2

30. Ni, J., Lin, X., & Shen, X. S. (2018). Efficient and Secure Service-Oriented Authentication Supporting Network Slicing for 5G-Enabled IoT. *IEEE Journal on Selected Areas in Communications*, *36*(3), 644–657. https://doi.org/10.1109/JSAC.2018.2815418

31. Phil Ling (2011). Traffic Monitoring System Uses Bluetooth Sensors over ZigBee. https://www.eenewsanalog.com/news/traffic-monitoring-system-uses-bluetooth-sensors-over-zigbee

32. Sabella, D., Serrano, P., Stea, G. et al. (2018). Designing the 5G Network Infrastructure: A Flexible and Reconfigurable Architecture Based on Context and Content Information. *Journal of Wireless Communications Network*, *2018*, 199. https://doi.org/10.1186/s13638-018-1215-1

8

Flood Management Policies in Megacities: A Case Study of Southern India

Y. Rekha

KCG College of Technology, Chennai, India

S. Suriya

Jerusalem College of Engineering, Chennai, India

Carolin Arul

Anna University, Chennai, India

CONTENTS

8.1 Introduction

Water is considered an elixir for human life. A country's socio-economic development is based on the availability and proper management of water resources [1]. Improper management of water resources leads to hydrologic extremes (floods and droughts). Flooding is considered most severe since it causes loss of life and damage to properties. Almost three quarters of the Chennai city was inundated and more than 50,000 people were affected due to the floods in 2008 [2]. Rapid urbanization, mass migration, industrialization and encroachment of the waterbodies aggravate flooding [1, 3].

One of the major causes for flooding is encroachment of wetlands and flood plains, which obstruct the floodways and hence the loss of natural flood storage space in urban areas [4, 5]. For example, primary and second-ary storm-water drainage system in Bangalore city, which often fails to carry the runoff due to silt and garbage causing blockage of the same [4]. Also, the degradation of urban wetlands due to the so-called developmental activities and encroachments had resulted in shrinkage of wetlands [6]. The spatial shrinkage of the water spread area in the Perungudi lake, the Pallikaranai lake, the Velachery Lake, the Taramani lake etc., which formed a part of Pallikaranai Marsh Land (PML) in and around the Chennai city is due to urban expansion [7, 8]. The allocation of lands for the Railways, the Chennai Corporation, the National Institute of Ocean Technology (NIOT), and the Centre for Wind Energy Technology (CWET), has resulted in the reduction of storage of the swamp [9, 10]. The retention time of floods in marshy lands has increased considerably due to occupation of the swampy areas as men-tioned above. Some experts suggested that construction of new waterways and storm-water drains would lessen the impact of flooding as the surface runoff would flow freely to the Buckingham canal via PML [11]. This urges to the formation of a public policy with an intention to preserve the wetland under the umbrella of environmental management. Therefore, this study on reviewing the existing water policy for flood management and its field impli-cation is essential to attenuate the severity of flooding.

The Water Policy of India, 1987, was formulated considering domestic, agriculture, hydropower, industrial, and other development activities. The policy was revised in the years 2002 and 2012. The Southern states of India like Karnataka, Andhra Pradesh, and Kerala have revised their respective water policies. Though, the Water Policy of Tamil Nadu was formulated in the year of 1994, it has not been revised so far. The revision of the State Water Policy is essential to address the issues like over exploitation of groundwater, degradation of surface water, encroachment of waterbodies, sea water intru-sion in the coastal areas, sand mining, siltation of tanks, and rapid increase of population, which causes urban flooding [12]. The aim of this study is to review and compare the water policies formulated by the Southern States with the National Water Policy, 2012, particularly in the category of flood management. This study tries to identify the gaps in policy implication in field level and provides suggestion for future revision of the State Water Policy.

8.2 Comparative Study on Water Policies

A comparative study between the Tamil Nadu and the other Southern states with respect to the National Water Policy, 2012, has been attempted, and is shown in Table 8.1. In the National Water Policy, 2012 the policy formulation includes aspects like water quality, information system, maximizing avail-ability of water, project planning, maintenance and modernization [7], safety of structures, groundwater development, water allocation priorities, water rates, water zoning, participatory approach, conservation of water, flood control and management, land erosion by sea or river, drought management, science and technology and training [13].

The Andhra Pradesh, Karnataka, and Kerala States have formulated their respective water policies. Some of the aspects were not covered in their water policies since these states have their own water resources management Acts like The Telangana Water, Land and Trees Act, 2002, The Karnataka Ground Water Act, 1999, The Kerala Ground Water Act, 2002, and The Kerala Irrigation and Water Conservation Act, 2006 [14].

The following inferences were made from Table 8.1:

- It was identified that the Tamil Nadu State should formulated its Water Policy in aspects like maximizing water availability, proj-ect planning, safety of structures, groundwater development and land erosion by sea or river [15]. The Andhra Pradesh state should strengthen the aspects like institutional mechanism, participatory approach, and governance [16].

- In NWP, 2012, there is a separate section which deals with manage-ment of flood. It focuses mainly on rehabilitation of natural drain-ages through structural and non-structural measures and flood forecasting for preparedness for floods [13]. The policy also insists about preparation of flood inundation maps [17] based on the flood recurrence in order to cope up the situation by providing access to shelter, food, and drinking water.

- The Andhra Pradesh State Water Policy, 2008, emphasizes the same aspects, which were mentioned in the NWP, 2002[14]. The policy includes carrying out flood protection works, soil conservation in the catchments, providing appropriate flood cushioning in water impoundment projects and giving top priority for flood forecasting for timely warning in the flood plains. There was a special mention about the necessity of regulation of settlements in the flood plain zones, protection of existing forest and increase in extent of forest coverage area as it plays a major role in flood management.

- The Karnataka Water Policy, 2019, was formulated in the year 2002, and revised in 2019 [18]. This state water policy was also in line with NWP, 2012. In the formulation year 2002, this policy gave a broader overview of disaster management strategy for both drought and flood management. But in 2019, it specifically indicated that flood-ing cannot be prevented but can be managed by both structural and non-structural measures, providing earlier warning system support

TABLE 8.1

Comparative Study on the Flood Control and Management Policies Existing in the Southern States of India

Sl. No.	National Water Policy (2012)	Tamil Nadu Water Policy (1994)	Andhra Pradesh Water Policy (2008)	Karnataka Water Policy (2019)	Kerala Water Policy (2008)
1	There should be a master plan for flood control and management for each flood-prone basin.	Same as NWP, 1994	Flood protection can be achieved by the use of a combination of structural and non-structural measures.	Disaster management strategy for drought and floods should be formulated.	—
2	Adequate flood cushion should be provided in water storage projects to facilitate better flood management.	-do-	This includes providing adequate flood cushion in water storage sites, identifying, zoning, and regulating flood risks, using current flood forecasting and communications systems, and ensuring flood preparedness in vulnerable communities.	—	—
3	Flood management should take precedence in reservoir regulation policies in flood-prone regions by even sacrificing the irrigation and power benefits.	-do-	—	—	—
4	Structural flood protection measures such as embankments and dikes are required, but non-structural interventions such as flood forecasting and warning, flood plain zoning, and flood proofing should also be prioritized in order to minimize damages and reduce recurring flood relief costs.	-do-	—	—	—

(Continued)

TABLE 8.1 (Continued)

Comparative Study on the Flood Control and Management Policies Existing in the Southern States of India

Sl. No.	National Water Policy (2012)	Tamil Nadu Water Policy (1994)	Andhra Pradesh Water Policy (2008)	Karnataka Water Policy (2019)	Kerala Water Policy (2008)
5	Loss of life and property due to flooding can be reduced by implementing strict regulation of settlements and economic activity in flood plain areas which includes flood proofing also.	-do-	—	—	—
6	Flood forecasting practices should be modernized, enhanced, and expanded to include uncovered areas.	-do-	—	—	—
7	Inflow forecasting to reservoirs should be instituted for their effective regulation.	-do-	—	—	—

during high rainfall periods and flood plain administration. There was a special mention about the urban flood which indicated about incipient risk of flood due to increase in impermeable area and encroachment of storm-water drainage. This gives attention to adopt proper storm-water management under an umbrella of Integrated Urban Water Management Programmes. This policy also insisted about providing support to all Urban Local Bodies (ULB) in eviction of encroachment from storm-water drains.

• The Kerala State Water Policy was formulated in the year of 2008. In this, there was no special mention about flood and disaster management [19].

• The Tamil Nadu State Water Policy, 1994, was in line with NWP, 1987, in the flood management category. This policy deals with the preparation of river basin plans and also it identifies flooding extent and intensity. Further it indicates about the watershed management, flood forecasting for reservoir operations, zoning of flood plains and restricting the encroachments in flood plain areas by both public and private in order to improve the carrying capacity of the river. Frequent increase in flood and non-implication of existing water policies due to lack of awareness it is necessary to revise the State Water Policy [15].

Based on the gaps identified, the revision was carried out in 2007 in the state of Tamil Nadu, but it is in draft mode [16]. There was a review report on the State Water Policy of Tamil Nadu in line with NWP, 2012, with regard to Climate Change prepared by Institute of Water Studies (IWS) and it was published in the year 2015, which mentions disaster mitigation measures and preparation of the masterplan for the flood prone basin, which includes basin plans, soil conservation, and catchment area treatment, adequate provision of flood cushion in water storage projects wherever possible, establishing the network of flood forecasting stations in the state, zoning the flood plains and also constructing the embankments and dikes [20]. A flooding hot spot area identified by CMDA, West Velachery was chosen for study to understand the implementation strategy.

8.3 Study Area Description

West Velachery, the 32nd flooding hotspot of Chennai [4] is situated at latitude 12°59′18.32″S and longitude 80°12′45.70″E. The line sketch of Water Resources Department (WRD) works carried out in the study area is shown in Figure 8.1 and also shows the direction of flood water carried by seven waterways to Bay of Bengal. The water spread area of Velachery Lake is found to be around 0.98 mm^2 was reduced to 0.49 mm^2 due to encroachment in 2019. This area is subjected to frequent flooding due to its low-lying nature and urbanization [8]. The WRD has constructed a storm-water drain in study area in order to minimize the inundation. Still, the flooding problem persists due to excess runoff from upstream of the study area and also an afflux created by the Pallikaranai Swamp.

The Pallikaranai Swamp is fed by 96 tanks out of which majority of the tanks are defunct. The surplus water from the Velachery Lake gets drained into the Pallikaranai Swamp since it is the last tank located in the cascading of 96 tanks. An afflux is created by the swamp due to the reduction of carrying capacity of the Buckingham Canal, through which the entire flood gets drained into the Bay of Bengal through Kovalam Creek.

8.4 Tools Available for Problem Identification

There are various types of the tools available to identify the problem in the study area are semi-structured interview, Focus Group Discussion (FGD), questionary survey, and stakeholder interview. In semi-structured interview,

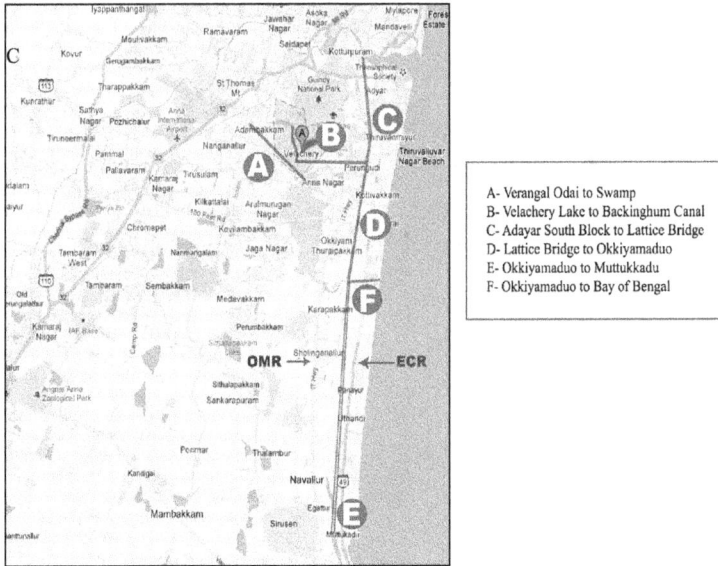

FIGURE 8.1
Line sketch of WRD works carried out in the study area.

(Source: Google Image, 2019.)

only predetermined questions were asked to the officials. The major disadvantage of the tool is the conversation will be between one to one. In FGD, a group of identified people (not more than 10) will be opted for to know about the views and perception of problems. As, it does not involve the participation of any other officials and the members are minimum it is difficult to relay on the opinion of target group members. Sometimes, there is a chance of hesitation from them to express their thoughts, particularly when their views differ with other participants. The Questionnaire survey method is used to collect the data which is related to cause and effect of problems using a set of common questions circulated among the affected people. The disadvantages in this method are limited response, sometimes lack of personal contact, poor response, incomplete entry, and people might be unaware of the seriousness of the problem. Hence, there may be difficulty in making out the ground reality of the affected area using the above methods.

Stakeholder interview was conducted with all the stakeholders (both officials and affected people) to identify the details about the specific issues. This method helps in collecting the relevant information, stakeholder's reaction, their suggestions, and also broader overview. Using this tool, key stakeholders who has specific knowledge about the cause of the problem can be targeted. This method also provides an overview of how to communicate with the public. Hence, a stake holder meeting (which consists of a gathering

FIGURE 8.2
Stakeholder meeting with the local resident association members.

FIGURE 8.3
Portion of seven-vent culvert carrying flood water to the swamp.

of around 10 to 12 members of local resident association) was conducted in the study area to identify the cause of flooding and its impacts shown in the Figure 8.2. The major causes for the flooding insisted by stakeholders are due to flat topography, minimal maintenance in cross drainage works available, insufficient number of culverts in swamp (Figure 8.3 depicts a portion of seven-vent culvert which is insufficient to carry flood water toward the swamp), defunct and encroached tanks in floodway and dysfunctional storm-water drains. The effects of flooding were severe in the study area during 2005, 2008, and 2015 floods. The people suffered with power interruption for more than five days and strived hard for the basic needs like milk, food, and drinking water, which were supplied by Government and Welfare Associations. Some people were forced to move to safer places. Flood relief was given to the affected people by the Government. Majority of the stakeholders are not aware of the flood mitigation policies.

8.5 Case Study Summary

From Tamil Nadu Water Policy, 1994, it was inferred that by constructing the structural measures like embankments and dikes flood inundation could be minimized. Through non-structural measures such as flood forecasting and warning would help to reduce the losses as well as recurring expenditure on flood relief. But the implementation of the State Water Policy is facing lot of hitches during execution at the field level due to social, economic, and political issues. Adding to that, improving the coordination among various Government and non-Governmental agencies at flooding site would make the situation better during floods. Hence, proper implementation of policies at ground level by the Government along with the support of stakeholders will reduce the impact of flood inundation in future.

8.6 Recommendations Based on Water Policy

The following recommendations were devised for the policy implication in the study area:

- The storm-water drains located in the study area should be connected to the macro drain in order to minimize the flood inundation.
- The existing micro- and macro-drainage system provided is not sufficient to carry the flood discharge in future due to rapid urbanization. Hence, proper mitigation works should be carried out.
- The South Buckingham Canal should be widened from Adyar to Muttukadu to carry flood discharge by the Pallikaranai Swamp. This will reduce the afflux produced by the swamp thereby the flooding impact be minimized.
- The encroachments along the banks of the lake and canal have to be removed for the free flow of flood water.
- The widening and desilting of waterways before onset of monsoon will minimize the flooding impact. These works should be done every year so that effect of flooding can be reduced.
- The seven-vent culvert located on the Velachery–Tambaram road shown in Figure 8.3 to drain the floods collected on the west of the swamp is quite inadequate at present and this bridge can be reconstructed considering future scenario of floods.
- The solid waste dumped in the water courses, canal and storm-water drain should be managed with utmost care to cope up with

the flooding situation and also for the improvement of ground water quality. The rainwater harvesting system should be implemented as per Government norms at every household level to reduce the surplus overflow of rainwater and also this would help in recharging the ground water. As per standards provided by Government, rainwater recharge pits should be constructed for each and every 30 m intervals for uninterrupted flood discharge.

- Bottom-up approach should be encouraged for implementation of policies related to flood management and mitigation.
- Awareness should be created among the stakeholders about the existing water policies.

8.7 Conclusions

From this study, it is understood that the water policies are existing and but there are difficulties in implementing them in field level. Proper awareness about the water policies should be created among public, private, and common people in order to get support from people during execution of mitigation work in future. The corresponding state Government should focus on revising the State Water Policy by understanding the priority and need for flood management and mitigation at the earliest. The State Water Policy should incorporate aspects like forest and wetland preservation and improvement of effective flood warning system through web-based information. The eviction of encroachment should be implemented in future properly. It should also focus on preserving the existing tanks from encroachment and rehabilitation of tanks like strengthening tank bunds, maintaining sluice gates, effective construction of check dams to improve the ground water level, so that the flooding effect can be minimized in future.

References

1. Rekha, Y., Arul, C., and Thirunavukkarasu, S. (2015), 'A Critical Review of Flood Control and Management Policies-An IWRM Perspective,' *International Journal of Science, Engineering and Technology Research (IJSETR)*, Vol. 4, No. 10, pp. 3427–3432, ISSN: 2278–7798.
2. Dorairaj, S. (2009), An Article in The Hindu, dated January 2, https://frontline.thehindu.com/the-nation/article30183301.ece

3. Suriya, S., Mudgal, B. V., and Nelliyat, P. (2012), 'Flood Damage Assessment of an Urban Area in Chennai, India, Part I: Methodology,' *Natural Hazards*, Vol. 62, No. 2, pp. 149–158.
4. Gupta, A. K., and Nair, S. S. (2011), "Urban Floods in Bangalore and Chennai: Risk Management Challenges and Lessons for Sustainable Urban Ecology," *Current Science*, Vol. 100, No. 11, pp. 1638–1645.
5. Gupta, A. K., and Nair, S. S. (2010), "Flood Risk and Context of Land-Use: Chennai City Case," *Journal of Geography and Regional Planning*, Vol. 3, No. 12, pp. 365–372.
6. Institute of Water Studies, Government of Tamil Nadu, Public Works Department, Water Resources Organisation, Taramani, Chennai-600113. http://www.wrd.tn.gov.in/IWS/IWS_activities.pdf
7. Kiruthika (2009), 'Study on Degradation of Waterbodies and the Field Implementation of Policies for Protection,' unpublished ME. Thesis, Centre for Water Resources, Anna University, Chennai 25.
8. Pareva (2006), Irrigation and Flood Control, Govt. of National Capital Territory (NCT) Delhi, http://nidm.gov.in/idmc/Proceedings/Flood/B2%20-%2036.pdf
9. Chandramohan, B. P., and Bharathi, D., (2008), "Role of Public Governance in the Conservation of Urban Wetland System: A Study of Pallikaranai Marsh," *Proceedings of the Indian Society for Ecological Economics (INSEE). 5th Biennial Conference*, Ahmedabad, India, http://www.ecoinsee.org/fbconf/Sub%20Theme%20B/Chandramohan%20and%20Bharathi.pdf
10. Ramsundram, N. (2008), 'Flood Auditing of an Urban Area Thirisulam Hill to Pallikaranai Marsh,' unpublished ME. Thesis, Centre for Water Resources, Anna University, Chennai 25.
11. Suriya, S., Mudgal, B. V., and Nelliyat, P. (2012), 'Flood Damage Assessment of an Urban Area in Chennai, India, Part II: Results and Discussions,' *Natural Hazards*, Vol. 62, No. 2, pp. 159–167.
12. Rekha, Y. (2018), 'Impact of Urbanisation on Waterbodies-Chennai,' *International Journal of Advanced Information and Communication Technology (IJAICT)*, Vol. 4, No. 11, pp. 1370–1375, ISSN: 2348–9928.
13. National Water Policy (2012), Ministry of Water Resources, Government of India, http://jalshaktidowr.gov.in/sites/default/files/NWP2012Eng6495132651_1.pdf
14. Andhra Pradesh State Water Policy (2008), Irrigation and Command Area Development Department, Government of Andhra Pradesh, http://www.ielrc.org/content/e0817.pdf
15. IELRC (1994), Water Policy of Tamil Nadu, 1994, Public Works Department, Government of Tamil Nadu, Institute for Water Studies, http://www.ielrc.org/content/e9424.pdf
16. India Water Partnership (2015), Review of Sate Water Policy of Tamil Nadu in Line with National Water Policy-2012 with Regards to Climate Change, http://cwp-india.org/wp-content/uploads/2018/03/Review-report-of-Tamil-Nadu-Water-Policy-with-respect-to-NWP-2012.pdf
17. Mesmer (2008), 'Evaluation of a Digital Supported System for the Flood Management in Chennai, India,' Unpublished Thesis, (B.Sc.) in Geography, Faculty of Forest and Environmental Sciences, Albert-Ludwigs-University, Germany.

18. Karnataka State Water Policy (2019), Water Resources Department, Government of Karnataka, https://karunadu.karnataka.gov.in/jnanaayoga/Other%20Reports/KJA%20Recommendation%20on%20KSWP.pdf

19. Kerala State Water Policy (2008), Water Resources Department, Government of Kerala, https://kerala.gov.in/documents/10180/46696/Water%20policy#:~:text=The%20State%20shall%20follow%20the,the%20various%20categories%20of%20users.&text=The%20necessity%20of%20conservation%2C%20development,11%20river%20basins%20of%20Kerala

20. Suriya, S., Gopinath, Mudgal B. V., and Karunakaran, K. (2009), 'Reconstruction of Volume and Spread of Flood Water to Velachery Area through Field Investigation,' *Malaysian Journal of Civil Engineering*, Vol. 21, No. 2, pp. 243–248, ISSN 1823–7843.

9

Maintenance Methodologies Embraced by O&M Department for Track Geometry at Kochi Metro Rail Limited, India: A Case Study

Priyanka Prabhakaran and S. Anandakumar
Kongu Engineering College, Erode, India

CONTENTS

9.1 Introduction

Modern transportation organizations have shifted their focus from erection and expansion of the transport infrastructure toward how to intelligently maintaining them [1]. Maintenance is the combination of the technical and associated administrative actions intended to retain an item or system in its state. Many papers in this stream address the problems in maintenance and operation. This study focuses on track maintenance methods of rapid transit systems (Metro) at Kochi, India. Increasing demand for railway transportation requires large amounts of money spent on track maintenance and renewal [2]. Track Sections need to be closed now and again for track maintenance and upgrades to ensure a satisfactory level of safety, comfort, and future availability [3]. Kochi is one of the most popular commercial hubs and a tourist paradise in Kerala. According to the estimated figures, the population of the city is 655,697 in 2016, increasing by over 3% every year as per the Handbook on Building maintenance, KMRL. The government has witnessed the growth in population over the last ten years and has put forth the metro project to improve mobility within city areas. Phase 1 of the Kochi Metro Rail Project (KMRL) is 25.612 km long on the double line elevated track. As part of Phase 1, 13 km with 11 stations from Aluva to Palarivattom became operational in the year 2017. Phase 1 of Metro comprises fully elevated 22 stations. At Muttom station, a platform face has crossover facilities as entry/exit to depot. A study based on Delhi metro ridership resulted in an average of 57,953 vehicles off the road each day [4]. The study by DMRC has further initiated the project at Kochi, which has many road commuters. The Operations and Maintenance department of KMRL has the following departments, namely civil and track (CTR), Power supply and Traction (PST), Signaling and control (S&T), Rolling stock, MEP (Maintenance of electrical and Plumbing works). The CTR wing consists of infrastructure maintenance and track maintenance. A detailed study on rapid transit systems is at stake, and thus the objective of this paper is to present insight into the track maintenance methods adopted at Kochi Metro Rail Limited for the past two years. Kochi Metro has rendered day and night service free of charge to the people of its native state during the most devastating floods the nation has ever faced. The paper does not portray a vast description of all its maintenance methods, but an overview of the main concepts and day-day strategies that have been practiced extensively for track maintenance and smooth operation.

9.2 Track and Traction

9.2.1 Types of Tracks

Kochi Metro rail Limited (KMRL) has two types of track: ballast and ballastless track. Ballastless track with "Fastening system Vossloh 336" has been provided over the mainline while in the maintenance depot. A track spacing of 4.2 m for at depot and 4.87 m for the viaduct was provided. The 110 kV

FIGURE 9.1
Ballasted track at Muttom depot ramp entry (left) and metro train stationed on Ballastless track in the stable line area within Muttom depot.

power supply from KSEB is stepped down to 750 V DC for traction and supplied through the third rail (bottom shoe collection type). The standard gauge adopted is 1435 mm. The minimum design radius of curvature the plan is 120 m for the mainline. The center to the center of the track is 4.2 m for the depot station and 4.87 for the viaduct. The center to the center of the track in scissors crossover in the depot is 5.0 m. Figure 9.1 illustrates the ballasted track at the depot and ballastless track in the stabling shed of Kochi Metro Depot.

9.2.2 Classification of Maintenance at KMRL

As per the standard clause of maintenance, works on tracks are subdivided as scheduled, non-scheduled, renewal. Scheduled maintenance is into preventive and regular maintenance. Maintenance of fixed railway infrastructure typically results in up to 70% of the total cost of the lifecycle of infrastructure assets divided [5]. Non-scheduled maintenance is divided into corrective and temporary maintenance. The scheduled maintenance of track should fit within clearly defined cycles and includes minor interventions such as lubrication, tightening of bolts, Cleaning, etc. The later mentioned aims to restore the track condition to utilize it within permissible tolerances. The above-mentioned consists of interventions in the track geometry or track component defects. This frequently involves correction of equipment position, localized renewals, fasteners tightening, etc. Predictive maintenance is a practice that allows for the identification of irregularities on the track before they reach a critical state. Non-scheduled maintenance accommodates the entire repair done whenever unexpected action arises due to the operation of the train, which includes rail fracture, broken fastening, or defective insulating joints. Renewals are considered to be the final hand phenomena when corrective maintenance is found to be technically ineffective or uneconomical. Renovation involves the replacement of rail pads, fastenings, ballast, buffer stops. Track geometry degrades with age and usage and can affect track performance and safety negatively. When track geometry degrades to an unacceptable level, it can lead to derailment [6]. They are planned well in advance when maintenance expenses become too high. Track maintenance is carried out usually in cycle's namely weekly inspection, fortnightly inspection, six months intervention, and yearly intervention.

9.3 P-Way and Traction Store

9.3.1 KMRL P-Way

The permanent way and traction store department are responsible for the day-to-day maintenance of the track activities that spawns between day and night. Kudumbasree workers are responsible for the extensive internal and external washing works. The two main elevated transit lines coming from the Muttom station have 16 transit lines, which end up at the stable area for inspection and cleaning works. Out of 16 transit lines, three belongs to the inspection area. The power supply to the track will be provided by the signaling department of the Alstom Company. The fleet of trains at KMRL is under DLP (Defect during the liability period) of the Alstom Company.

9.3.2 Methods of Track Maintenance

The value of entry and exit line curves gradients are maintained and revised once in 3 or 6 months, or whenever the need arises. The standard values are noted in the abstract of the curves register. The schedule of inspection is done as planned by the SE. Table 9.1 shows the parameters incorporated in the construction of rail at the metro depot and mainline. Kochi metro is designed to provide double track on the main lines and is designed to accommodate the max design speed of 90 kmph and maximum operating speed of 80 kmph. The mainline rail type is 60 E, 1080 grade HH rails conforming to IRS-T-12-2009 supplied in length. The maximum cant is 125 mm.

The rail seat is designed to carry a 60 kg UIC rail. The length, width, and depth of the sleeper are 2500 mm, 280 mm, 220 mm, respectively. Insulated rail pads are provided in between the rails to keep the track insulated with an adequate power supply. The curve registers for entry and exit lines of railway along with the turnouts holds a record of the values to differentiate

TABLE 9.1

Track Parameters

S. No.	Criteria	Dimension
1	Gauge	1435 mm
	Design Speed	MPH
2	Maximum operating speed – Mainline	80 kmph
	Maximum operating speed – Platform mainline	50 kmph
	Maximum operating speed – Depot and Non-running lines	35 kmph
	Maximum permissible gradient compensated for curvature	
3	Desirable	2.5%
	Exceptional	4.0%
	Horizontal curves minimum permissible radius	
4	On main lines other than platform lines	120 m
	On platform lines without cant	1000 m
	On non-running lines including depot tracks	100 m
5	Vertical curves minimum permissible radius	1500 m
6	Inclination of rail	1 in 20

the variations observed in the values in comparison with that of the ideal values (designed values).

9.4 Track Gauge and Versine

9.4.1 Gauge Measurements along Tracks

The track consists of a stock rail and tongue rail. The stock rail remains stable throughout the transit, whereas the tongue rail supports the movement of the train from one track to another. The tongue rail restricts longitudinal movement. The values recorded from the first, second, third, and fourth inspections are compared with that of the ideal values. The values of versine, cant, and gauge are recorded to observe the difference in various points at curves. Superelevation of the track is also referred to as cant. versine is said to be the lateral difference of the mid-ordinate of two rails or curves. Turnouts are said to be the cross levels from one track to another. Track gauge is defined as the distance between the gauge points on the face of each rail. The reference value of the standard gauge is 1435 mm with 600 mm sleeper spacing [7]. A similar value of the standard gauge is adopted at KMRL is 1435 mm with 650 mm sleeper spacing at the depot and 600 mm sleeper spacing at the station track. Maximum variations for ballastless and the ballasted track are ±2 mm, ±6 mm at the maintenance stage, and +6 mm/−3 mm

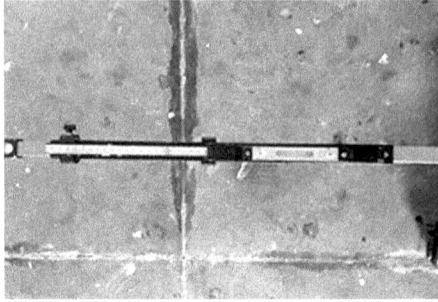

FIGURE 9.2
Gauge used to measure values along the mainline.

maintenance at turnouts. Layout tolerances on versine's in lead rail in the middle of diverted track are +5 mm/−10 mm. The extent of irregularities is categorized as A (up to +3 mm), B (±3 mm to +6 mm), C (above ±6 mm). The values were taken from the Handbook for Track Maintenance, Indian railways Institute of Civil Engineering, Pune, 2016. Figure 9.2 illustrates the equipment with which a gauge is recorded along the mainline. The gauge is operated manually by maintainers during daytime. The track at stations is maintained during nighttime by maintainers who walk along the stretch for about 2–3 hours using a night lamp. In Singapore, the maintenance activities are carried for 3–5 hours at night, whereas significant work is limited to 1 or 2 days, depending on how busy the track operation is during weekends [8].

The federal railroad administration (FRA) funded the development of the Gauge Restraint Measurement System (GRMS) vehicle, which is used to collect gauge measurement data [9]. The gauge consists of a metal scale attached to a frame along with a water level in the middle. The water level is adjusted after placing the bottom screws available on the gauge to the rail. The deviation from the ideal value is noted down along the straight track as well as the crossing level.

9.4.2 Gauge along Mainline and Turnout

Figure 9.3(a) indicates the variation in the gauge readings observed from the site during the first inspection. The gauge readings are found to be 0 at stations 5, 6, 7, 9, 12, which indicates that there is no difference in the values observed in comparison with that of the standard gauge values. The positive and negative value of the gauge indicates the correction values from the standard gauge values. Figure 9.3(b) shows the variation in the gauge readings observed from the site during the second inspection. The gauge readings are found to be 0 at stations 5, 6, 8, 9, 11, which indicates that there is no difference in the values observed in comparison with that of the standard gauge values.

Figure 9.3(c) shows the variation in the gauge readings observed from the site during the third inspection. The gauge readings are found to vary due to the massive monsoon floods that let the station depot submerged. The above readings were observed after the floods, and hence the values have highly deviated. The sections with higher variations suffer more considerate deterioration along the straight sections [1]. Figure 9.3(d) indicates the variation in the gauge readings observed from the site during the fourth inspection. The gauge readings are found to be 0 at stations 1, 2, 5, 6, 8, 9, 11, which indicates that there is no difference in the values observed in comparison with that of the standard gauge values. If the distance of the gauge (1435 mm) varies more than expected, it may be due to weak gauges, clusters of weak ties, and broken fasteners that affect the movement of trains.

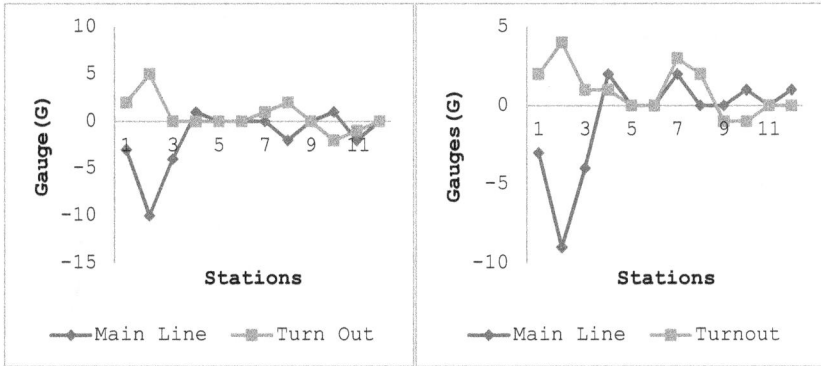

9.3 (a)   9.3 (b)

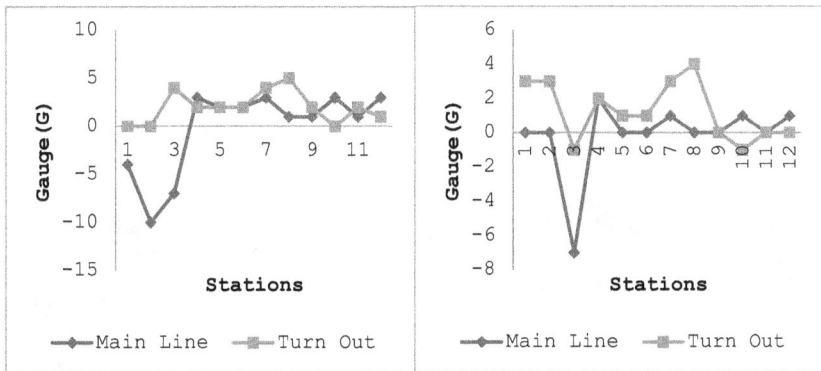

9.3 (c)   9.3 (d)

FIGURE 9.3
Variations in gauge level observed along mainline and turnout.

9.4.3 Variations in Versine along Stock Rail and Tongue Rail

The versine is measured using a 6 m chord on-site. At stations, versine is measured for every 3 m interval. A Chord is placed along with points zero and two, and the value at one is recorded. The formula used for the calculation of versine is $C^2/8R$, where C denotes the chord length, and R denotes the radius at turnout i.e., 190 m. The standard value of versine is 23 mm. At turnout, the versine of stock rail and tongue rail is measured at every 6 m interval. Figure 9.4(a) indicates the values recorded for versine (mid-ordinate) of stock rail and tongue rail recorded during the first inspection. From the above figure, it is observed that station number 4, 9, 10 is out of the standard versine range along the stock rail. The rails are subjected to lateral movements. Figure 9.4(b) indicates the values recorded for versine (mid-ordinate) of stock rail and tongue rail recorded during the second inspection. From the above figure, it is observed that station number 4 is out of standard versine range along the stock rail as well as the tongue rail.

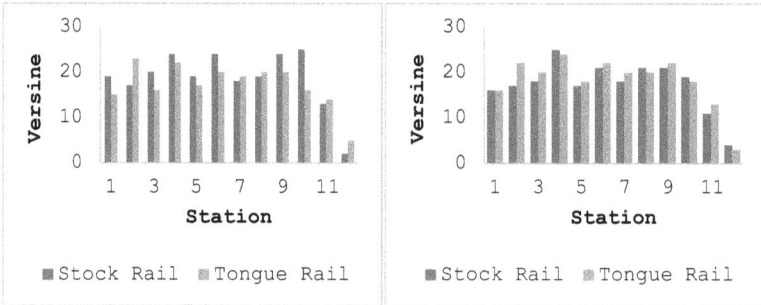

9.4 (a)   9.4 (b)

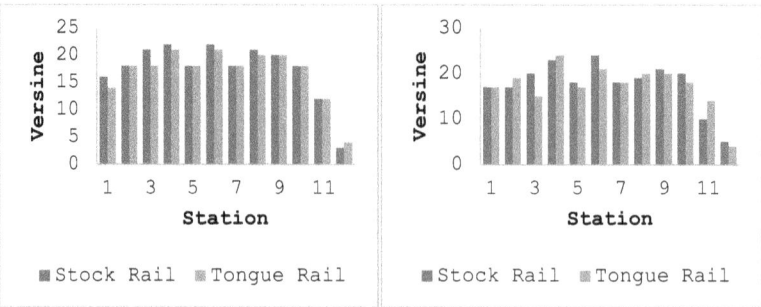

9.4 (c)   9.4 (d)

FIGURE 9.4
Variations in versine along stock rail and tongue rail.

Figure 9.4(c) indicates the values recorded for versine (mid-ordinate) of stock rail and tongue rail recorded during the third inspection. Figure 9.4(d) shows the values recorded for versine (mid-ordinate) of stock rail and tongue rail recorded during the fourth inspection. The above figure shows all stations fall under the standard versine range.

9.5 Cross Levels along Mainline and Turnout

The Figure 9.5(a) and (e) indicate the variation in the cross-level readings along the mainline and turnout observed from the site during the first inspection. It is a well-known fact that the outer rail is raised than the inner rail at the curves, and they are referred to as super elevation, or "cant." If the elevation difference is low, the speed of the trains should be reduced to avoid derailment. LL values show that the right rail is lower than the left rail in mm; LC indicates the level is correct; LL indicates that the left rail is lower than the right rail in mm. At the station, zero 1RL indicates the right rail is 1 mm low than the left rail. The gauge is set in such a way that only the lower value in comparison with the right or left rail can be found. The rail doesn't usually tend to rise. Rails always settle down, and hence, the gauge is set to measure the difference low in mm. RL2, LC2, LL2 indicate the value of cross-level at the turnout. Station number 5, 6, 12 is found to be of the same level along the mainline and turnout. Figure 9.5(b) and (f) indicates the variation in the cross-level readings observed from the site during the second inspection. LL values indicate the right rail is lower than the left rail in mm; LC indicates the level is correct; LL indicates that the left rail is lower than the right rail in mm.

The values at stations 3, 5, 7 against the RL values were found to be 0 not because the values are 0 but those stations were also under the LC (level correct) area. It is observed from the above that at no station the rail is found lower than the right rail which indicates that the rails were maintained at correct geometry when laid at curves. RL2, LC2, LL2 suggest the value of cross-level at the turnout. Station number 3 is found to be of the same level along the mainline and turnout. The above Figure 9.5(c) and (g) indicates the variation in the cross-level readings observed from the site during the third inspection. LL values indicate the right rail is lower than the left rail in mm; LC indicates the level is correct; LL indicates that the left rail is lower than the right rail in mm. The values at stations 3, 6, 7, 12 against the RL values were found to be 0 not because the values are 0, but those stations were also under the LC – level correct area. It is noted from the above that at no station the rail is found lower than the right rail. In addition, it is noted that station number 6, which was 2 mm low during the second inspection, has been leveled in the

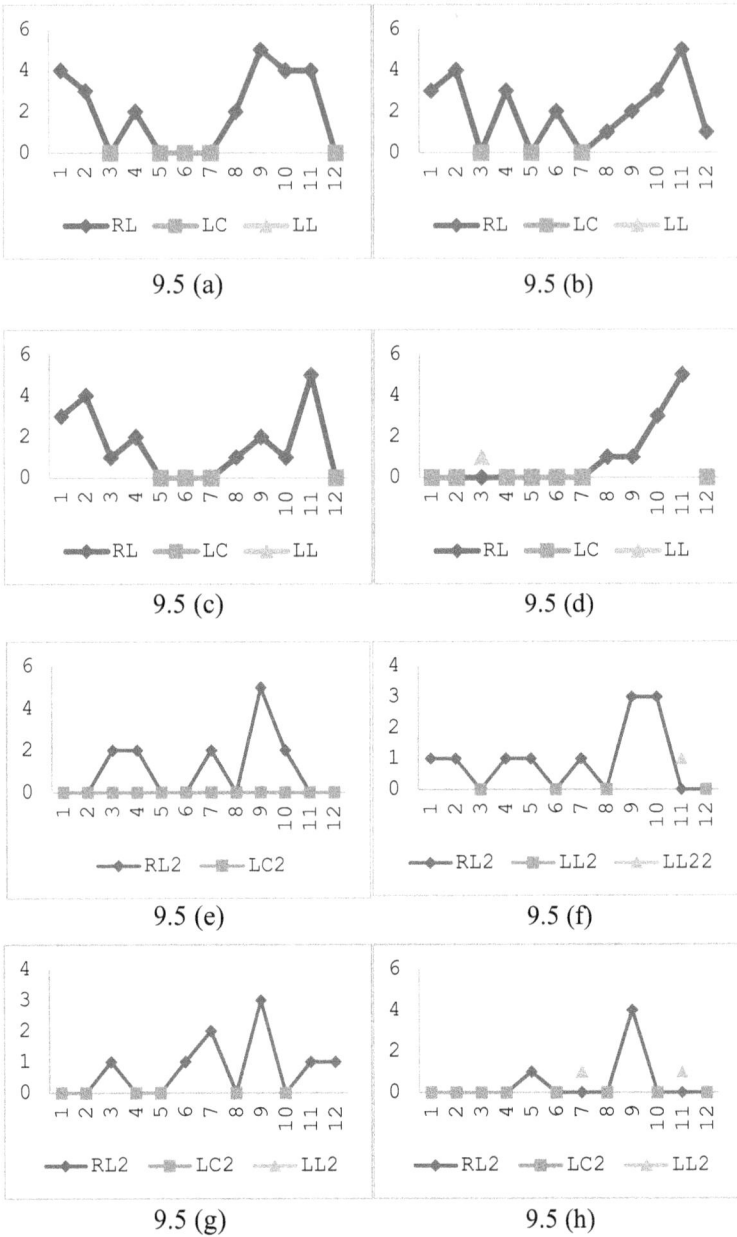

9.5 (a)

9.5 (b)

9.5 (c)

9.5 (d)

9.5 (e)

9.5 (f)

9.5 (g)

9.5 (h)

FIGURE 9.5
Rail geometry at cross-level along mainline and turnout.

third inspection and found to be under the LC area. Station number 3, 7 has remained to be under LC area throughout the first, second, and third inspection. RL2, LC2, LL2 indicate the value of cross-level at the turnout. Station number 5 is found to be of the same level along the mainline and turnout. The above Figure 9.5(d) and (h) indicate the variation in the cross-level readings observed from the site during the fourth inspection. The values at stations 1, 2, 4, 5, 6, 7, 12 were found to be level correct along the mainline as well as the turnout. The investigations show the tracks at the depot were maintained to be at a certain level after the monsoon. Station 11 tends to be at the same level throughout the second, third, and fourth inspection, and indicates the need for immediate attention.

9.6 Variations of Gauge

9.6.1 Comparison of Variations in Gauge Levels along Mainline

Figure 9.6(a) compares the values recorded along with the gauge levels at the mainline during all four inspections. The values after station four have maintained a constant decorum throughout whereas the values along with

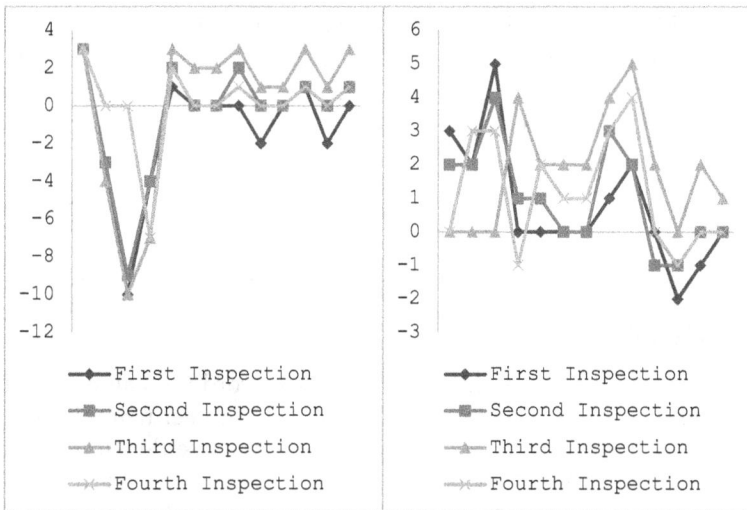

9.6 (a)   9.6 (b)

FIGURE 9.6
Comparison of variations in gauge levels along mainline and turnout.

station zero to station three, which indicates these stations require more rectification. Figure 9.6(b) compares the values recorded along with the gauge levels at the turn out during all four inspections. The values were scattered, and, hence, there is a need for stringent maintenance rectification at the turnout. From the above tables and figures, it is clear that there is a difference in values recorded for one observation to another. The creation of an integrated maintenance management system is the cornerstone of enhanced maintenance [10]. Before the fourth inspection, the areas of the Muttom depot were partially submerged due to the floods caused by the heavy monsoon.

Thus, we can observe that until the third inspection the right rail was found to be lower than the left rail, but in the case of the fourth inspection after the floods, the left rail was found to be low due to the improper settlements of ballast track as the soil beneath was washed away. Another observation that the researcher could observe was that the construction of the Muttom Depot lies in the area where agricultural practices were carried out. After the land acquisition, red soil was dumped, and the check rail and maintenance depot were constructed. This serves to be the critical reason behind the variation in values of the track. The gauge error is due to the calibration error and at times, manual error, which are further rectified in the day-to-day inspection by the maintainers. A separate checklist for joint inspection of points and crossings is recorded under the O&M department in association with the maintenance and signaling department to ensure both wings are working intact. The overall track maintenance cost for the year 2018 is 44 lakhs in INR (e-tender, KMRL). The ballasted track maintenance is, in any case, costly: 42% of its operating budget on track maintenance and renewal [11]. The maintenance operations are not at negligence because of the cost incurred, whereas the advantages offered by MRTS (Mass Rapid Transit System) are always into consideration.

9.7 Conclusions

The study focuses on the maintainer's duty to record the values noted during the inspection works manually. The section engineer heuristically takes decisions based on the experience to go for renewal or settle within the maintenance measures available. The study identified the variation in the quoted amount is due to the unexpected climatic rage, which is still under rehabilitation. The O&M has also further initiated their automation systems to be on the second floor of the depot unit. The above shift can prevent the systems from being deteriorated and also reduce maintenance costs annually. Rather than entrusting the yearly maintenance through an external agency, the work is to be entrusted among the maintenance employees. Manual maintenance puts forth errors in recorded inspection values, and there is an urgency to digitalize and calibrate the available tools and equipment. Operation and maintenance is viewed as an integral part of the planning process where the staff communities are planned to train for a regular maintenance schedule with the collected revenue. This practice is done very well in advance within the KMRL Community, forecasting the further maintenance cost. Though the maintenance department has taken ample time for getting back to normal after floods, it is believed that with strong support from the staff community and further expansion of stretch, the revenue can be increased in the coming years with an increase in commuter's access to KMRL.

The paper further needs a large number of references that can serve as a support system in the development of better maintenance models, software, and tools. The further study will include all the boundaries of engineering with new comprehensive approaches that can also serve as a basis for the regular maintenance of upcoming rapid transit systems.

References

1. Karimpour M, Hitihamillage L, Elkhoury N, Moridpour S, Hesami R. Fuzzy approach in rail track degradation prediction. *Journal of Advanced Transportation*. 2018; 2018:3096190.
2. Guler H. Decision support system for railway track maintenance and renewal management. *Journal of Computing in Civil Engineering*. 2012 April; 27(3):292–306.
3. Forsgren M, Aronsson M, Gestrelius S. Maintaining tracks and traffic flow at the same time. *Journal of Rail Transport Planning and Management*. 2013 August; 3(3):111–23.
4. Sharma N, Dhyani R, Gangopadhyay S. Critical issues related to metro rail projects in India. *Journal of Infrastructure Development*. 2013 June; 5(1):67–86.
5. Heyns FJ. Construction and maintenance of underground railway tracks to safety standard of SANS: 0339. *Journal of the Southern African Institute of Mining and Metallurgy*. 2006 December; 106(12):793–8.
6. Soleimanmeigouni I, Ahmadi A, Kumar U. Track geometry degradation and maintenance modelling: A review. *Proceedings of the Institution of Mechanical Engineers, Part F: Journal of Rail and Rapid Transit*. 2018 January; 232(1):73–102.
7. Kaewunruen S. Monitoring of rail corrugation growth on sharp curves for track maintenance prioritization. *International Journal of Acoustics and Vibration*. 2018 March; 23(1):35–43.
8. Dao C, Basten R, Hartmann A. Maintenance scheduling for railway tracks under limited possession time. *Journal of Transportation Engineering, Part A: Systems*. 2018 May; 144(8):04018039.

9. Higgins C, Liu X. Modelling of track geometry degradation and decisions on safety and maintenance: A literature review and possible future research directions. *Proceedings of the Institution of Mechanical Engineers, Part F: Journal of Rail and Rapid Transit.* 2018 May; 232(5):1385–97.

10. Karaa FA. Infrastructure maintenance management system development. *Journal of Professional Issues in Engineering.* 1989 October; 115(4):422–32.

11. Abadi T, Pen LL, Zervos A, Powrie W. Improving the performance of railway tracks through ballast interventions. *Proceedings of the Institution of Mechanical Engineers, Part F: Journal of Rail and Rapid Transit.* 2018 February; 232(2):337–55.

10

Smart Lights for Smart City

Maheswaran Shanmugam, Indhumathi Natarajan,
Balasubramaniam Vivek, R. D. Gomathi, and Sathesh Shanmugam
Kongu Engineering College, Perundurai, India

CONTENTS

10.1 History of Streetlights

Streetlights are the most important tool of the city to provide safety for pedestrians, motorists, and emergency services. So, it is necessary to install and maintain the streetlights. During the eighteenth century, gaslit streetlights were used in many countries. The first electric streetlight was installed in Paris, France. After a year, arc electric lamps were used. Arc lamps were placed on towers 60 to 150 feet tall. In arc lamps, carbon rods were used, which conducted high current between the electrodes. These was used mainly for high lumen light situations such as lighthouses. In later days, the xenon arc lamp was used for greater brightness, and because it occupied a smaller area. Sometimes, it was used for car headlamps. After 20 years, incandescent lights were introduced. These were the first low-electric lamps, and made use of tungsten filaments for lighting. The color rendering index of the lamp is rated at 100. Naturally, tungsten-halogen incandescent lights are efficient and brighter. Due to the short lifespan, it was not widely used in streetlighting. In traffic signals, however, these lights were commonly used.

In the late 1930s, the fluorescent lamp was introduced. It needed a small current to make the tube glow. The glow was powerful in ultraviolet and dim in visible light. To make the visible light strong, the glass is coated with a mixture of phosphors. It is used widely because of its efficiency and newness value. The major problem with this is that it can produce non-directional diffused light. It should be mounted not more than 20–30 meters above the walkway to produce an acceptable level of light. In 1948, the mercury vapor lamp was introduced. Over the course of a year, it uses same energy, and produces dimmer light even when they can burn out over time. For this reason, the mercury lamp glasses are coated with a material made of phosphors that helps to increase the color rendering index (CRI). The ultraviolet light excites the specially coated material to produce furthermore white light. Mercury lamps are also called color corrected lamps. Due to the power factor, the mercury vapor lamp is replaced by high-pressure sodium (HPS) lamp and LED. Around the 1970s HPS lamp was invented, and has been used commonly since 1980. There are two types of Sodium lamps that are available in the market, high-pressure and low-pressure. Low-Pressure Sodium (LPS) lamps are more efficient than HPS lamps, but it has the ability to produce a single wavelength of yellow light. LPS is used for low height mounting applications due to less intense light than HPS.

HPS lamps have slightly added electrical components than MV lights [1]. Transformer and ballast are used to change the voltage and regulate the current for both the lamps. A starter circuit is needed for HPS lamps. HPS and MV have the same lifespan and give more efficiency in lower wattage. For example, 175 Watts MV is replaced by 100 Watts or 150 Watts HPS. In recent years, the metal halide (MH) lamp has been introduced. It produces truly white light. The average lifespan of MH is 10,000 to 12,000 hours. At the end of its lifecycle, it tends to dim or flicker. Due to the high cost and low lifespan, it does not tend to be in use widely. After that, ceramic MH has been introduced for pure white light with 78–96 CRI. Since the development of LED, it has diminished. During the twentieth-century induction, the lamp was introduced for the streetlighting system. It was more efficient, had high

CRI, and a long lifespan. These induction lamps were controlled and monitored through the internet. Due to high installation and maintenance costs, it was not highly recognized. After certain years, CFL (Compact Fluorescent Lamp) has come into existence. This is used still in municipal walkways and streetlights, it has high efficiency, and better CRI. It faces some issues such as low lumen, low lifespan, and burnout if the lights are frequently ON/OFF. Now, LED (Light Emitting Diode) is being used for streetlighting and traffic light signals and pathway lighting [2]. In comparison with the traditional streetlighting system, it is more energy efficient, and only half of the power is required to produce the same lighting level. Figure 10.1 shows the history of streetlights.

10.1.1 Advantages of LED

- Low energy consumption.
- Predictable and long lifetime compared to traditional streetlights.
- Accurate color rendering.
- Immediate on and off.
- Does not release any gases if damaged.
- Less attractive to insects.
- Minimum glare.
- Higher light output even at low temperature.

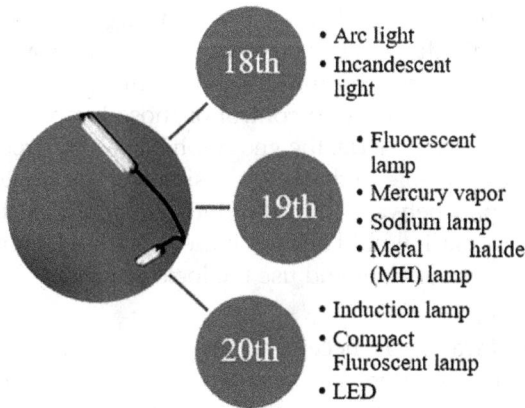

FIGURE 10.1
History of streetlights.

10.2 Smart Lighting

Smart lighting is the technology that is mostly aiming at the conservation of energy. The world is spending around 19% [Wikipedia] of the produced energy for lighting purposes and accounts for 6% [Wikipedia] of the greenhouse gas emissions. Energy conserved is the energy saved. The world governments are working toward saving the environment. They try to do so by reducing the amount of energy consumed. How can energy consumption be reduced? The answer to this question is by using smart systems, whether it is for light, or any other system concerned. As said earlier, more energy is spent on lighting than any other purpose. In such a scenario, it is the best to go with smart lighting for the conservation of energy.

10.2.1 Features Needed for Smart Lighting

a) Individual and group Control.

b) Communication.

c) Security.

d) Brightness control.

e) System integration.

a) Individual and Group Control
Individual lights are the lights that function as a single unit; for example, light installed in a kitchen, hall, or bedroom belongs to the individual lights. Group lights include the lights that are installed on the perimeter of the house, that are used for aesthetic purposes, or in industry. These lights form a group, and function as a single unit. The smart lighting concept should be able to work for individual and group control of lighting systems.

Considering the scenario, normal incandescent lights are being used at homes. The individual and group control of those lamps can be done with a single switch. On the flip side, the geographic span of streetlights is more. Therefore, it is not possible to do control using a single switch, because the streetlight may span between two different electricity distribution transformers. In such a case, it would be best, if the lighting systems get command from a common control center and use the local power.

b) Communication
Effective communication is necessary for controlling and securing data transfer. The communication can be wired or wireless. Wired communication involves connecting each device by wires and communicating to the smart lights using a predefined protocol. Wireless communication uses RF,

Bluetooth, etc. The purpose of communication is to control the lights effectively. It would be good to turn on and off the streetlights at a single point rather than switching on and off individual streetlights.

c) Security

The smart lighting systems at one point may be connected to the internet and shall function as IoT devices. This poses a great deal of risk since these devices may be penetrated and may not function as expected. The flaw of Philips Hue smart lighting systems and Kaspersky Research tells that attack on smart home devices has gradually been increased to 700% during the last 12 months.

d) Brightness Control

Most energy is spent on lighting at homes. Different amounts of light intensities are required for various tasks at home. The main activities in the home include watching TV, reading, cooking, showering, washing, and working out (see Table 10.1 [3]). But the public does not consider this as a big concern. They tend to use the same amount of lighting for all the activities except sleep. Five countries, the United States of America, Europe, Malaysia, Japan, and Britain [3] have set the required light intensity levels needed for different activities. Refer to Table 10.1 for the standard values provided by the five countries.

e) System Integration

Ideally, the smart lighting system would be integrated with an existing system. For example, there are many video surveillance systems installed that monitor traffic congestion and home security.

TABLE 10.1

Intensity Levels Needed for Different Activities

Activity	America (IES)	Europe (EN 12464)	Malaysia (JKR)	Japan (JISZ9110)	British (BSEN 12464)	Work [6]
Engaging in TV	50	300	50	—	100	50
Comforting	50	100	50	50	200	50
Book read	300	500	300	500	500	300
Preparing food	—	500	—	—	400	450
Showering	100	200	100	100	150	100
Washing	300	300	200	100	300	200
Taking food	100	200	300	300	—	300
Working out	500	300	300	—	—	300
Slumbering	0.1	—	—	2	—	0

10.3 Smart Lights on Smart City

A major part of the city is streetlighting, where the main function is to brighten the streets during the night. Saving power is very important. It should be switched off to avoid unnecessary waste. In many cities, streetlight accounts for as much as 25% of the total energy in use. During the morning, street-lights were on, leading to much wasted energy. There were fewer towns and cities long ago. So, the streetlights were controlled and monitored manually. But now, the number of streetlights has increased rapidly in line with high traffic density. Using different technologies, these systems are proposed to control, monitor, and reduce the energy consumption of public lighting sys-tems [4]. Nowadays, led streetlights are widely used, but to make the system smart, the LoRa module is fixed along with the led streetlight. Controlling of streetlights is based on two major factors that are light intensity during the daytime, and object detection during the night [5–9].

10.3.1 Working of Smart Streetlight System

An intelligent streetlight system shown in Figure 10.2 consists of LED, Sensors, GPS, and Lora module. The traditional streetlights are replaced with an array of LED lights. The sensor network is composed of the cur-rent, voltage sensor, LDR sensor, and movement sensor. To avoid the lights be on during daytime, an LDR sensor is used. If the LDR resistance is low, it sends the low voltage to the lights and it automatically goes to OFF state. There is no need for manpower to switch on and off. To detect the object during the night, a movement sensor is used. The object is detected before it approaches the lamppost and sends the information to the controller where it increases the intensity of the lights, if the object crosses the lamppost, it automatically reduces the intensity during the night. This helps to save energy consumed by the streetlights and increases the lifespan of LED [10, 11]. The current and voltage sensors are used to calculate the power con-sumed by a particular array of lights. This information is sent to the central control. If there is any drastic change in the power consumption, such as the lamppost having a faulty light, it is identified by the GPS. This infor-mation is sent to the central controller through the LoRa module. A smart streetlight system doesn't need any manpower to monitor and control the system [12–14].

10.3.1.1 Advantages

- Power saving: in a smart streetlight system, during nights and based on the traffic density on roads, the power consumption of the lights is comparatively lesser than traditional streetlights.

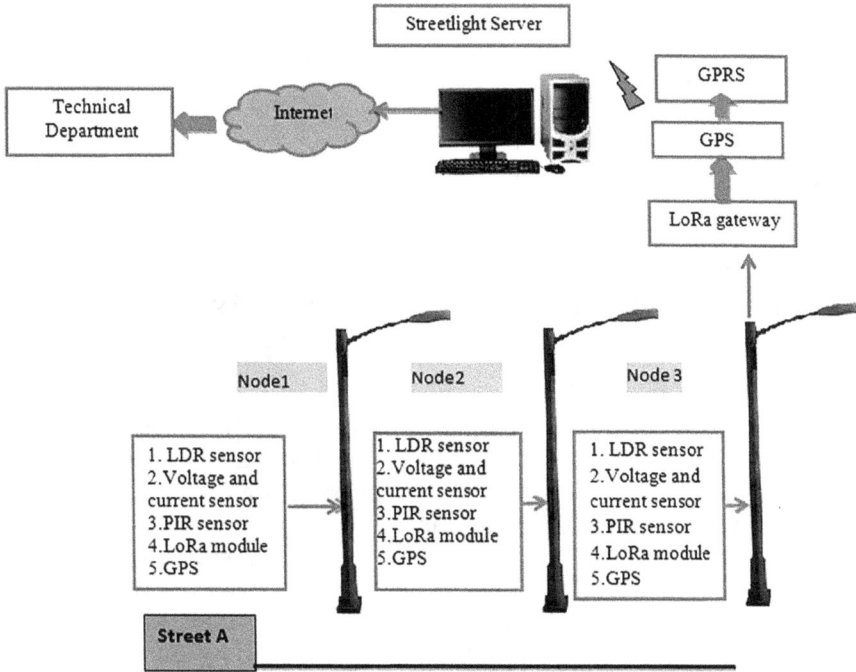

FIGURE 10.2
Smart streetlight system.

- Automatic monitoring and control: this system does not require the manpower to switch lights on/off and identify faults.
- LoRa module is able to communicate up to 10 km and consumes low power.
- The location of faulty light is identified by the GPS and the corrective of lights will be taken faster.

10.3.2 Requirements of Smart Light

10.3.2.1 LDR

The light sensor (LDR) is used for the detection of daytime and nighttime. The LDR is working under the principle of photoconductivity. If the LDR is placed in the dark light, the resistance of the LDR gets increased and if it is in the light, the resistance gets decreased. when sunlight falls on it, it will consider as daytime, the resistance of LDR is low if the resistance is low, the controller sends the low voltage to the LED, so the light will be off and when there is no sunlight on it, it will be considered as night, it produces high

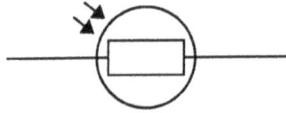

FIGURE 10.3
LDR sensor.

resistance if the resistance is high, the controller sends the voltage to the LED, so the light will be on. Figure 10.3 illustrates the symbol of LDR.

10.3.2.2 Movement Sensor

Figure 10.4 shows that the motion sensor is used to detect the object and it is a small electronic device. There are two types of motion sensors available, one is an active motion sensor, and the other is a passive motion sensor. The active motion sensor has a transmitter and a receiver, where it uses the maximum ultrasonic sound to detect the object. The passive motion sensor doesn't have a transmitter to detect infrared waves [15], such as heat dissipated from a moving object. If there is any change in the temperature, it sends the signal to the controller before it reaches the streetlight at nighttime, and increases the intensity automatically. After the object has passed, the temperature returns to normal, and it automatically reduces the intensity level.

10.3.2.3 Voltage Sensor

Power usage is evaluated by voltage and current sensor and is fixed to the module that is subjected to register the power consumption. The current sensor is a device that is used to calculate the current through the conductor. If the current is passing through the conductor, it creates the magnetic field around that. The magnetic field is directly proportional to the current in the conductor. Using this principle, the current sensor calculates the current consumed

FIGURE 10.4
Movement sensor.

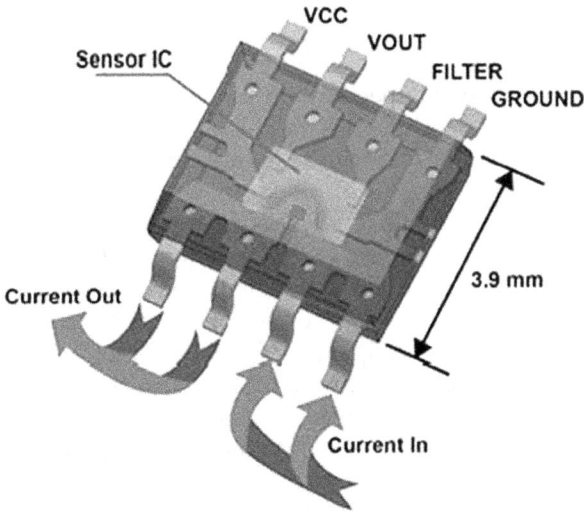

FIGURE 10.5
Current sensor.

by the lights. If there is any disfigurement in the streetlight, it is easily identified by fluctuations of the current value which is measured through the sensor. Figure 10.5 shows the current sensor and its pin representation.

10.3.2.4 LoRa Module

These controlled and monitored information are sent to the central server unit using LoRa module. LoRa is a low power wide area network. It is able to pass the information up to 10 km of distance. It is a physical silicon layer device and it uses LORAWAN protocol for communication [16]. The device is configured with a transceiver with end node or sensor devices. The sensed data is transmitted through the gateway at the destination. The symbol of LoRa is shown in Figure 10.6.

10.3.2.4.1 The Contrast between LoRa with Other Technologies

a) LoRa versus Cellular Network
GSM, 2G, 3G, 4G cell innovations are renowned, and widely available. These developments are firmly rooted in high knowledge of flow, the force efficiency in these advancements is not considered. While the amount of information that is transmitted is less constant and moderate, these organizations consume a lot of intensity. At the underlying level, the cost of constructing this type of enterprise is high. With the introduction of 5G technology,

multiple cell managers suspended 2G administrations and vaguely inoperative IoT gadgets.

b) LoRa versus LAN

The machine management specification for the LAN is widely received. It is mostly used for connecting PCs to a standard organization, in schools, universities, and at work. The wired organization is as available in LAN as a remote organization. The most popular inventions are Ethernet and Wifi. Remote constancy of Wifi. The Wifi works are of a small area, such as a home office structure. There is just 1 km of Wifi in length. LoRa invention however offers a long range of opportunities. Wifi management is bad. The Wifi is continuously overflowing with a huge volume of details, the information is hard to identify, and the correct recipient does not get a lot of information over time. In contrast, the LoRa breakthrough has a sensitive administrative character. The remote Wifi is therefore very secure. During transmission, anyone will interfere with the content. This was quite helpless and cannot be used as stable encryption. LoRa's safety is incredibly important. The breakthrough in LoRa relies on CSS, which is deeply imperceptible to multiply and blur. LoRa is modulated using RF and refers to the physical layer of the reference model of Open Systems Interconnection (OSI).

c) LoRa versus ZigBee

ZigBee is a high-level international conference on free communications. ZigBee is generally used to render local organizations. A network of ZigBee includes various hubs that are specialized radios with low power. These mailings from ZigBee are the most suitable applications for small scales that require information to be moved around small sections. It has a range of 10–100 m. The hubs of the ZigBee network structure provide long range messages through several hubs. In the middle of the road, the hubs are used for communications, ZigBee's correspondence should not be used by this burning force and consequently by low force conditions [17]. LoRa relies on the star geography whereby the information is sent to the center gadgets along with these lines, which decreases the use of force in general. Working networks are not suitable for correspondences of short to medium-range distance and do not have LoRa innovation's long-term capability.

FIGURE 10.6
LoRa.

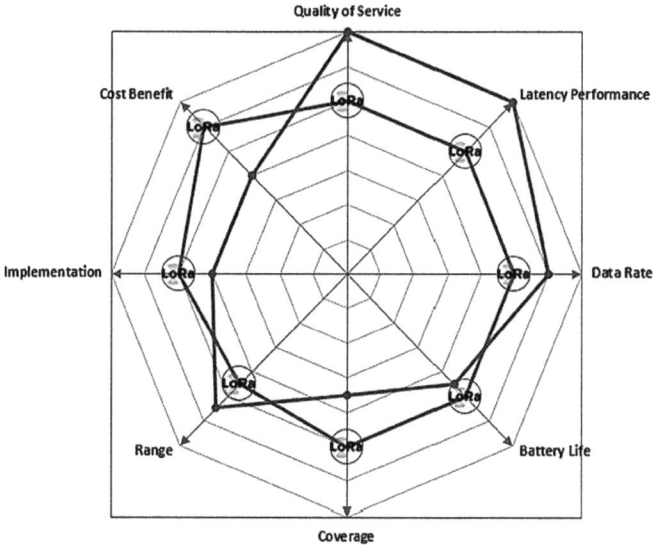

FIGURE 10.7
The contrast between LoRa and NB-IoT.

d) LoRa versus NB-IoT

Figure 10.7 shows the performance comparison of LoRa and NB-IoT. Each breakthrough has its favorite location and disadvantage. Both rely on their highlights for different applications. LoRa's invention is based on ALOHA, while NB-IoT relies on FDMA and hence leads to a higher power use than LoRa's. In contrast to LoRa, low density and the speed of knowledge of NB-IoT is sufficient. Those programs are in need of great inactivity and a high information rate can also use NB-IoT and those that require a lower information rate should choose LoRa. LoRa innovations are commonly suitable for all IoT-based communications.

10.3.2.5 GPS (Global Positioning System)

Global positioning systems are a satellite-driven radio map-reading structure. They offer the geo-position data and time information to the receiver (GPS) through GNSS (Global navigation satellite system). The location of something on earth is determined by GPS. It consists of satellite groups and transceivers for communications. The location is provided in Lat (latitude), Long (longitude), and Alt (altitude). GPS also provides the exact time is also provided by GPS. Twenty-four satellites are there which surround the earth at correct orbits. Within 12 hours, every satellite covers a complete earth orbit. Constant radio signals were sent out by these satellites. The receivers of GPS are automatically (through the program) set to gather data on each and every satellite

FIGURE 10.8
GPS.

at any point in time. The receiver of the GPS computes the location of its own by calculating the period. The calculated timings are based on the signals received from a minimum of four satellites. Since the RF (Radio Frequency) waves travel with steady speed, and the distance from that of other satellites is calculated using time measurements. Irrespective of climatic conditions, the GPS will accurately function. GPS is operated independently through the handler's applications through wired networks or mobile networks. By various satellites, GPS provides the information more precisely. When a distance is calculated from one satellite, it is the accurate distance regardless of the direction from the satellite. At any instant of time, four satellites (GPS) are in a straight line which is nothing but LoS (Line of sight) to the receiver of GPS on earth. At regular intervals of time, GPS satellites send the position and time to the receiver. The collected data is communicated as signals to the receiver. The distance (satellite and GPS receiver) is found by the time difference amid the signal that was initiated from the GPS satellite and the signal that was received by the receiver of GPS. It requires at least a signal from three satellites to analyze the positioning of 2-D, for 3-D it requires a signal from at least four satellites. Figure 10.8 shows the symbolic representation of GPS.

10.4 Smart Streetlight System (SSL)

10.4.1 Dimming of Streetlight Based on Time

Figure 10.9 illustrates that if it is daytime, the light will be in the off condition based on LDR output. From 6 pm to 10 pm, the lights will glow at maximum intensity. From 10 pm to 3 am, the lights will glow at a low-intensity level, and after 3 am the lights will automatically return to their normal intensity level until 6 am [18]. This helps in saving energy. The figure shows the day-time lights off mode. LDR senses the sunlight and based on that it sends the voltage output to the controller. The voltage level is compared with the

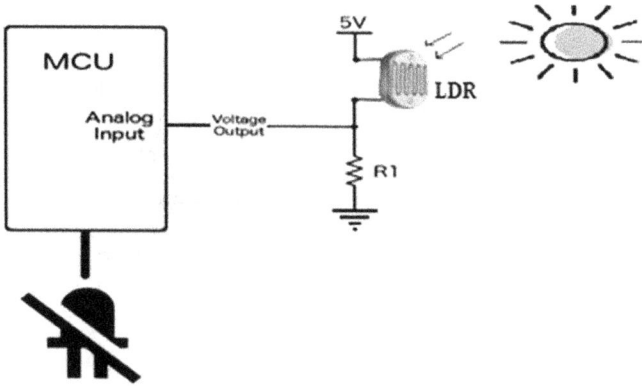

FIGURE 10.9
Dimming of streetlight based on time.

threshold level, i.e. stored in the controller based on that, it will deactivate or activate the streetlight.

Figure 10.10 confirms the pictorial representation of the dimming of lights based on time. The light intensity automatically varies based on the timer unit. From 6 am to 6 pm, the lights will be switched off. Moreover, 6 pm to 10 pm, the lights will glow at full intensity. From 10 pm to 3 pm the lights will glow at 2/3 intensity level [19].

10.4.2 Increasing Intensity Based on Movement Sensor

Figure 10.11 shows the block representation of increasing intensity based on the movement sensor. The lights are glow at a low-intensity level between

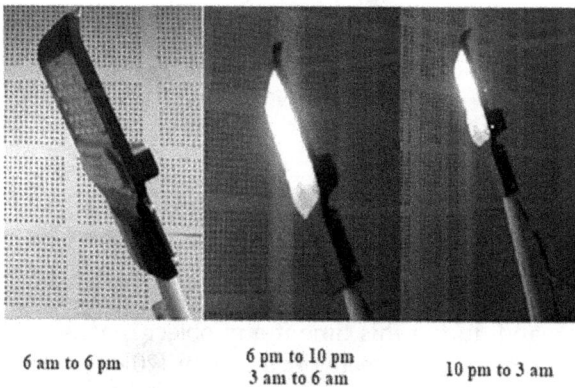

6 am to 6 pm   6 pm to 10 pm / 3 am to 6 am   10 pm to 3 am

FIGURE 10.10
The output of street light according to timings.

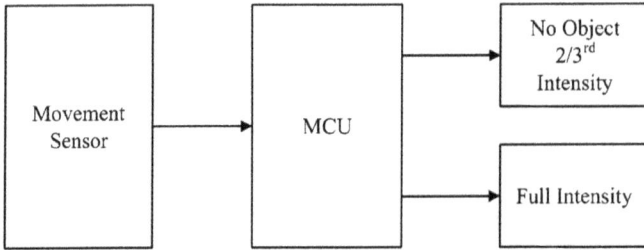

FIGURE 10.11
Block diagram of increasing intensity based on movement sensor.

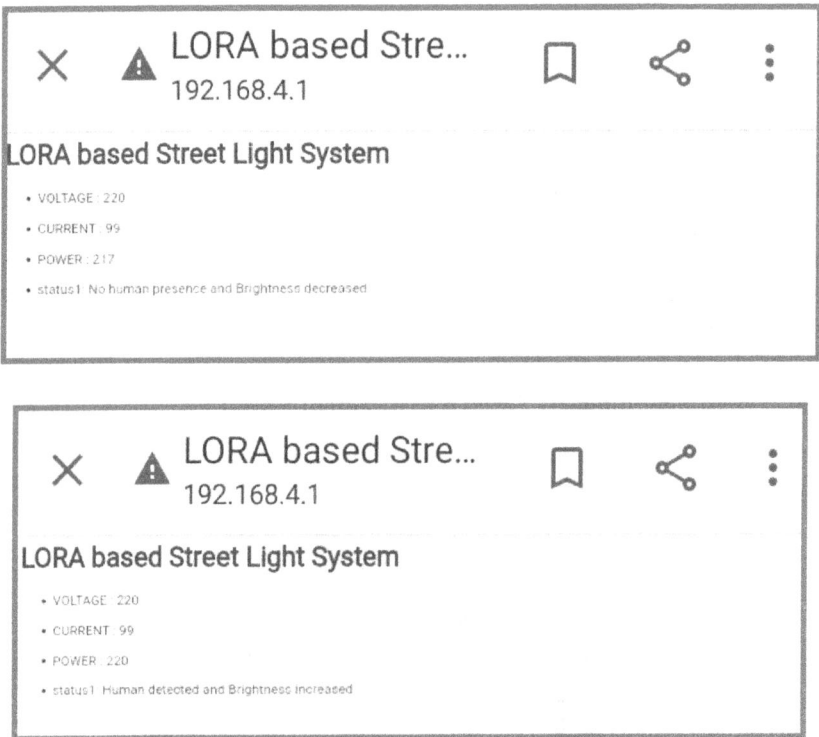

FIGURE 10.12
Movement sensor output status.

10 pm to 3 am, and during this time if any object is sensed on the road, it automatically increases the streetlight intensity [20]. This is used to avoid accidents at night.

Figures 10.12–10.14 show the movement sensors that detect the temperature variation in the environment and it sends the information to the

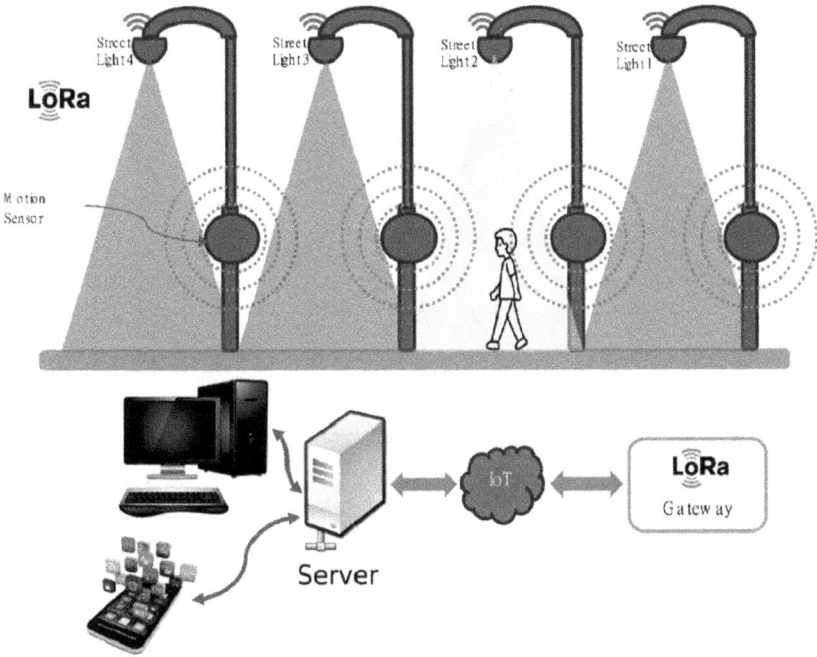

FIGURE 10.13
Increasing intensity based on movement sensor.

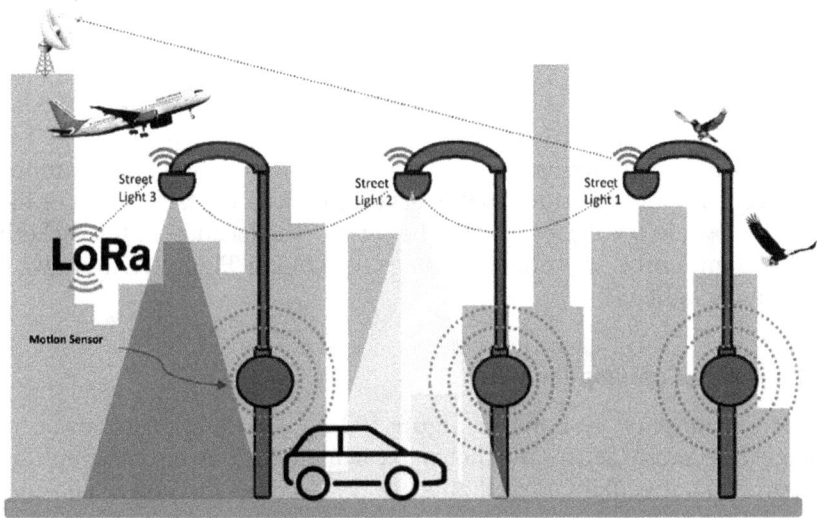

FIGURE 10.14
Based on object detection intensity control.

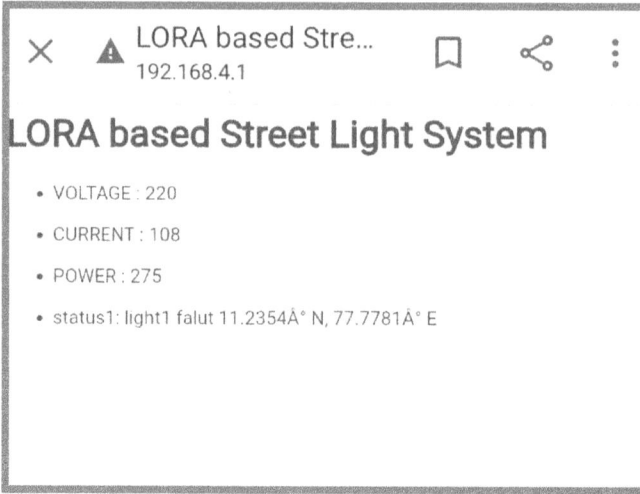

FIGURE 10.15
Faulty light detection.

controller before the object is arriving nearer to the lamppost [21, 22]. The streetlights are glowing at 2/3 of the intensity level, if the motion sensor sends the signal to the MCU, the MCU automatically increases the intensity level of the streetlights.

10.4.3 Faulty Light Detection

If there is any fault in the array of LEDs, it will be identified and the location of the faulty light or lights is sent to the central controller via the LoRa module. Each LoRa module has an IP address based on this, each lamppost power consumption is loaded in the server. If there is any faulty light or flickering it shows in the light status in the server. The location of the flickering or faulty lamppost is identified by using GPS [23, 24]. Figure 10.15 shows the status of the light.

10.4.4 Power Consumption Rate

Power consumption is shown in Figure 10.16. Every lamppost consists of a voltage and current sensor that helps to calculate the power consumption of the particular streetlight. Streetlights are IP-based, so the power consumed by all the streetlights is stored in the central server based on the IP address of the post. If there are any fluctuations in the consumption, it should be identified and rectified quickly.

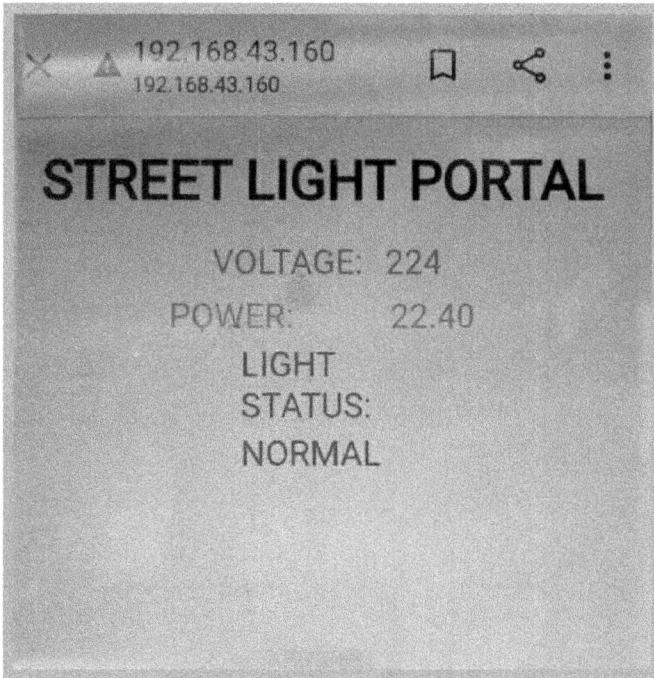

FIGURE 10.16
Power consumption rate.

The central server is loaded with all the IP addresses of the streetlights. It periodically checks the received information and takes action based on the received data [25–29].

10.4.5 SSL Workflow

Figure 10.17 illustrates the workflow of SSL. The smart street system initially checks the LDR output and based on this, and the lights may be turned on or off. The timer unit calculates the time if the time reaches evening 6.00 pm, the lights start glowing at full intensity level up to 10 pm, after 10 pm to 3 am the lights glow at 2/3 of intensity level. Here the movements sensor is fixed with the smart system between 10 pm and 3 am. If any of the objects approaches the street it sends the signal to MCU, the MCU sends the command to increase the intensity. From morning 3 am to 6 am, the lights will glow at full intensity. If the timer reaches 6 am, the streetlights go to the off state.

Figure 10.18 shows the status of the central server, the voltage sensor output recordings, and the same is compared and the light status is sent to the central unit. If there are any changes in the output, there will be flickering in the status of the light.

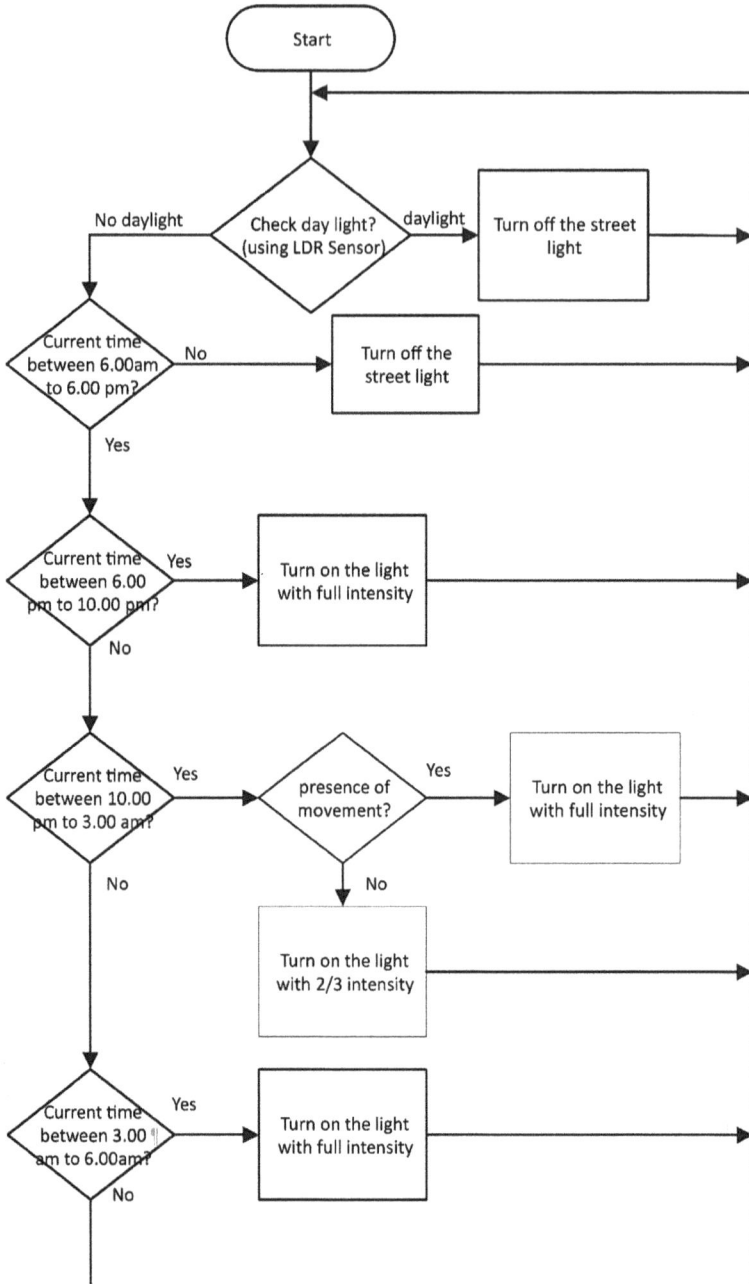

FIGURE 10.17
Workflow of SSL.

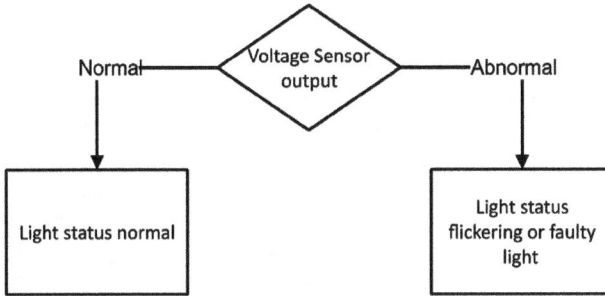

FIGURE 10.18
Flowchart for light status.

10.5 Summary

This chapter has focused on the evaluation of streetlights, features of smart lights and implementation of smart lights for the streets of smart cities. The first section discussed various factors like time, intensity level, power factor, faulty light detection, and object detection.

The second section of this chapter described the workflow of the smart streetlight system corresponding to the discussed factors, the power consumption and faulty light detection techniques were also discussed.

References

1. A. Chammam, W. Nsibi and M. Nejib Nehdi, "Behaviour of a high-intensity discharge lamp fed by a high-frequency dimmable electronic ballast," Sage J., vol. 49, no. 2, pp. 277–284, 2017.
2. A. Iorkyaa, A. I. Richard and A. N. Amah, "The efficacy of light-emitting diode (led) lamps used in rural communities of Nigeria," Energy Environ. Res., vol. 2, no. 1, pp. 121–127, 2012.
3. N. Adnan, N. Kamal and K. Chellappan, "An IoT-based smart lighting system based on human activity," In *2019 IEEE 14th Malaysia International Conference on Communication (MICC)* (pp. 65–68). IEEE, December 2019.
4. A. S. Jalan, "A survey on automatic streetlightning system on Indian streets using Arduino," Int. J. Innovative Res. Sci. Eng. Technol., vol. 6, no. 3, pp. 4139–4144, 2017.
5. S. Srivastava, "Automatic streetlights," Adv. Electron. Electr. Eng., vol. 3, no. 5, pp. 539–542, 2013.

6. A. Rao and A. Konnur, "Streetlight automation system using Arduino Uno," Int. J. Innovative Res. Comput. Commun. Eng., vol. 5, no. 11, pp. 16499–16507, 2017.

7. M. Abhishek, S. A. Shah, K. Chetan and K. A. Kumar, "Design and implementation of traffic flow based streetlight control system with effective utilization of solar energy," Int. J. Sci. Eng. Adv. Technol., vol. 3, no. 9, pp. 195–499, 2015.

8. C. Bhuvaneswari, R. Rajeswari and C. Kalaiarasan, "Analysis of solar energy based streetlight with auto tracking system," Int. J. Adv. Res. Electr. Electron. Instrum. Eng., vol. 2, no. 7, pp. 3422–3428, 2013.

9. D. K. Rath, "Arduino based: Smart light control system," Int. J. Eng. Res. Gen. Sci., vol. 4, no. 2, pp. 784–790, 2016.

10. S. Kim et al., "Networked smart LED lighting system and its application using Bluetooth beacon communication," In *2016 IEEE International Conference on Consumer Electronics-Asia (ICCE-Asia)*. IEEE, 2016. DOI:10.1109/ICCE-Asia.2016.7804837

11. A. Ikpehai et al., "Smart streetlighting over narrowband PLC in a smart city: The Triangulum case study," In *2016 IEEE 21st International Workshop on Computer Aided Modelling and Design of Communication Links and Networks (CAMAD)*. IEEE, 2016. DOI:10.1109/CAMAD.2016.7790365

12. I. Oditis and J. Bicevskis, "The concept of automated process control," Sci. Pap., vol. 756, pp. 193–203, 2010.

13. E. Adetiba, V. O. Matthews, A. A. Awelewa, I. A. Samuel and J. A. Badejo, "Automatic electrical appliances control panel based on infrared and Wifi: A framework for electrical energy conservation," Int. J. Sci. Eng. Res., vol. 2, no. 7, pp. 1–7, July 2011.

14. S. A. E. Mohamed, "Smart streetlighting control and monitoring system for electrical power saving by using VANET," Int. J. Commun. Network Syst. Sci., vol. 6, pp. 351–360, 2013.

15. G. Benet, F. Blanes, J. E. Simó and P. Pérez, "Using infrared sensors for distance measurement in mobile robots," Rob. Auton. Syst., vol. 40, no. 4, pp. 255–266, 2002.

16. F. Viani et al., "Evolutionary optimization applied to wireless smart lighting in energy efficient museums," IEEE Sensors J., vol. 17, no: 5, pp. 1213–1214, 2017

17. M. Srikanth and K. N. Sudhakar, "Zigbee based remote control automatic streetlight system," Int. J. Eng. Sci. Comput., pp. 639–643, 2014. DOI: 10.4010/2014.208

18. H. Satyaseel, G. Sahu, M. Agarwal and J. Priya, "Light intensity monitoring & automation of streetlight control by IoT," Int. J. Innovations Adv. Comput. Sci., vol. 6, no. 10, pp. 34–40, 2017.

19. N. Xavier et al., "Design, fabrication and testing of smart lighting system," In *Future Technologies Conference (FTC)*. IEEE, 2016. DOI:10.1109/FTC.2016.7821690

20. P. C. Cynthia, V. A. Raj and S. T. George, "Automatic streetlight control based on vehicle detection using Arduino for power saving applications," Int. J. Electron. Electr. Comput. Syst., vol. 6, no. 9, pp. 291–295, 2017.

21. K. S. Sheela and S. Padmadevi, "Survey on streetlighting system based on vehicle movements," Int. J. Innovative Res. Sci. Eng. Technol., vol. 3, no. 2, pp. 9220–9225, 2014.

22. P. Mestry, I. Darekar, A. Adurkar and S. Ojha, "Vehicle movement based streetlights with external light sensing," Int. J. Adv. Res. Eng. Sci. Technol., vol. 4, no. 2, pp. 2394–2444, 2017.
23. K. H. S. D. Abhishek and K. Srikant, "Design of smart streetlighting system," Int. J. Adv. Eng., vol. 1, pp. 23–27, 2015.
24. Y. Chunjiang, "Development of a smart home control system based on mobile internet technology," Int. J. Smart Home, vol. 10, no. 3, pp. 293–300, 2016.
25. L. A. Akinyemi, O. O. Shoewu, N. T. Makanjuola, A. A. Ajasa and C. O. Folorunso, "Design and development of an automated home control system using mobile phone," World J. Control Sci. Eng., vol. 2, no. 1, pp. 6–11, 2014.
26. K. P. Shinde, "A low-cost home automation system based on power-line communication," Int. J. Creative Res. Thoughts, vol. 5, no. 3, pp. 20–24, 2017.
27. P. C. Joshin, M. Joseph, S. James and V. Sasidhara, "Automation using power line communication with web-based access," Int. J. Adv. Res. Electr. Electron. Instrum. Eng., vol. 4, no. 1, pp. 229–234, 2015.
28. S. Escolar, J. Carretero, M. Marinescu and S. Chessa, "Estimating energy savings in smart streetlighting by using an adaptive control system," Int. J. Distrib. Sens. Networks, vol. 10, no. 5, pp. 1–17, 2014.
29. BEEP, Building Energy Efficiency Project (BEEP) database, Available: http://beepindia.org/beep-commercial

11

Flexible Communication Technologies Utilized in Developing Smart Cities

Pooja N. Kakani and Logesh Rajendran

L&T Smart World, Chennai, India

CONTENTS

DOI: 10.1201/9781003287186-11

11.1 Introduction

In the last few decades, the world population has increased rapidly at an average rate of 1.2% per year. For millions of people worldwide, migrating to cities has become synonymous with opportunities and prosperity, and urbanization is expected to be multifold by 2050. Urbanization has its own merits and demerits, with increases to the pressure for an optimized solution to manage the limited resources available such as Water, Electricity, and Transport to meet the citizen needs. Therefore, sensors, IoT devices, field elements, and technology such as Machine Learning, Artificial intelligence, and Information and Communications Technology (ICT) applications are implemented to handle smart city infrastructure. This fusion of technology with the available resources is the base of smart cities. In the intervening decades, communities have begun to adopt technology for communication to improve services and the quality of life of civilians, which implies that they are evolving from "wired cities," through "virtual cities," to "information cities, "intelligent cities," "digital cities," "environmentally sustainable cities, and smart cities" In the future, ICT is effectively linked as "smart sustainable cities." We could witness the cities with "smart urban sustainability" in the future that has been linked with ICT effectively. Communication has distinct definitions, theoretical models, and perspectives are available in the research-based literature. However, the most general is the layered architecture model where Networking and Communication Technologies acts as a backbone.

Walkthrough the overview of smart city technologies to understand the potential and scope of communication and the Strategy adopted for selecting adequate communication technology. The available technologies are classified based on the mode of transmission, network classes, IEEE standards, objectives, mode of operations, future and emerging technologies. A comprehensive study of different communication and networking technologies based on essential factors characterizes facilitated communication in the smart city ecosystem. Ultimately, various research challenges, such as scalability of wireless solutions, mobility management, interference management, high-energy consumption, and interoperability are discussed as a future research direction.

11.2 The Smart City: A Modern Vision of Urban Living

Technology Innovations help to improve the life quality and develop a sustainable plan to utilize the available resources. Smart cities are stepping-stones toward efficiently manage the physical infrastructure resources by connecting them to technology.

Multiple Research definitions are cited for a smart city and ICT (Information and Communications Technology) is commonly used. The sector like the International Telecommunication Union (ITU) and Institute of Electrical and Electronics Engineers (IEEE) are communication standardization bodies that drive new initiatives and governance. The ITU-Telecom Standardization community focuses on sustainable smart cities and encompasses the concept: "A smart, sustainable city maintains operational performance, services efficiency, and competitiveness while at the same time maintaining for the economic and environmental needs of present and future generations." The ITU-Telecom Standardization community focuses on sustainable smart cities and encompasses the concept: "A smart, sustainable city maintains operational performance, services efficiency, and competitiveness while at the same time maintaining for the economic and environmental needs of present and future generations" [1]. According to the IEEE, a smart city can easily handle and process networked information to optimize urban management impacts in the cities. Such operations include a wide range of activities such as surfacing information. government, businesses, and individuals and enhancing the reliability of energy and water supply or use, traffic control, public protection, and emergency services [2].

11.2.1 Attributes of Smart City

There are several defining characteristics of smart cities available that provide a broader spectrum of functions. The core support functions of a smart city are Economy, Governance, Environment, Society, Technology and Infrastructure, as illustrated in Figure 11.1.

- Economy: ability of the city to thrive and provide opportunities for living. It must maintain the economic balance in terms of employment, market values, GDP (Gross Domestic Product), global and local viability, investments, innovations, and compensations.
- Governance: a robust administration maintaining all the policy and integrated all the essential components such as process, transparency, structure, communication, standards, compliance, and citizen services.
- Environment: the smart city must have a sustainable plan of operation to preserve the resources such as land, biodiversity, water, air, and intact from any form of pollution.
- Society: the city ensures the flexibility of the citizens. It integrates all the social networks, cultures, and community needs. A smart city provides a better user experience and quality of life with equal access to all.

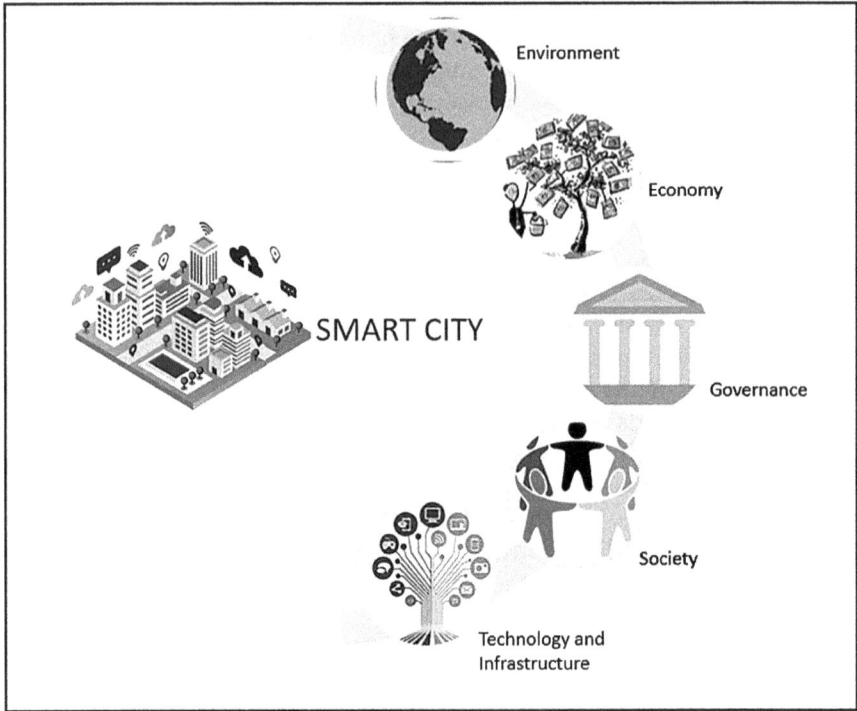

FIGURE 11.1
Characteristics of smart city.

- Technology and Infrastructure: the city combines available physical infrastructure and technology services to make the system smart energy, smart water, smart healthcare, smart education, and smart transportation.

11.3 Prime Applications of Smart City Infrastructure

The fusion of technology with the traditional infrastructure makes the fundamental services smarter. Services such as water, energy, health, and education over time have evolved to be reliable and widely available, with the help of communication technology. Applications of smart cities are illustrated in Figure 11.2. A brief discussion on several smart city applications is presented below.

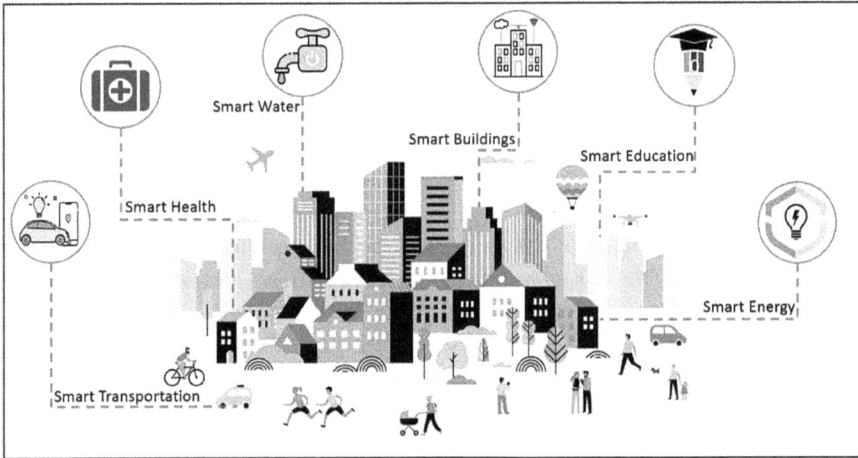

FIGURE 11.2
Smart city applications.

11.3.1 Smart Energy

Energy is one of the essential natural available resources. Households and Industries consume a large amount of electricity on a daily and regular basis. To regulate and monetize the usages, a smart metering system is adopted in the cities. Traditionally, power line consumption, or RF mesh was used for communication. Cellular-based communication such as ZigBee, LoRa WAN, and Wifi has gained popularity recently to collect meter data for monitoring purposes and to regulate usages. The meter data is drawn and used for consumption analytics and automatic billing, which helps in improving energy consumption forecast, track power generation, improve customer satisfaction by introducing transparency in the system and prevent energy theft. Besides, the current major goal is to integrate renewable energy into the ecosystem. Thus, developing a sustainable plan for energy consumption plays a vital role.

11.3.2 Smart Water

Smart water involves managing the water resource efficiently as it is priceless, and the infrastructure systems are surrounded to help water sourcing, water treatment, and water delivery. The real-time monitoring of the water distribution and management system requires IoT sensors, communication infrastructure, and central command control. Several IoT devices, such as water meters, a leakage detection module, and water level sensors are integrated into the water management system to monitor the water supply systems. GPS/GPRS-based location tracking is used to monitor water leakage. These devices connect wirelessly to the central network to monitor the statistics and maintain balance.

11.3.3 Smart Transportation

Intelligent transportation system employs technologies to monitor, evaluate, and manage the transportation system to enhance efficiency and safety. It also makes moving around the city convenient, safer, and more cost-effective. The adaptive traffic management uses the Vehicle to Internet method for communication for location tracking, traffic flow data, accident-prone areas prediction, and parking management. The light poles integrated with AI-enabled CCTV cameras help in detecting the traffic density and peak hours. The vehicles are integrated with wireless proximity sensors with the ability to communicate V2I, V2V to ensure safety, location-based services, and traffic updates. Today, parking management faces a major problem as the number of vehicles is increasing. RFID-based vehicle recognition enables easy parking and toll collection. Besides, vehicles are GPS enabled, which facilitates navigation, speed control, and emergency services. These connect wirelessly through a cellular network to the central command and control. ICCC plays the third eye in monitoring the whole ecosystem. They also control the traffic lights to avoid traffic congestion based on data collected. Thus, communication technologies make mobility in cities smoother and safer.

11.3.4 Smart Health

Healthcare is an essential service to the inhabitants. With the advancement in biomedical devices and communication accuracy, there is a shift from the traditional hospital-focused method to a distributed patient-focused manner. Smart healthcare enables the delivery of information regarding medical conditions and their possible solutions. It facilitates remote monitoring services that reduce the treatment cost and provides healthcare solutions across the vast geographical limits with much faster response in critical emergencies. IoT sensors play an essential role in revolutionizing healthcare. Based on the range, the sensors are connected through ZigBee, Lora WAN, Z-wave, or Wifi. The health vitals are tracked on applications or central consoles to take necessary measures. The heartbeat-monitoring sensor uses Bluetooth and Wifi to transfer the data to the device for continuous evaluation. Applications such as oxygen level detection, remote surgery, temperature monitoring are support by modern communication technologies and wearable devices.

11.3.5 Smart Buildings

The ambiance and building sophistication are directly connected to an individual lifestyle. Safety, Security, utility, and comfort are key features of smart buildings. Building safety and security must be ensured from natural, human, and all forms of threats. Hence, the integration of multiple IoT sensors and the trigger of an early warning notification to the user mitigate the chances of accidents and threats. The automated fire alarm systems alert residents and rescue departments for immediate aid. Incorporation of monitoring and access control systems ensuring security and prevents human attacks 24/7. Building resources such as light, water, and electricity are automatically monitored and regulated as per requirements. All the smart buildings components are connected either over Ethernet, fiber backbone, and wireless to provide faster communication.

11.3.6 Smart Education

Mobility in education and smart ways of teaching are essential factors of smart education. In growing, education cannot be limited to classroom and books. The digital classroom and massive open online courses have started a new wave in education. It provides abundant resources and flexibility to learners. The interactive online courses make learning more efficient. Internet connectivity plays a crucial role in the smoother learning experience. The real-time motion sensor backed by AI improves the quality of learning. 4G and 5G have provided high-speed data rates required for functioning. Thus, with the support of high-speed connectivity traditional classrooms are transformed into smart classrooms equipped with smart boards, automated attendance systems, and behavior monitoring.

For smart city applications, ICT has a vital role as it collects and aggregates information from the ground to enhance the interpretation of the system. Thus, ICT enables information and knowledge sharing, forecasting, and integration for achieving smart city goals.

11.4 IEEE Standards

While designing and developing materials, products, methods, and services for general people's use, a set of procedures and specifications are followed, referred to as standards to ensure reliability. These standards specify various protocols that maximize product compatibility, interoperability, and functionality. In addition, ensures quality, and consumer health and safety. The standards are the pillars for product development and consistency. It simplifies product development by ensuring compatibility and interoperability as they are universally adopted. These standards are globally accepted making product comparison and selection easier. Several

TABLE 11.1

IEEE Standards for Communication Technologies

Standards	Description
IEEE 1901	Power line communication
IEEE 802.3	Ethernet
IEEE 802.8	Fiber optic
IEEE 802.11	Wireless LAN physical and MAC layer
IEEE 802.15	Wireless Personal Area Network (WPAN)
IEEE 802.16	Wireless Metropolitan Area Network (WMAN)

IEEE standards for communication technologies and smart cities are proposed and adopted. They specify protocols and procedures to build the technologies.

11.4.1 Standards for Communication Technologies

Communication technologies are used to interconnect different devices in the ecosystem Table 11.1. Contains essential IEEE standards frequently used in smart city applications. Commonly used IEEE standards in smart cities for wired communication are IEEE 1901, IEEE 802.3, and IEEE 802.8. On the other hand, IEEE 802.11, IEEE 802.15.1, IEEE 802.15.3, IEEE 802.15.4, IEEE 802.15.6, and IEEE 802.16. IEEE 802.11 are wireless communication standards used in several smart city applications [3]. For applications such as smart lighting and health monitoring, where a relatively short range of coverage is suitable IEEE 802.15.1, IEEE 802.15.3, IEEE 802.15.4, and IEEE 802.15.6 are used. IEEE 802.16 defines standards for long-range communication technologies often found in smart energy applications.

11.4.2 Standards for Smart Cities

IEEE defines serval other standards related to critical smart city factors such as healthcare, security, IoT frameworks, and energy efficiency. IEEE 1609, IEEE 602, and IEEE 139 state the medical devices and RF emission from ISM equipment used in healthcare applications. The standards related to IoT architecture are stated in IEEE p2413 and IEEE P1451-99. For data privacy process and ePrivacy IEEE, P7002 and IEEE P802E are established. Besides, for physical security IEEE 1402 can be found.

Generally, used IEEE standards for energy management are IEEE 1801 and IEEE P1889.

11.5 Classification of Existing Communication Technologies

Smart cities handle a huge population and unevenly distributed city infrastructure. This results in a large number of applications ending, and communication tightly coupled the applications and physical resources. Hence, it is probable that various communication technologies are used. Communication technologies are classified based on several parameters such as mode of operation, mode of transmission, network classes, and objectives [4]. Desired parameters for a communication model are error-free communication links, high data rates, and minimal power consumption. The frequently deployed technologies that impact ICT are presented in Figure 11.3. They are classified based on the range and data rates. Nowadays, it is evident that wireless communications are the most used technology due to advantages such as ease of deployment and mobility. The Internet is accessed chiefly via either 3G/4G or Wifi networks. Nonetheless, wired communications are used in the landscape, which requires high speeds [5].

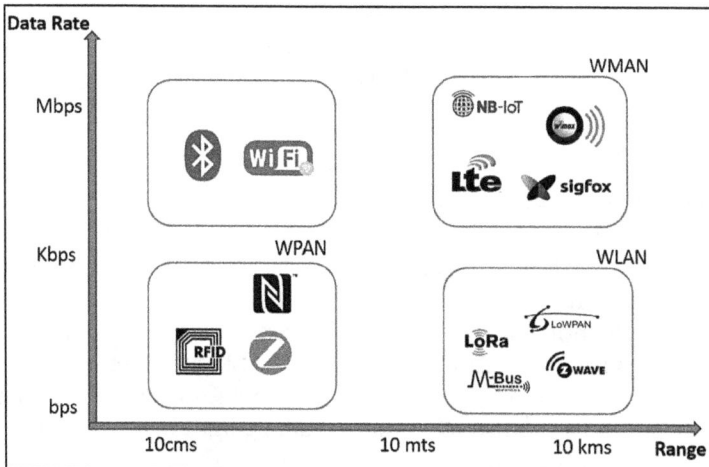

FIGURE 11.3
Essential technologies to enable smart city applications.

11.5.1 Wired Technologies

Wired communication is an essential component of communication technologies in smart cities. It covers high data rates, no channel interference, stable bit error rates (BER). Whereas wireless communication suffers channel loss due to interference, path loss, reflections, refractions, and fading. This reduces the overall signal strength and performance of the system. The wired technologies involve laying cables on the ground, civil works, high investments, etc. Wired communication aims to have full control over communication infrastructure. Copper Ethernet cable networks and fiber optic cables are widely used physical mediums. Wired communication is used for multiple purposes like voice, Internet, digital subscriber line (DSL) solutions, power line communication (PLC), digital subscriber line, and cable TV (CATV). Fiber optic communication provides ultra-broadband and low latency compared to the traditional copper Ethernet cable network. The high bandwidth fiber leads smart city-wide networks to be available for improved road safety through robust traffic signals, smart mobility, public services like information signages, smart infrastructure, and optimizing traffic flow by connecting various signals, and all these are to the benefit of citizens.

11.5.2 Wireless Technologies

In recent years, wireless technologies have seen tremendous growth. Over the period, several new standards, coverage range, and data rates are proposed. Based on the range of communication, the wireless technologies are categorized into three: Wireless Personal Area Networks (WPAN), Wireless Local Area Networks (WLAN), and Wireless Metropolitan Area Networks (WMAN) [6]. Figure 11.4 illustrates the network topologies of wireless communication. An overview of area networks and technologies is presented below.

11.5.3 WPAN

The devices connect and communicate through the personal network for sharing information in Wireless Personal Area Network (WPAN). Through wireless communication, the individual devices connect and access the data over the Internet. It is also known as a short-range wireless network. NFC and RFID are technologies having a range from 10 cm to 3 m. A few longer-range technologies include Bluetooth, Z-wave, Blue Low energy (BLE) and ZigBee having a 10–15 m range. These technologies provide data rates up to 250 kbps and are used in smart building, smart lightings, and smart health for short-range applications.

a) Bluetooth

Bluetooth is a low-power, high-speed wireless technology, designed to connect portable equipment. It follows IEEE 802.15.1 and uses low-power radio

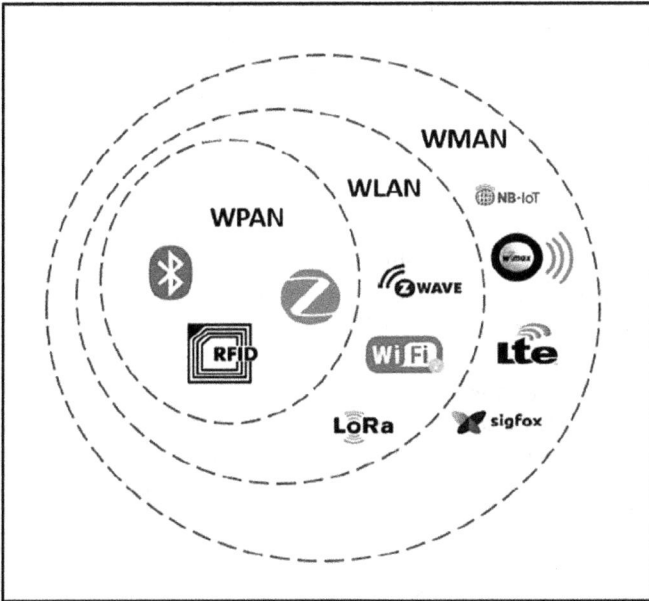

FIGURE 11.4
Network topologies for smart city technologies.

links to connect sensors, phones, and other network devices for short distances without wires. The typical range of wireless communication is 30 feet (10 meters). Low-cost transceivers are embedded into the devices to achieve the same. It works on a 2.45 GHz frequency band along with three and has a speed up to 721 kbps. This frequency band is only for the use of industrial, scientific, and medical devices (ISM) as per international agreements.

b) Z-wave

Z-wave is designed and developed by Sigma Designs Inc., which integrity, authentication, and encryption. It uses a low-power mesh network with a single controller designed to support up to 232 repeater nodes. It employs BFSK modulation and operates in frequency bands of 868/915 MHz. It applies the carrier sensing multiple access techniques combined with collision avoidance (CSMA/CA), a network multiple access method. Furthermore, 9.6 kbps and 40 kbps are the available data rates.

c) Zigbee

ZigBee is a wireless technology designed as an open global standard that addresses the need for low-power, low-cost wireless IoT networks. This standard operates on the IEEE 802.15.4 specification and operates in 2.4 GHz, 900 MHz, and 868 MHz including unlicensed bands. It is used in sensor nodes, suitable for creating WPAN such as medical device collection, and home automation. The application includes wireless light switches, traffic management, and electric meters. Generally, most suitable for limited but can connect billions of devices. However, to deploy ZigBee, additional equipment is involved such as routers, coordinators, and ZigBee end devices.

11.5.4 WLAN

A local area network formed by connecting two or more devices in the wireless network is termed as Wireless Local Area Network. WLAN provides a medium range of communication, from 100 m to 300 m radius. Medium range communications technologies like ZigBee, Wifi, Z-wave, wireless M-bus, and 6LoWPAN (IPv6 over Low-power Wireless Personal Area Networks) are used in situations such as smart home automation, personal area network, smart lightings, smart traffic management, and smart waste management, etc. These communications are used highly by consumers on a wider scale and technologies are easily scalable and can connect to multiple devices for a large and effective communication.

a) Wifi

Wireless Fidelity is a wireless local area network technology allowing the users/devices to access or connect to the network using radio waves. The IEEE developed 802.11 standards for Wifi. The bandwidths are range from 2.4 and 5 GHz. It uses carrier sense multiple access with collision avoidance (CSMA/CA) as a network access protocol along with Rivest Cipher 4 (RC-4) based stream encryption algorithm for security. The data rate varies with the version of IEEE standard deployed such as 54 Mbps for 802.11a and up to 5.5 Mbps for 802.11b.

b) LoRa

LoRa stands for long-range working in the unlicensed band below 1 GHz. It is a low-power proprietary design owned by Semtech Corporation which uses a spread spectrum modulation technique. It provides long-range transmission with improved immunity to channel interferences at low cost, thus employed for IoT solutions. It utilizes 868 MHz and 915 MHz and provides a vast coverage area of up to 15 km in suburban and 5 km in urban areas. LoRa architecture makes the deployment easy and the flexible gateway enables thousands of end devices to connect in the network. It provides a data rate varying from 0.3 kbps to a maximum of 27 kbps for 125 kHz of bandwidth. The low-power transmission with high-cost efficiency makes it suitable for deploying M2M of IoT applications.

c) 6LoWPAN

The 6LoWPAN stands for the combination of Internet protocol and low-power WPAN. Over time, due to exhaustion of address blocks in IPv4, and the inability to address billions of devices, which is a dominant characteristic of IoT, addressing technology for an Internet host, is being replaced by IPv6. The 128-bit address provided by IPv6 solves the shortage of nodes but creates the problem of compatibility. This limitation is resolved by the compression technique used in the address format by 6LoWPAN.

11.5.5 WMAN

A Wireless Metropolitan Area Network (WMAN) provides point to point or point to multiple point connection through a wireless connection. WMAN has a lesser coverage area than wireless wide area network (WWAN) but more significant than WLAN. It provides a coverage range of approximately 30 miles, such as a city's size. It not only provides a larger coverage area but supports backhaul applications in metropolitan environments. Few examples of these technologies are LoRaWAN, SigFox, LTE, LTE-A, WiMAX, and NB-IoT which provides flexibility for communication as well as collecting the data from all the IoT devices situated in various locations of the city.

- LTE

Long-term evolution (LTE) is broadband communication technology introduced in the 3GPP radio interface. Works on the mobile network-based technologies and has IP-based network architecture providing higher data rate and channel capacity. It employs advanced modulation techniques which support both time and frequency division multiplexing where carrier bandwidth varies from 1.4 MHz to 20 MHz. Besides, it facilitates smooth handoff for voice and data to older cellular towers because of negligible transmission delay of fewer than five milliseconds. On average, LTE offers a download and upload speed of 77.8 Mbps and 26.9 Mbps, respectively (see Table 11.2).

a) LTE-A

Long-term evolution is broadband communication technology introduced in the 3GPP radio interface. Works on the mobile network-based technologies and has IP-based network architecture providing higher data rate and channel capacity. It employs advanced modulation techniques which support both time and frequency division multiplexing where carrier bandwidth varies from 1.4 MHz to 20 MHz. Besides, it has a transmission delay of fewer than five milliseconds, facilitating handoff for data and voice to older mobile towers. Compared to the previous LTE version, LTE-A provides higher upload (1.5 Gbps) and download (3 Gbps) data rates. LTE-Advance supports multiple antenna schemes and new transmission protocols to enable smoother handoff between different cells. It has better bits per second across the spectrum and improved data throughput at the edges. Thus, enabling more consistent connections, higher network capacity, and cheaper data rates.

b) SigFox

SigFox uses Differential binary phase-shift keying modulation (BPSK) techniques and transmits in Ultra-Narrow band spectrum (UNB), generally less than 2.5 KHz. It uses a unique transmission method where the device sends the data to the base station by three uplink packets in the order of three

TABLE 11.2

Overview of Wireless Communication Technologies

Communication Technology	Standard/ Governing Bodies	Frequency	Range (App.)	Data Rates
NFC	ISO/IEC – ISO 13157 etc.	(HF) 13.56 MHz	10 cm	106 kbps, 212 kbps, 424 kbps
RFID	ISO/IEC – ISO/IEC 18000	(LF) 125–134 kHz (HF)13.56 MHz (UHF) 856–960 MHz	3–10 m (active up to 100 m)	40 kbps–640 kbps
Bluetooth/ Bluetooth Low Energy (BLE)	IEEE 802.15.1/ Bluetooth SIG	2.4 GHz	10 m typical 30–50 m (BLE)	1–3 Mbps 1 Mbps (BLE)
Z-wave	Z-wave alliance	900 MHz	30 m (ind.) 100 m	96 kbps–100 kbps
ZigBee	ZigBee alliance	2.4 GHz	10–100 m	250 kb/s
Wifi	IEEE 802.11 (a/b/g/n)	2.4/3.6/4.9/5/5.9 GHz	100 m	1–54 Mbps
6LoWPAN	Internet Engineering Task Force (IETF) IETF RFC4944	868 MHz/915 MHz/2.4 GHz	200 m	250 kbps (2.4 GHz) 40 kbps (915 MHz) 20 kbps (868 MHz)
Wireless M-BUS	EN 13757-4	169/433/868 MHz	300 m	2.4 kbps–100 kbps
Wifi low power (802.11ah)	IEEE 802.11 working group	Sub-1 GHz	1 km	150 kbps ~ 346,666 Mbps
LTE-M release 12/13	3GPP release 12	700–900 MHz	2.5–5 km	200 kbps (DL) 200 kbps (UL)
LoRaWAN	LoRa alliance	433, 863–870 MHz (EU) 902–928 MHz (US)	2–15 km	250 bps–50 bps (UL EU) 250 bps–50 kb/s (DL EU)
SigFox	SigFOX	868–902 MHz	10–50 km	256 b/day (DL), ≤100 bps (UL)
LTE	3GPP	2.5/5/10 GHz	30 km	300 Mbps (DL), 75 Mbps (UL)
LTE-A	3GPP	2.5/5/10/15/20 GHz	30 km	1 Gbps (DL), 500 Mbps (UL)
WiMAX	3GPP	3.5 GHz	50 km	75 Mbps
NB-IoT	3GPP Release 13	Can be deployed in 2G/3G/4G spectrum (e.g. 450 MHz to 3.5 GHz),	1 km (urb.) 10 km (rur.)	234.7 kbps (DL) 204.8 kbps (UL)

(Continued)

TABLE 11.2 (Continued)

Overview of Wireless Communication Technologies

Communication Technology	Standard/ Governing Bodies	Frequency	Range (App.)	Data Rates
EC-GSM	3GPP Release 13	800–900 MHz	15 km	~300 kbps (DL) 10 kbps (UL)
CS IoT	–	700–900 MHz	–	200 kbps (DL) ~48 kbps (UL)

different frequencies of the carrier. Sigfox evolved to support bidirectional connection from original uplink communication. It provides a maximum throughput of 100 bps and a coverage range of 3 km to 10 km in urban areas while 30 km to 50 km in rural environments.

11.6 Communication Enablers

A plethora of communication activities takes place in smart cities involving a vast range of technologies such as sensing and actuation, edge and cloud computing, data analytics, and data management. Machine-to-Machine and the Internet of things efficiently unite these activities under a single platform. Often, IoT is confused with M2M communication as they both connect sensors, actuators, and devices to the communication network via wired or wireless communication. However, IoT spreads across the entire paradigm, including design principles and technologies associated with systems and emerging Internet-connected things. IoT is expected to build an ecosystem like that of today's Internet with the support of M2M acting as enablers. The basic working methodology of M2M communication can be depicted in Figure 11.5.

The fundamental operations of IoT-based applications are M2M-based data collection from sensors, generate high-level abstraction using data processing to provide intelligence, and the ability to react and actuate based on derived intelligence [7]. Hence, IoT is an expansion of the present Internet, with control, supervision of physical resources, and automatic data collection, through remote control and monitoring. In contrast to M2M, IoT connects the system and sensors through Internet-protocols (IP) networks as shown in Figure 11.6. This ensures the interoperability and connectivity of devices from distinct manufactures while opening opportunities for mass production and deployment. In the future, the devices will act as intelligent creatures interacting, exchanging information with humans, and supporting business processes. Through the IoT ecosystem, real-world objects will be able to connect, interact and communicate with each other the same way people do via the web today.

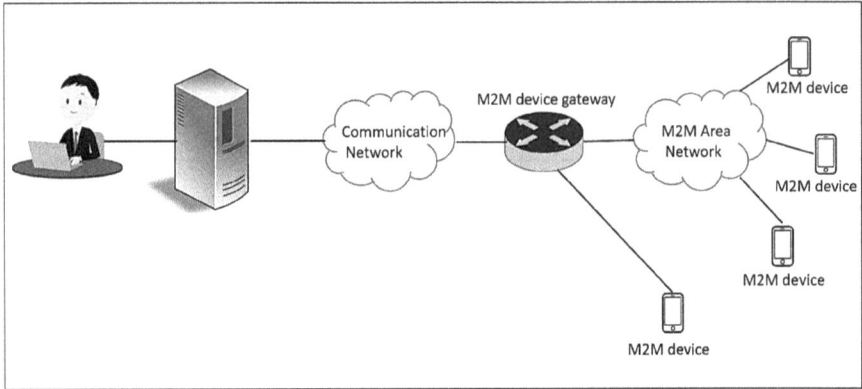

FIGURE 11.5
A generic model of M2M communication.

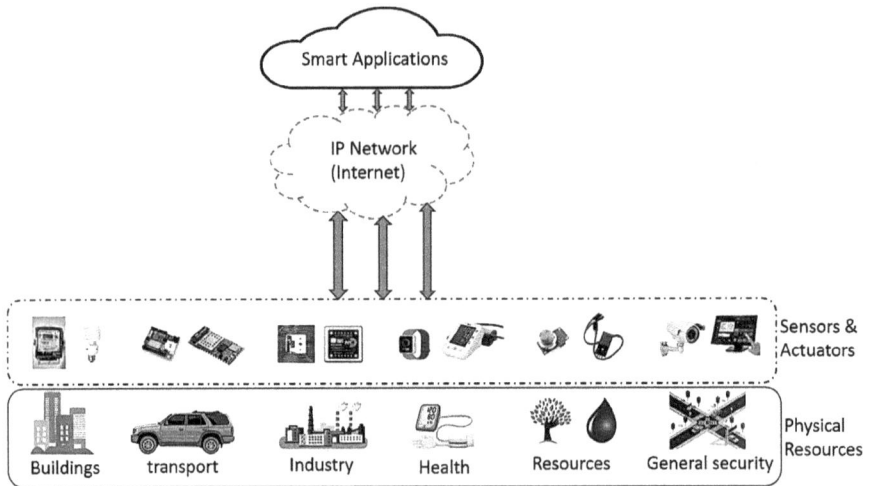

FIGURE 11.6
Overview of IoT ecosystem.

11.7 Smart Industry 4.0 (IIoT 4.0)

Industry technologies evolved a lot in a few decades from Industry 1.0 to Industry 4.0 as depicted in Figure 11.7. Considering industries connect multiple IoT devices so it is also termed IIoT (Industry Internet of things) or smart manufacturing. Industry 4.0 is a new phase of the industrial revolution that interconnects and integration of digital technologies like IoT, physical

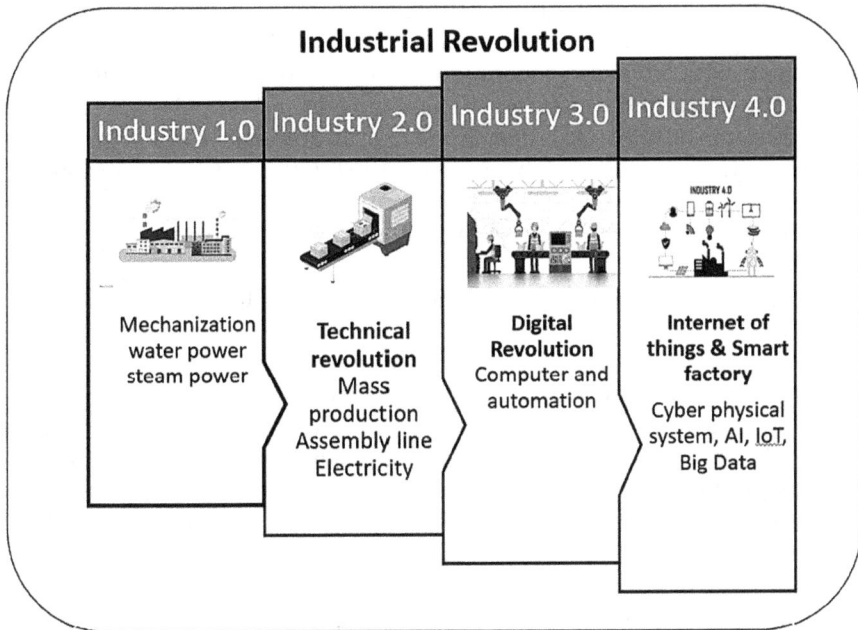

FIGURE 11.7
Stages of the industrial revolution.

devices, machine learning, big data analytics, Communication technology, automation, and real-time data with physical production and operations to achieve a connected ecosystem for industries focusing on manufacturing and supply chain management. While every organization and company faces various challenges today, the most common difficulties are connectivity and access to real-time data across processes, products, partners, and people. Industry 4.0 transforms the way business functions and grows than just integrating technologies and tools to improve industry manufacturing efficiency.

The industrial revolution began decades ago, with the invention of water and steam power, steam-powered machines, and other sorts of machinery lead to the first industrial transformation of society with the mechanization of manufacturing, trains, and production of smog. With inventions in manufacturing technologies and the use of electricity, the second industrial revolution commenced. These enabled improved industrial output with the help of assembly-line production, mass production, and basic automation. The ascent of electronics, computers, robotics, network connectivity, and the Internet embarked on the third industrial revolution. A new wave of e-everything hit the industries and changed the way of information handling and sharing with better automation. With the growth of the Internet, industries moved toward bridging the digital and physical environment, often referred to as Cyber-physical systems (CPS) leaving the traditional models. The convergence of information technology (IT) and operational technology (OT) along with advanced robotics, IoT, and big data facilitates automation and optimization in the industries. These modern technologies lead the way for huge opportunities and innovations to fully automate the industry to the next level. Industry 4.0 mainly consists of IoT, CPS, and cloud computing-based manufacturing [8]. IoT aids the transformation of solutions for the operations and promotes the existing industrial systems to tomorrow's digital industrial era.

IoT enables the construction of virtual networks to aid smart factories in the Fourth industrial revolution. Thus, it is an essential technology for upcoming manufacturing industries. Cloud computing-based manufacturing contributes significantly to the revolution of industries. Cloud manufacturing uses a highly distributed approach to form a network of resources effectively. Service manufacturing is also known as, manufacturing as a service has predominantly become popular, as it reduced the cost of manufacturing for the industries. Industry 4.0 foundation relies centrally on CPS. These systems collaborate computational entities connected intensively with ongoing processes and the physical world. To achieve the above functioning, CPSs utilize the services available on the Internet for data accessing and processing services. Physical hardware and software layers are immensely intertwined in CPS, each functioning in distinct planes and scales but interacting in innumerable ways. With the introduction of CPS, Machine-to-Machine communication cab be enabled and central control can be decentralized to optimize the production. Generally, Industry 4.0 encompasses several complex elements and had large applications in the industrial sector. Communication technologies play the role of binding agent in the whole Industry 4.0 ecosystem.

The fundamentals of IoT are considered, as worldwide infrastructure for network constitute of innumerable connected devices that bank on sensors, networking, information processing, and communication technologies. The essential devices for IoT networks are RFID and wireless sensors. Services such as track, distinguish, and monitoring objects are easier with RFID tags. This technology can be found in a variety of industries such as healthcare, defense, package delivery, transportation, retailing, materials management, etc. By contrast, for sensing and monitoring, WSN applies the interconnection of intelligent sensors.

The cyber-physical system is a product of tight integration between the physical world and connected computing devices. With the advancements in CPSs, capability, scalability, adaptability, resiliency, security, safety, and usability will be further elevated than simple embedded systems present today. CPSs provide a combination of sensing, central control, computation, and network functions that improves the overall productivity of the industry. These employ sensors to transmit and receive information regarding physical parameters, data of actuators, and devices to engage in control over physical processes. The real-time interactions with the physical world involved in CPS distinguish it from conventional IT and ICT systems.

11.8 Security for Smart Cities

Although there are advancements in smart cities that improved the whole ecosystem, smart applications are vulnerable to hacking threats. Security attacks such as spamming, denial of service, unauthorized access, Sybil, outside attacks, and identity attacks can decrease the quality of services [9]. All the devices in the smart city ecosystem are collecting and processing the data in one form or another. Thus, sensitivities of the data must be considered. Smart healthcare and smart home service providers can have access to customer's sensitive data. Besides, the huge amount of data captured by smart transportations applications can be used to predict the mobility and location of the users. Hence, by considering the properties of devices deployed infield, the security and privacy threats and the complex ecosystem of smart cities, this section highlights basic and necessary specifications to secure smart cities [10].

a) Authentication

Authentication is the fundamental necessity for different levels of a system. It ensures that services are only available to authorized clients across the system. Specifically, IoT devices connected to the network can authenticate and send alerts from the management stations in smart cities. Besides, since authentication data is snowballing due to a variety of applications in smart cities, it is essential to create improved and advanced technologies to ensure precise real-time client authentication.

b) Confidentiality

Information confidentiality ensures that the security of data from passive attacks and vulnerability to illicit sources. In IoT-based applications, the invader may have the capability to eavesdrop on networking and communication devices. Hence, encryption-based technologies are deployed to construct reliable storage, network, and communication system to secure the information during transmission between the nodes. The reliability and transparency make the design of the authentication and identification process difficult.

c) Availability

The availability signifies that functions and devices are available when required. Here, smart applications or systems must have the potential to function effectively even under threats or attacks. Also, the smart system should have the capability to detect abnormal activities and protect the system from additional damage. Protection mechanisms must have the ability to continuously learn and adapt to handle the more and more intelligent attacks.

d) Integrity

Apart from connectivity between the IoT devices, ensuring integrity between devices, exchanged information and the cloud is critical. The data while transmission can easily tamper, if not well protected. IoT devices can manage the data traffic flow using protocols and firewalls because of the low compute power but cannot ensure integrity end-to-end.

e) Privacy protection

Security and privacy are interconnected. Every smart application layer requires privacy protection. In several instances, the privacy of the user can be breached even through a secure system. Intelligent data algorithms are one of the commonly used techniques. With the data mining tools, few third-party companies and service providers can fetch the consumer's data. In addition to the technical solutions, measures such as governance, education, and policies will improve awareness for data security.

To subdue, the security issues faced by smart applications have different solutions are incorporated into the system. A few of the current common methods are cryptography, blockchain, biometrics, data mining, and machine learning. The backbones of privacy protection for smart city applications are cryptographic algorithms. They secure the data lifecycle of storing, processing, and sharing by preventing access by distrusted parties. As IoT devices have limited computing power, lightweight cryptographic technologies are used for authentication and protect the network from DDoS attacks. Besides, blockchain works on the framework level that guarantees reliable communication as well as security of the devices. The wireless sensor networks the key component of smart city applications are strength by Machine learning techniques to improve the network handling capability and protect the network from attacks. These are some of the examples of security solutions for smart applications.

11.9 The Necessity of Communication Technologies for Smart City Infrastructure

A smart city is an absolute change of paradigm experiencing in daily life incorporating new technologies, especially ICT, to achieve efficient results. Sensors, communication providers, service providers, equipment providers,

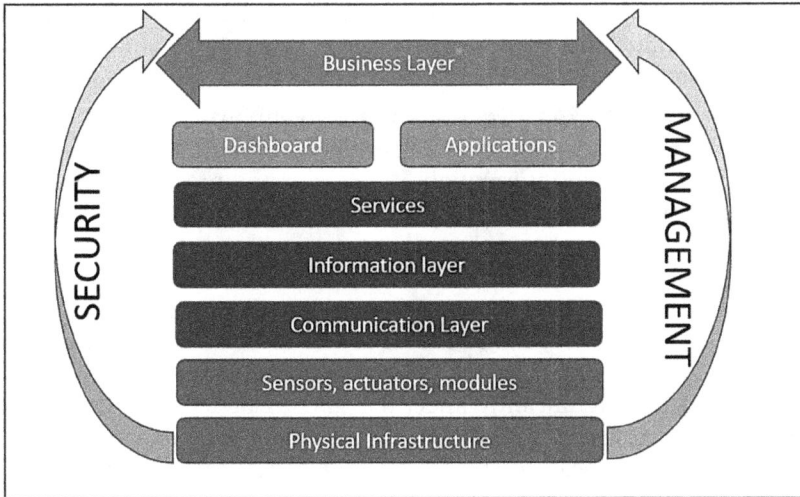

FIGURE 11.8
The smart city application framework.

business innovation are some of the components and players involved in smart city realization. These elements can be arranged in a layered framework, as represented in Figure 11.8. As the architecture depicts, ICT is the critical element in building smart cities by coupling the city infrastructure with applications and services. Currently, most of the focus is on smart cities to solve the issues of data security, sharing, privacy, APIs, standardization, and commercialization. Thus, several opportunities are accessible for different players to understand what data they have and use it efficiently for benefit.

However, while considering the significance of data, the layer collecting the data becomes much more important as communication networks are the central layer. It defines and determines how interrelated objects communicate efficiently with each other, arrange systems, and provide a higher quality of service. In another word, efficient deployment of ICT makes the designing of smart cities easier and provides a better understanding of gathered information. The essential functions of ICT that help achieve the goals and multiply the productivity of the smart city are:

- Forecast: preparing for adverse events such as natural disasters requires a significant amount of data to recognize patterns, recognize the style, threat areas, and forecast potential events. With the help of ICT, information can be gathered and managed efficiently to improve the preparedness and response abilities of the city.

- Integration: access to relevant information at regular intervals of time facilitates a better understanding of the city's strengths and vulnerabilities.
- Information and knowledge sharing: conventional cities may not be capable enough to respond and resolve a problem instantaneously due to insufficient information. With faster and precise information sharing cities can obtain insights on issues much quicker and take preventive actions before it escalates.

Thus, strategically design a system that fulfills all the functional requirements and specifications. As the number of interconnected devices is increasing with the expansion of smart city applications, the technical capabilities of communication significantly affect the same. Due to the unprecedented increment of wireless devices coexisting in the same plane, interference management is a critical design challenge. Besides, scalable wireless solutions are needed to connect the immense number of devices embedded in the system. As the ecosystem is connected using a wide range of technologies, interoperability support among the heterogeneous. Wireless networks become essential. Above all, energy consumption by the devices deployed in the infrastructure is significantly growing.

11.10 Conclusion

The need for building a future sustainable city as the concept of the smart city emerged as a solution that benefits the populations in a subject of urban growth, environmental constraints, and resource scarcity. Smart cities are a combination of processes, services, and applications backed by engineering and technological advancements. In addition, they ensure optimal use of available resources and foresee future needs. A tremendous volume of raw data and information is collected, shared, and processed by smart city applications. Therefore, reliable networking and communication infrastructure should be established to make a smooth and secure data transmission. Various applications build on smart city architecture have their ICT expectations and requirements. Hence, various technologies are selected based on applications, system specifications, and deployment environments to build ICT infrastructure. Wireless technologies have gained more attention because of several benefits, especially ease of deployment and flexibility. However, there are specific challenges such as power consumption, interoperability, and security threats faced by communication technologies. Combined efforts of key players such as equipment manufacturers, framework designers, and mobile network providers to create and implement a new range of communication technologies, named secured low-power communication networks, can overcome the challenges.

References

1. ITU-T Focus Group on Smart Sustainable Cities (FG-SSC). An Overview of Smart Sustainable Cities and the Role of Information and Communication Technologies Focus Group Technical Report. 2014.
2. Dameri, R, and Rosenthal, Sabroux C. Smart City—How to Create Public and Economic Value with High Technology in Urban Space, Springer International Publishing, New York, NY; 2014.
3. Kuzlu, M., and Pipattanasomporn, M. "Assessment of communication technologies and network requirements for different smart grid applications." *2013 IEEE PES Innovative Smart Grid Technologies Conference (ISGT)*, Washington, DC, 2013, pp. 1–6, doi: 10.1109/ISGT.2013.6497873
4. Yaqoob, Ibrar, et al. "Enabling communication technologies for smart cities." *IEEE Communications Magazine* 55.1 (2017): 112–120.
5. Khorov, E., et al. "A survey on IEEE 802.11 ah: An enabling networking technology for smart cities." *Computer Commun.* 58 (2015): 53–69.
6. Talari, Saber, et al. "A review of smart cities based on the Internet of things concept." *Energies* 10.4 (2017): 421.
7. Höller, J., Tsiatsis, V., Mulligan, C., Karnouskos, S., Avesand, S., and Boyle, D. From Machine to- Machine to the Internet of Things—Introduction to a New Age of Intelligence, Elsevier, Amsterdam; 2014.
8. Alcácer, Vitor, and Cruz-Machado, Virgilio. "Scanning the industry 4.0: A literature review on technologies for manufacturing systems." *Engineering Science and Technology, an International Journal* 22.3 (2019): 899–919.
9. Ismagilova, E., Hughes, L., Rana, N. P., and Dwivedi, Y. K. (2020). Security, Privacy and Risks Within Smart Cities: Literature Review and Development of a Smart City Interaction Framework. *Information Systems Frontiers*. doi: 10.1007/s10796-020-10044-1
10. Cui, L., Xie, G., Qu, Y., Gao, L., and Yang, Y. "Security and privacy in smart cities: Challenges and opportunities." *IEEE Access* 6 (2018): 46134–46145, doi: 10.1109/ACCESS.2018.2853985

12

Sewage Management: Sources, Effects, and Treatment Technologies

A. S. Ramya and C. Maheswari

Kongu Engineering College, Erode, India

CONTENTS

12.1 Introduction

Wastewater, as the name suggests, is a really important topic to remember. In both its analysis and existence, it is unavoidable. It is tied to water, which is essential to humans and their environment's long-term survival. Human health is essentially determined by their environment. Indeed, many human health problems can be traced back to negative environmental causes like water contamination, soil pollution, air pollution, bad living conditions linked to hunger, animal reservoirs, and pathogenic organisms, which all pose a persistent danger to human health [1]. Such human activities include urbanization, and industrialization. Without a shadow of a doubt, the world is critical and complex. As a result, it is important to protect and conserve the atmosphere from harmful effects that could further damage it. This is made possible by using engineering expertise and scientific knowledge to solve a variety of environmental issues. It's worth noting that engineering and the environment are inextricably linked.

The vast majority (99.9%) of sewage is water, and 0.1% solids, often referred to as "sludge." One-tenth of one percent (0.1%) of solutes in sewage is composed of organic and inorganic substances. Furthermore, they are partially suspended and partially dissolved. The presence of organic compounds in sewage contributes to its aggressive nature. Sewage comprises a number of living species originating through feces, and some of those are disease agents. Furthermore, one gram of feces is believed to contain 1,000 million E. coli and 10 to 100 million fecal streptococci. An average human ejaculates around 100 grams of waste a day. Sewage contains a significant volume of contaminants accumulated as a result of agricultural, consumer, and domestic applications of water, in relation to impurities present in the water supply. Sewage can be characterized by physical, chemical, and biological properties. Solid content, color, odor, and temperature are the most essential fundamental properties of waste [2]. It is evident that sewage, if not adequately monitored, treated, and handled, poses various risks to public health, posing a danger to humans living in a given climate. Many communities have also been influenced by sewage and its implications on the ecosystem as a result of poverty and indifference. As a result, there is a need to resolve the issues that have arisen as a result of the detrimental effects of sewage on the environment. The aim of this paper is to reveal the potential (both favorable and unfavorable) effects of sewage on the environment as well as potential solutions to the issues.

12.2 Waste Production

Globally, day by day, the generation of waste is increasing due to human activities. Annually, 1.3% of waste production is increased per capita and total waste generation is increased to 5% annually.

As seen in Figure 12.1, in India, around 60,000 MLD of sewage is generated, more than 10,000 MLD of sewage is produced by metropolitan cities and around 8,000 MLD of sewage is produced on the river Ganga due to human habits. Further, only 23,000 MLD of sewage is treated and remaining

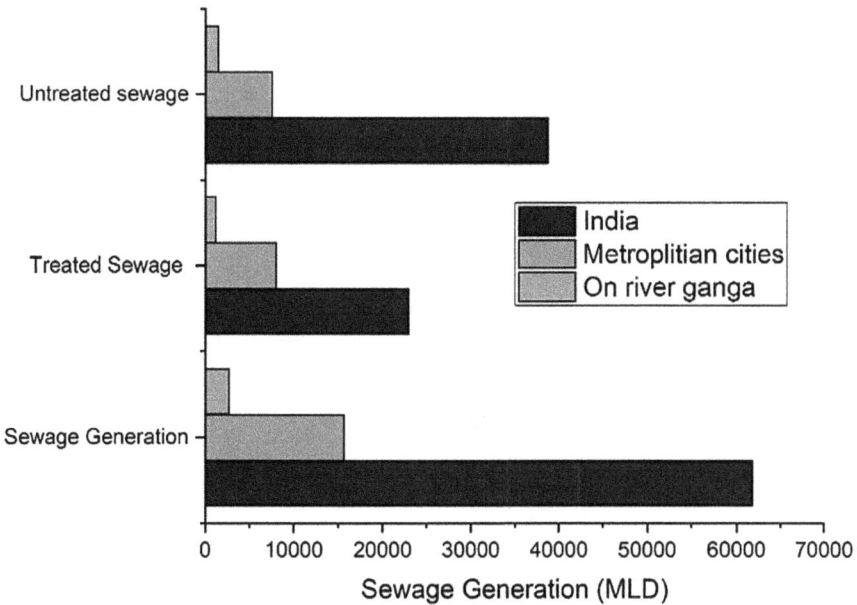

FIGURE 12.1
Current scenario of sewage pollution in India (MLD).

sewage is discharged into land or river bodies without any prior treatment, which leads to environmental degradation.

In the environment, sewage can be generated in various kinds of activities involving domestic, industrial, municipal, agricultural, biomedical and nuclear as are described in Figure 12.2 and its detailed explanation is given in the following sections [3].

12.2.1 Domestic Waste

Wastes, in particular, are a major threat to rural ecology because of the enormous quantities of waste dumped and the poor management of waste. Domestic waste is composed of two major fluxes: one would be gray water from kitchen sinks, laundry sinks, washing-up, douches, kitchens, etc. and another is black water from toilet and urinary items. It is in the form of cartons, yard cuts, timber, and other sources of trash that is excessive substance. The products normally lie on land and in residential properties. Aerosol latches, ammunition, engine oil, paints, and solvents are some other types of household waste. In 2010, Americans generated 250 million tons of household waste, including lawn clippings, clothing, packaging and food waste, equipment, and newspapers. Every single citizen receives 4.3 pounds of waste.

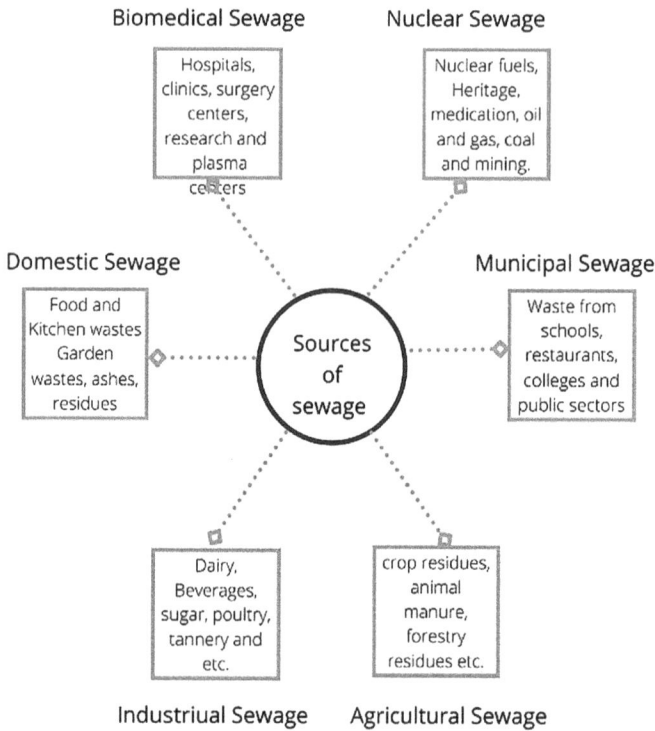

FIGURE 12.2
Sources of sewage.

Household waste also includes products considered to be dangerous, which include radioactive, flammable, and corrosive waste. Paints and chemicals are other articles requiring special disposal methods [4]. Drainage of these threats is an inappropriate form of disposal that presents a danger to human health and the environment. These products often constitute a hazard to animals and can destroy fish in rivers as toxicity. Septic and sewage treatment plants may also be damaged by harmful materials which are incorrectly discarded [5].

12.2.2 Industrial Waste

Industrial waste (IW) is referred to as substance from manufacturing processes such as chemical and cement factories, textile, food processing, petroleum industries, power plants, construction, paint and metal/steel industries (Table 12.1). IW comprises solvents, dirt, scrab chamber, chemicals, oil, scrap metals, weed grass, trees, trash, and concrete. Besides, the everyday introduction of new products such as drugs, medicines, paints, dyes, textiles, plastics and computers into the environment is also a source of the IW [6].

TABLE 12.1

Different Types of Pollutants Generated by Various Types of Industry

Industry	Chemical	Painting	Petroleum	Steel	Food	Construction
Pollutants	Acids and bases Solvents Organic materials and Reagents Reactive products	Rags, masking, floor sweepings, paints, resins, solvents, and brushes	Benzene, Hydrocarbons, Sludges, Oils and Refining products	Cyanide, Steel, Reinforcement bars, Ion, Stainless steel, Copper and Aluminum,	Vegetables, Fruits, Fats, Oil, Carbohydrates, Milk products, Meat, Beverages, and cans, etc.	Sawdust, concrete, cement, chip wood, shavings, asbestos, bricks, mortar, reinforcement, scrap metals, electric wiring, etc.

IW could be in the form of liquid or solid. Depending on the industry, manufacturing process, and products IW are toxic, reactive, corrosive, ignitable, and explosive. For example, the steel industry consumes more water for cooling, washing and organic matter removal and it is also subjected to various chemical and physical processes, including gasification and blast furnaces. During this process, water from this industry comprises toxic chemicals including cyanide, ammonia, phenols, cresols, naphthalene, and benzene, which are a heavy threat to the environment. Besides, sewage from the chemical industry comprises organic and inorganic molecules, including pesticides, paints, dyes, detergents, paper, solvents, resins, and petrochemicals, causing water pollution. So, proper and new treatment technology is required to treat IW before its disposal into the environment.

12.2.3 Agricultural Waste

Agricultural waste is is produced by agro-based industrial activities, such as animal and plant waste. Agricultural wastes include pesticides, herbicides, fertilizers, palm oil, animal manures, poultry and slaughterhouse wastes, livestock, tobacco processing, forestry residues, horticultural plastics, rubber, wood wastes, and crop residues. About 65.7% of agricultural pollution is caused by the agro-based industrial sectors like manufacturing and processing of agricultural products and livestock waste, which is the major backer of Biological Oxygen Demand (BOD). Among the agro-industries, the pig farm is the one that generates more than 82% of waste which is directly discharged into freshwater bodies without any prior treatment. This pig farm waste causes oxygen depletion and eutrophication in rivers because it pollutes the water with bacteria and parasites. Other than that, the application of a surplus amount of pesticides, chemicals, and fertilizers on agricultural land leads to pollution of water sources and land/soil. The excess use of pesticides such as parathion, lindane, DDT, endosulfan, BHC, malathion and dieldrin are toxic because they are absorbed by the soil, which leads to hindering the growth of crops or spoiled. Animal waste and some other yard residues should not be used in ways that establish new pathways for bacteria and infectious diseases among livestock, people, or the environment [7].

12.2.4 Municipal Waste

At present, India generates around 42 MT of Municipal Waste (MW) annually and about 200–600 g of sewage is generated per capita per day. MW is referred to as the rubbish/refuse/garbage that is collected from the streets, markets, schools, institutions, offices, residential buildings, hotels, restaurants, and public sectors. Generally, it consists of a mixture of organic matter, clothes, paper, wood, glasses, metals, and plastics. Depending on the living standard, lifestyle, food habits, industrial, and commercial activities, the constitution of waste varies from season to season. MS has a great venomousness in the environment. Generally, MW contains 5–10% of suspended solids and it looks like grease in nature and settleable solids. In developed countries, the MW can be treated by using biological, chemical, and physical methods. Typically, flotation, and sedimentation techniques are implemented to separate settleable and un-settleable solids in cities. MW is extensively treated before its discharge. Nonetheless, both the above treatment methods are costly and need further post-treatment. Sludge and municipal effluent after the treatment might be converted into valuable products/sources by reuse of water, and recovery of nutrients and organic matter which are used as fertilizer in agricultural land.

12.2.5 Biomedical Waste

Biomedical waste may be in liquid or solid which comprises blood, tissues, cells, organs, syringes, scalpels, bandages, plaster casts, dressings, and fetuses. These wastes are generated during the treatment, diagnosis, surgeries, research activities, biological, or in health camps. "The certain solid or liquid waste, such as its container and any medium substance, produced during inspection, treatment, or immunization of humans or animals, or research practices relating to or in the processing or evaluation of biological or in camps" is what biomedical waste refers to. As per statistical analysis [Figure 12.3(a)], in 2020, due to corona in September, there is a large quantity of bio waste is produced which could increase by 30.56% respectively the previous months.

12.2.6 Nuclear Waste

The nuclear power sector, as the most liquid sector, needs vast volumes of water each day, mostly for cooling. The polluted heated discharges, which

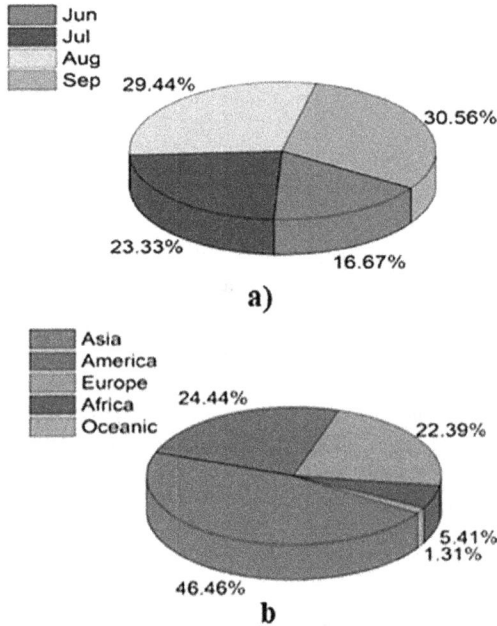

FIGURE 12.3
Statistical analysis of (a) Bio waste and (b) E-waste.

have been used in different applications, are drained back into rivers, streams, and seas, causing significant environmental risks. The extraction and refining of uranium and thorium, as well as the fission reaction used for the manufacturing process, are the primary sources of these radioactive pollutants in nuclear plants. The front part of the nuclear fuels tends to produce chromatin contaminants through uranium mining. Fission compounds that emit α and β radiation, as well as lanthanides which produce alpha radiation, such as neptunium, uranium, americium, and plutonium are contained in the rear end, which is mainly discarded nuclear fuel. In 2018, 2.7 million tons of selective electronic products were produced, accounting for less than 1% of all MSW produced. Televisions, VCRs, remote controls, video recorders, stereo equipment, mobile phones, and computing equipment are examples of listed electronic products which are otherwise called E-waste. A certain E-waste (such as televisions) includes toxic elements such as lead, arsenic, and cadmium, which are dangerous to both individuals and society. Asia is the world's largest continent, and it generates more E-wastes of about 46% than other countries like America, Europe and Africa as illustrated in Figure 12.3(b). Atomic weapons, disposal of radioactive material, and nuclear explosions are also forms of radioactive waste.

12.3 Types of Sewage

12.3.1 Biodegradable Waste

A material that could be decomposed by bacteria or other species and cannot be added to pollutants can be identified as a biologically degradable material. Biodegradable waste is waste that can be degraded by natural factors such as microbes (for example, bacteria, and fungi) and abiotic elements such as sunlight, ultraviolet light, oxygen, and so on. It includes foodstuffs, cooking waste, and other natural waste. Complex compounds are broken down into basic organic matters, which eventually stop and dissipate in the environment. It is normal to proceed faster or slower. Environmental problems and dangers from biodegradable waste are also poor. Biodegradable waste is a form of waste that can be destroyed by other living species, and usually originates from plant or animal sources. City waste, such as green waste, fruit, paper waste, and biodegradable plastics, is widely found as biologically degradable waste.

12.3.2 Non-Biodegradable Waste

A non-biodegradable material could be characterized as a kind of matter that is not distinct and separate, and acts as a source of contamination for natural organisms. Non-biodegradable waste cannot be readily handled, unlike biodegradable waste. The waste that cannot be decomposed or dissolved by natural agents is non-biodegradable. They have been without decay on the planet for decades. The threat they pose is therefore even more serious. Plastics, a common material in virtually any area, are a remarkable example. In order to achieve a long-term effect on these plastics, suitable plastics standards are used. This makes them resistant to temperatures and longer lasting after use. Such types are agriculture and retail cans, tools, and chemicals. They are the primary sources of contamination of air, water, soil, and disease. As waste that is not biologically degradable is not environmentally safe, it must be substituted. In developing substitutes, scientists have developed several ideas, such as biodegradable plastics. They added some biodegradable materials and made them conveniently and quickly degradable. It's a pricey operation, though.

12.3.3 Hazardous Waste

Commercial items such as cleaning fluids, lacquers, or pesticides discarded by commercial institutions or individuals can also be classified as hazardous waste. The Environmental Protection Agency's (EPA) definition of hazardous waste is that of non-hazardous industrial waste, and that is not urban waste. Hazard waste is categorized based on its biological, chemical, and physical characteristics. These characteristics produce poisonous, volatile, inflammable, corrosive, contagious, or nuclear materials. Toxic waste, whether in limited quantities or trace concentrations, is radioactive. They can have immediate consequences, killing or aggressive diseases, or cumulative effects, which may cause irreparable damage steadily. These are cancerous and after several years of exposure they cause cancer. Others are mutagenic, which cause significant biochemical modifications in exposed human and wildlife offspring.

12.4 Waste Management Techniques

Sewage has an unavoidable and irreversible effect on the setting. It's all pessimistic, with a little positive thrown in for good measure. The beneficial effect can be seen in several countries around the world, such as India and Nigeria, where sewage can be used for sewage farming, also known as Large Irrigation, which is a practice in which waste is fed into furrows and crops are cultivated on the ridges. Crops that do not interact with waste and are susceptible to contamination, such as fodder grass, tomatoes, and plantain, have been found to be suitable for cultivation. Sugarcane, vegetables, and onions, for example, cannot be cultivated. However, strict caution must be applied, as the field should be operated by a trained agricultural specialist.

A sewage treatment scheme that includes waste generation, storage, disposal, and treatment has been put in place. In the first place, an attempt to control excessive waste disposal and clean up the waste that has been accumulated along the canals and streets. In addition, a home garbage disposal network was developed that involves a number of waste collection points and transfer stations. For a final reason, waste reclamation technologies such as methane fermentation, composting, and the development of compost-based fertilizers, as well as industrial-scale fertilizer technologies, such as synthesized nitrogen, were created and placed to use. The waste management includes liquid and solid waste as shown in Figure 12.4. Liquid waste can be treated by biological, chemical, physical, and tertiary methods whereas solid waste can be managed by incineration, composting, landfill, and recycling.

12.4.1 Liquid Waste

Sewage from domestic, commercial, and manufacturing areas of cities and towns makes up the bulk of liquid waste. Many soluble unused and rejected compounds are present in this sewage. Wastewater is conveyed through a sewerage system, which consists of a series of tunnel pipes known as sewers,

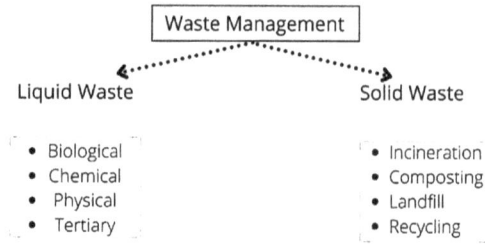

FIGURE 12.4
Waste management techniques.

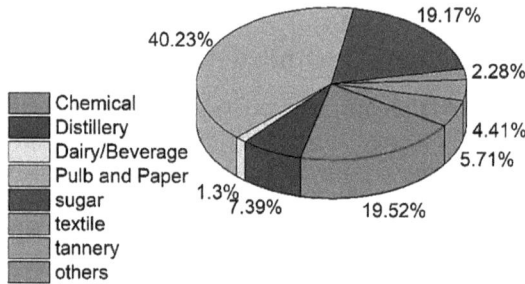

FIGURE 12.5
Types of industries that generating sewage.

in towns and cities. As shown in Figure 12.5, different industries consume more water for industrial processing, such as chemicals, distilleries, dairy, pulp, sugar, textiles, and so on. The pulp and paper industry consumes the most water (40.23%), followed by the tannery and distillery industries, which consume 19.52% and 19.17%, respectively.

Sedimentation of household waste inside the water is the first step of physical wastewater treatment. This is achieved after the water has been filtered to remove larger toxins. Wastewater is filtered and separated from toxins in a series of tanks and filters. The resultant "sludge" is then pumped into a digester, where it is further processed. This first batch of sewage comprises almost half of the wastewater's dissolved solids.

The biological process seeks to greatly decrease the bacterial content of water, such as that present in industrial waste, food scraps, soaps, and softener. The majority of urban and industrial facilities use aerobic processes to treat settled sewage liquor [8]. The biota, in order to be efficient, requires both oxygen and a living substrate. This can be accomplished in a variety of ways. Bacteria and protozoa absorb biodegradable soluble organic toxins (e.g. carbohydrates, fats, and nutrients) and attach most of the less soluble fractions into floc particles in both of these processes. Fixed film and suspended growth biological systems are the two types. Biomass grows on media in fixed film structures, including roughing filters, and sewage passes over the surface [9]. The biomass is well combined with the sewage in suspended growth processes, like activated sludge. Fixed film systems often provide smaller footprints than comparable suspended growth systems; nevertheless, floating growth systems seem to be more able to cope with biological loading shocks and have higher BOD and suspended sediment removal speeds over fixed film systems. Aeration is then applied to the activated sludge. The organic matter in waste liquid is oxidized to carbon dioxide, water, and nutrients by aeration. In this stage, the elimination of disease-causing species such as typhoid and cholera occurred [10, 11].

To speed up disinfection, chemicals are used in various procedures during water purification. Chemical unit processes are chemical processes that cause chemical reactions which are used in combination with physical and biological cleaning practices attaining different water standards. Chlorine, sodium chlorite, hydrogen peroxide, and sodium hypochlorite are examples of specialized substances that are used to clean, sanitize, and aid in the purification of waste at treatment facilities. Chemical coagulation, chemical oxidation, chemical deposition and ion exchange, accelerated oxidation, and chemical neutralization and stabilization are some chemical unit systems that can be used to clean wastewater [12, 13, 14].

Advanced oxidation processes (AOPs)/Tertiary processes have proven to be highly effective in the treatment of naturally occurring pollutants, organic and inorganic chemicals, pesticides, and other hazardous contaminants such as nanomaterials in water and wastewater [15]. AOPs must contain enough hydroxyl radicals to have an effect on water wastewater treatment. Since the 1990s, many methods for creating hydroxyl radicals and other ROS, such as superoxide anion radical, singlet oxygen, and hydrogen peroxide, have been classified and produced [16]. UV/H_2O_2, UV/O_2, sonolysis, photo-plasmas, photocatalysis, radiolysis, and nonthermal and supercritical water oxidation processes are all examples of AOPs. Because of radiolysis, photolysis and sonolysis several AOPs are initiated [17, 18]. In aqueous media, both sonolysis and radiolysis create $OH^{\bullet}$ radicals without the need for chemical oxidants, while photocatalytic methods require the use of a catalyst or precursor. Hydroxyl radicals are nonselective, highly reactive species that are used to dissolve toxic pollutants in a medium [19, 20]. The $OH^{\bullet}$ radicals have a strong oxidative potential ($E_0 = 2.8$ V) and can react with a wide variety of organic substances, meaning that they are mineralized entirely [21, 22]. This means that hydroxyl radicals can degrade organic molecules into CO_2, water, salts, or less harmful chemicals. The hydroxyl radicals target organic pollutants using a four-step reaction mechanism: hydrogen extraction, radical mixture, electron transfer, and radical formation. As hydroxyl radicals react with organic materials, they produce carbon centered radicals that can be converted to peroxyl radicals in the presence of oxygen [23, 24, 25].

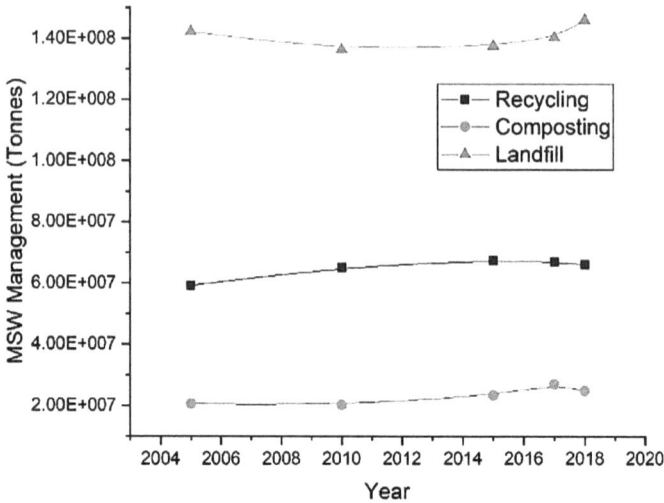

FIGURE 12.6
Municipal solid waste management.

12.4.2 Solid Waste

Solid waste management refers to the collection, treatment, and disposal of solid waste that has fulfilled its function or is no longer required. Inadequate sanitation can arise from insufficient urban solid waste management, which can lead to contamination of the atmosphere and outbreaks of disease. Solid waste disposal tasks pose a number of logistical difficulties. They also bring with them a slew of logistical, fiscal, and social issues that must be addressed. From Figure 12.6, in India mostly municipal solid wastes are simply dumped into land and as per analytical data day by day, it has increased consecutively. Less than 2 ML of waste is subjected to composting techniques.

The incineration process requires the high-temperature combustion of solid wastes until they are reduced to ash. When destroying solid wastes, incinerators are designed in such a manner that they don't produce a lot of heat. Waste-to-energy plants are incinerators which recycle heat energy using a furnace and boiler. Since they need specialized machinery and controls, highly trained technical staff, and auxiliary fuel tanks, waste-to-energy technologies are much more costly to set up and maintain than standard incinerators. Individuals, counties, and even institutions may use this form of waste disposal. The advantage of this approach is that it decreases the amount of waste by up to 20% or 30% of the initial volume. Burning waste decreases the amount of waste material substantially. It is an effective approach to eliminate scavenging in the situation of the dumps.

Biodegradable backyard garbage is permitted to biodegrade in a medium developed for the purpose due to a shortage of landfill space. Composting uses only degradable byproducts. It is a biological mechanism in which degradable waste material is converted into humus by microorganisms, especially fungi and bacteria. The carbon and nitrogen content of this end product, which resembles soil, is high. Compost is used to make high grade eco-friendly compost, and is a perfect place for plants to grow and is used for agriculture activities. But it reduces the efficiency of treatment due to large-scale projects which necessitate extensive supervision and skilled staff.

Landfill is the most common method of solid waste management today. Trash is laid out in small sections, crushed, and then coated with dirt or rubber foam. Traditional landfills are constructed with an impermeable lining, normally composed of many layers of dense plastic and sand, covering the base of the landfill. This liner prevents contamination of the groundwater due to leaching or filtration. To avoid water seepage, the landfill is filled with sandstone, mud, topsoil, and gravel until it is complete.

The method of repurposing valuable yet unused objects is known as recycling or resource recovery. Plastic bags, tins, bottles, and pots are frequently recycled immediately because they are probably rare items in certain cases. These products are usually washed and washed before being recycled. The method seeks to cut down on oil waste, new material use, and landfill waste. To reduce waste levels, most industrialized nations have a good recycling culture.

12.5 Impacts of Sewage

Most of these pollutants have virtually obscure human health and environmental effects. Animals and humans, particularly farmworkers and oil and gas workers who are constantly exposed to certain waste streams, have encountered high levels of toxic pollutants. Manufacturing waste and chemical activities lead to water pollution in industries. Relevant and easily detectable chemical compounds are typically present in industrial wastewater. In certain sub-sectors, water contamination is primarily concentrated in the form of hazardous wastes and chemical contaminants. A substantial part of this can be traced to modern chemical manufacturing and to the food industry. While many big factories have industrial effluent treatment facilities, this is not the case with small-scale enterprises, which cannot afford massive investment in emission control equipment because their profit margin is rather slender. Not only humans, but also cattle, animals, fish, and birds suffer from the effects of water contamination. The contaminated water is inappropriate for bathing, leisure, farming, and industry. The aesthetic standard of lakes and rivers has decreased. More seriously, polluted water kills and reduces the reproductive capacity of marine organisms. It's a threat to human wellbeing, finally. The consequences of water contamination cannot be escaped by anyone.

Many that are subjected to untreated waste damage will suffer serious consequences. The EPA estimates that 7 million people are sickened per year as a result of exposure to raw sewage. Only 7% of the 7 million people became seriously or critically sick. Though untreated sewage could be present in water and causes certain illnesses, the majority of cases are triggered by sewage or polluted water exposure contamination. Gastroenteritis can be caused by a number of viruses, including rotavirus, norovirus, and Norwalk virus. Gastroenteritis does not cause the same body and headaches as typical influenza. Diarrhea, stomach pain, soreness, coughing, and vomiting are all symptoms of gastroenteritis. Due to flood destruction, Hepatitis A can be present in raw sewage. Adults are more likely than infants to develop symptoms, such as jaundice, vomiting, stomach pain, lack of energy, fatigue, diarrhea, and flu. E. coli seems to be a more common bacterium attributed to sewage contamination, and it can lead to the fatal hemorrhagic systemic corticosteroids, which can lead to brain damage or even death if not treated properly. Nausea, fatigue, stomach pain, and flu are some symptoms of E. coli infection.

12.6 Conclusions

Billions of tons of waste were generated every year at a global level, considering the increased industrialization and urbanization in developed countries. In certain countries like China and India, the health concerns associated with waste disposal were rising. In order to minimize the health effects of inadequate waste disposal practices, massive investment in waste treatment systems, skills training was required. While numerous treatment technologies that meet these strict discharge requirements were currently on the market, the industry will benefit from research into lightweight and ecologically sustainable treatment systems. Furthermore, lifecycle evaluations should be performed in the valorization chain to analyze and compare different industry-generated waste management practices.

References

1. Adami L, Schiavon M. From circular economy to circular ecology: A review on the solution of environmental problems through circular waste management approaches. *Sustainability*. 2021;13:925.

2. Sharholy M, Ahmad K, Mahmood G, Trivedi RC. Municipal solid waste management in Indian cities – A review. *Waste Management*. 2008;28:459–67.

3. Vaneeckhaute C, Fazli A. Management of ship-generated food waste and sewage on the Baltic Sea: A review. *Waste Management*. 2020;102:12–20.

4. Melvin SD, Leusch FDL. Removal of trace organic contaminants from domestic wastewater: A meta-analysis comparison of sewage treatment technologies. *Environment International*. 2016;92–93:183–88.

5. Giusti L. A review of waste management practices and their impact on human health. *Waste Management*. 2009;29:2227–39.

6. Saad A, Hegazi N. Domestic Waste: Sources, Effects, and Management. Cairo, Egypt: Egypt. Atomic Energy Authority; 1999.

7. Wallace T, Gibbons D, O'Dwyer M, Curran TP. International evolution of fat, oil and grease (FOG) waste management – A review. *Journal of Environmental Management*. 2017;187:424–35.

8. Danalewich J, Papagiannis TG, Belyea RL, Tumbleson ME, Raskin L. Characterization of dairy waste streams, current treatment practices, and potential for biological nutrient removal. *Water Research*. 1998;32:3555–68.

9. Qazi JI, Nadeem M, Baig SS, Baig S, Syed Q. Anaerobic fixed film biotreatment of dairy wastewater. *Middle-East Journal of Scientific Research*. 2011;8:590–93.

10. Shete BS, Shinkar N. Anaerobic reactor to treat dairy industry wastewater. *International Journal of Current Engineering and Technology*. 2013:2277–4106.

11. Karadag D, Koroglu OE, Ozkaya B, Cakmakci M, Heaven S, Banks C, Serna-Maza A. Anaerobic granular reactors for the treatment of dairy wastewater: A review. *International Journal of Dairy Technology*. 2015;68:459–70.

12. Zakeri HR, Yousefi M, Mohammadi AA, Baziar M, Mojiri SA, Salehnia S, Hosseinzadeh A. Chemical coagulation-electro fenton as a superior combination process for treatment of dairy wastewater: performance and modelling. *International Journal of Environmental Science and Technology (Tehran)*. 2021. doi:10.1007/s13762-021-03149-w

13. Chezeau B, Boudriche L, Vial C, Boudjemaa A. Treatment of dairy wastewater by electrocoagulation process: Advantages of combined iron/aluminum electrodes. *Separation Science and Technology*. 2020;55:2510–27.

14. Alimoradi S, Faraj R, Torabian A. Effects of residual aluminum on hybrid membrane bioreactor (Coagulation-MBR) performance, treating dairy wastewater. *Chemical Engineering and Processing*. 2018;133:320–24.

15. Loures CCA, Samanamud GRL, de Freitas A, Oliveira IS, de Freitas LV, Almeida C. The use of advanced oxidation processes (AOPs) in dairy effluent treatment. *American Journal of Theoretical and Applied Statistics*. 2014;3:42–6.

16. Coha M, Farinelli G, Tiraferri A, Minella M, Vione D. Advanced oxidation processes in the removal of organic substances from produced water: Potential, configurations, and research needs. *Chemical Engineering Journal*. 2021;414:128668.

17. Pourgholi M, Jahandizi RM, Miranzadeh M, Beigi OH, Dehghan S. Removal of dye and COD from textile wastewater using AOP (UV/O_3, UV/H_2O_2, O_3/H_2O_2 and UV/H_2O_2/O_3). *Journal of Environmental Health and Sustainable Development*. 2018:621–29.

18. Rekhate CV, Srivastava JK. Recent advances in ozone-based advanced oxidation processes for treatment of wastewater – A review. *Chemical Engineering Journal Advances*. 2020;3:100031.

19. Suresh R, Rajoo B, Chenniappan M, Palanichamy M. Experimental analysis on the synergistic effect of combined use of ozone and UV radiation for the treatment of dairy industry wastewater. *Environmental Engineering Research.* 2020;26:44–54.
20. Lucas MS, Peres JA, Li Puma G. Treatment of winery wastewater by ozone-based advanced oxidation processes (O_3, O_3/UV and $O_3/UV/H_2O_2$) in a pilot-scale bubble column reactor and process economics. *Separation and Purification Technology.* 2010;72:235–41.
21. Wardenier N, Vanraes P, Nikiforov A, Van Hulle SWH, Leys C. Removal of micropollutants from water in a continuous-flow electrical discharge reactor. *Journal of Hazardous Materials.* 2019;362:238–45.
22. Sathya U, Keerthi, Nithya M, Balasubramanian N. Evaluation of advanced oxidation processes (AOPs) integrated membrane bioreactor (MBR) for the real textile wastewater treatment. *Journal of Environmental Management.* 2019;246:768–75.
23. Suresh R, Rajoo B, Chenniappan M, Palanisamy M. Feasibility of applying nonthermal plasma for dairy effluent treatment and optimization of process parameters. *Water and Environment Journal.* 35(3), 1038–1050. doi:10.1111/wej.12696sci-hub.se/10.1111/wej.12696
24. Ranade VV, Bhandari VM. Chapter 1 – Industrial Wastewater Treatment, Recycling, and Reuse: An Overview. In: Ranade VV, Bhandari VM, editors. Industrial Wastewater Treatment, Recycling and Reuse. Oxford: Butterworth-Heinemann; 2014. pp. 1–80.
25. Sgroi M, Anumol T, Vagliasindi FGA, Snyder SA, Roccaro P. Comparison of the new $Cl_2/O_3/UV$ process with different ozone- and UV-based AOPs for wastewater treatment at pilot scale: Removal of pharmaceuticals and changes in fluorescing organic matter. *Science of the Total Environment* 2021;765:142720.

13

Fabrication of Mullite Ceramic by Using Industrial Waste

Romit Roy, Dipankar Das, and Prasanta Kumar Rout
Tripura University (A Central University), Agartala, India

CONTENTS

13.1 Introduction

Fly ash (FA) is the unburned residue of coal produced in thermal power plants. The rapid growth of coal-based thermal power plants in India provides electricity in industries and the agriculture sector to achieve the economic goal. Hence, a large quantity of FA is now being generated by the thermal power plants that required sufficient land for disposal, one of the great sources of environmental pollutions [1, 2]. Therefore, it is crucial to increase the utilization of FA to ensuring life. FA consists mainly, oxides of silicon, aluminum, iron, and calcium. Some other oxides are present in FA in lesser amounts, such as magnesium, potassium, sodium, titanium, and sulfur. FA mainly consists of crystalline phases: quartz, mullite, amorphous silica glass, and some other phases such as hematite, ferrite spinel, merwinite periclase, and lime, etc. [3]. FA has utilized cement manufacturing, a green building material, chemical industry, waste stabilization, solidification, sinter glass, and ceramics preparation [4]. Mullite is the most significant material for structural applications. However, mullite can be prepared by fly ash with a small amount of alumina and other clay minerals to enhance the property. The use of mullite ceramics has increased tremendously over the past two decades due to its high-temperature stability, good mechanical strength, low thermal expansion, low heat capacity, magnificent thermal shock resistance, and outstanding creep behavior [5]. Nowadays, researchers fabricate mullite ceramic by FA, mainly mixing the raw materials and pressed into the desired shape, and then sintered. Most fabrication methods involve pressing and molding techniques [6, 7], which carry some obstructions. The fabrication process leads to high-cost equipment and gives an improper density, generates crack in the green body, variable composition throughout the sample, and very difficult to fabricate complex shapes. The gel casting process can overcome all these obstructions, and it involves an essential ceramic fabrication process to produce complex and oversized shapes, machinable green body, provide proper density with near-net accuracy. At the present investigation, mullite ceramic is fabricated through the gel casting process, using fly FA and kaolin, and alumina.

13.2 Experimental Procedure

13.2.1 Raw Material

Present investigation, FA obtained from north-east India (NTPC-Bongaigaon, Assam, India) and kaolin (make: Loba Chemie Pvt Ltd, India) and alumina (make: Loba Chemie Pvt Ltd, India) used to fabricate mullite ceramic. FA's representative chemical composition was determined by X-ray fluorescence technique (XRF: model AXIOS, make: PANalytical) shown in Table 13.1. Minerorolical phases of FA were analyzed by X-ray diffraction technique (XRD-Model: X'pert PRO, make: PANalytical), the morphology was analyzed by scanning electron microscope (model: Sigma-300, make:

TABLE 13.1

Chemical Composition Analysis of FA

Oxides	SiO_2	Al_2O_3	CaO	Fe_2O_3	TiO_2	MgO	K_2O	Na_2O	MnO	SrO	ZnO	SO_3
%	55.6	29.8	1.59	5.91	1.63	1.08	1.94	0.23	0.05	0.04	0.03	0.45

TABLE 13.2

Batch Compositions of Raw Materials

Batches	Raw Materials		
	FA	Kaolin	Alumina
MC1	60	20	20
MC2	50	20	30

Carl Zeiss). The particle size distribution (PSD) of FA was done by laser diffraction technique (model: MASTERSIZER, make: Malvern, UK).

13.2.2 Batch Compositions

The weight percentage of powders (FA, kaolin, and alumina) are detailed in Table 13.2.

13.2.3 Slurry Preparation and Processing

In this present work, a slurry made up of 50 vol.% of solid loading, 20 vol.% egg white and the remaining are distilled water. 1-Octanol and ammonium polyacrylate are used as an antifoaming and dispersant agent, respectively. The powders were weighed according to the batch composition and add egg white, distilled water, additives, as per calculations, and then mixed in a pot mill (Horizontal Pot Mill, make: Ants Ceramics Pvt Ltd.) for 24 hours using zirconia milling media (∅3). The milled ceramic slurry was cast into a plastic mold, where petroleum jelly was used as a mold release agent. Mold covered with aluminum foil and then placed in a preheated hot air oven at 80°C for an hour to gelation [8]. After gelation, the mold was left to cool for the whole night, and then samples dried up to 100°C. All the green bodies were subjected to binder burnout in the muffle furnace at 900°C with a heating rate of 3.5°C/min and a soaking period of 2 hours. Finally, the sintering operation was carried out at 1,300°C with a heating rate of 3°C/min and a soaking period of 2 hours.

13.2.4 Characterizations

The crystalline phases produce in the sintered samples were analyzed by X-ray diffraction technique [model: Xpert [9] MRD, make: PANalytical] using Ni-filtered Cu-Kα radiation at 40 Kv accelerating Voltage with 40 mA current, the 2θ value obtained in the range 10° to 80° (0.5 s scan time). The microstructure of sintered samples was analyzed by scanning electron microscope (model: Sigma-300, make: Carl Zeiss).

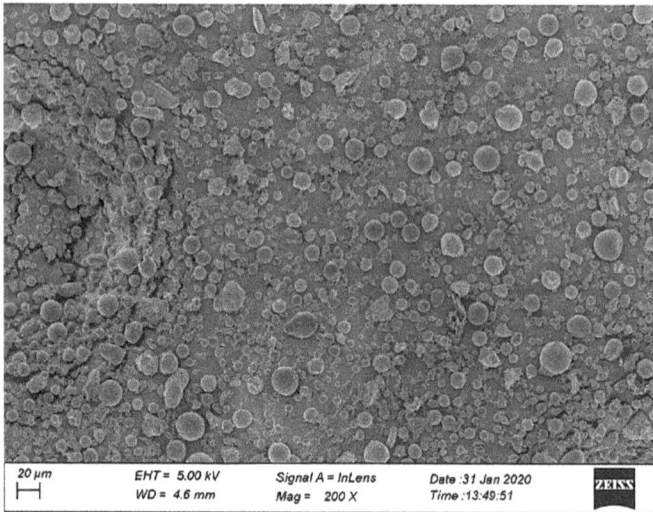

FIGURE 13.1
Scanning electron microscope (SEM) images of FA powders.

13.3 Results and Discussions

13.3.1 Characterization of Raw Material

Figure 13.1 shows the morphology of FA powders. Most of the particles were round and spherical with different sizes, and some irregularly agglomerated particles were also present [9, 10]. Figure 13.2 presents the XRD pattern of FA powders. The pattern shows that quartz, mullite is the major crystalline phases and amorphous silica glassy phases present in FA powder [3]. Figure 13.3 display particle size distribution curve (PSD) of FA powder; it shows that all the particles are less than 100 μm, where 90 vol.% of particles (d_{90}) is less than 13.62 μm and 50 vol.% (d_{50}) and 30 vol.% (d_{30}) particles size is less than 2.55 μm and 0.65 μm respectively.

13.3.2 Characterization of Sintered Sample

Figure 13.4 displays the XRD diffractogram of sample MC2, sintered at 1,300°C. It can be seen, mullite is the main crystalline phase formed at 1,300°C. Meanwhile, a small number of corundum and quartz phases present at 1,300°C sintering temperature. Figure 13.5 shows the microstructure of sintered samples MC2. Some porosity was observed in the specimen sintered at 1,300°C. Formation and growth of needle-like crystals were observed in MC2 sintered samples due to the solid-solid diffusion reaction of alumina and silica.

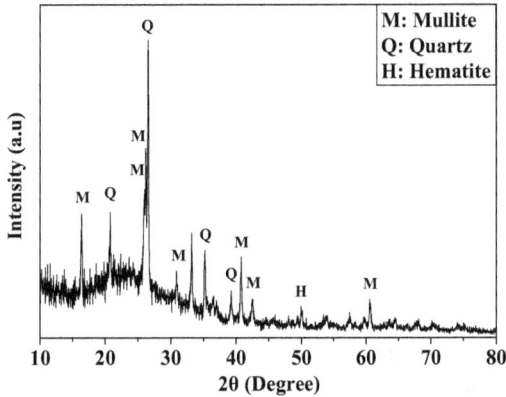

FIGURE 13.2
X-ray diffractogram of FA powder.

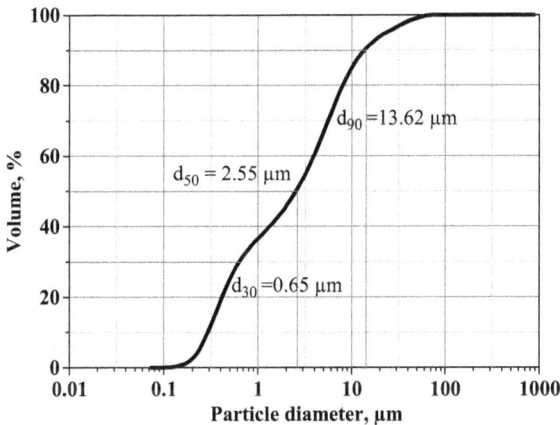

FIGURE 13.3
Particle size distribution of FA powders.

13.4 Conclusions

In summary, the mullite crystals were successfully fabricated using FA, kaolin, and alumina through the gel casting process, sintered at 1,300°C. XRD Figure 13.4 shows mullite is the main phase produced for the batch composition MC2, along with some quartz and corundum phase present. SEM microstructure shows needle-shaped mullite crystals are formed in batch MC2.

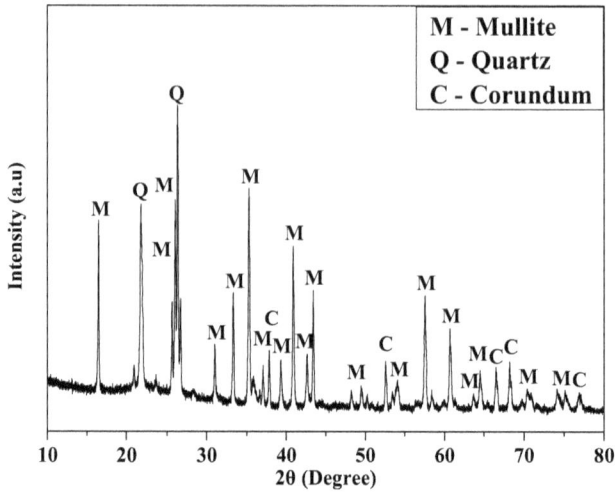

FIGURE 13.4
X-ray diffractogram of sample MC2, sintered at 1,300°C.

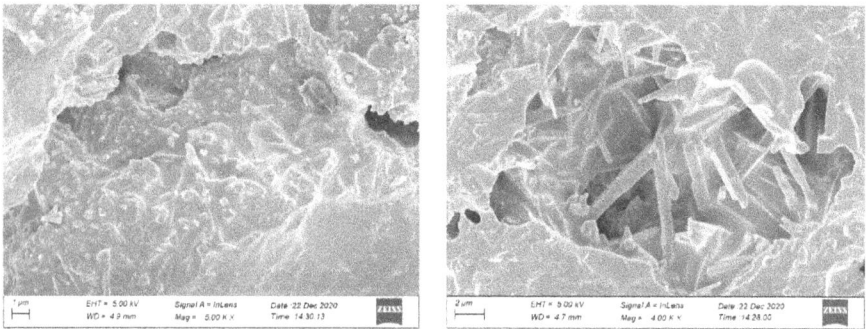

FIGURE 13.5
Scanning electron micrographs of the fracture surface of batch composition MC2.

References

1. Ahmaruzzaman, M. 2010. "A Review on the Utilization of Fly Ash." *Progress in Energy and Combustion Science* 36 (3): 327–63. https://doi.org/10.1016/j.pecs.2009.11.003
2. Chancey, Ryan T., Paul Stutzman, Maria C. G. Juenger, and David W. Fowler. 2010. "Comprehensive Phase Characterization of Crystalline and Amorphous Phases of a Class F Fly Ash." *Cement and Concrete Research* 40 (1): 146–56. https://doi.org/10.1016/j.cemconres.2009.08.029

3. Das, Dipankar, and Prasanta Kumar Rout. 2019. "Utilization of Thermal Industry Waste: From Trash to Cash." *Carbon – Science and Technology* 11 (2): 43–48.

4. Das, Dipankar, and Prasanta Kumar Rout. 2021a. "Synthesis, Characterization and Properties of Fly Ash Based Geopolymer Materials." *Journal of Materials Engineering and Performance*. 1–19. https://doi.org/10.1007/s11665-021-05647-x

5. Das, Dipankar, and Prasanta Kumar Rout. 2021b. "Synthesis and Characterization of Fly Ash and GBFS Based Geopolymer Material." *Biointerface Research in Applied Chemistry* 11 (6): 14506–19. https://doi.org/10.33263/BRIAC116.1450614519

6. Dhara, Santanu, and Parag Bhargava. 2001. "Egg White as an Environmentally Friendly Low-Cost Binder for Gelcasting of Ceramics." *Journal of the American Ceramic Society* 84 (3–12): 3048–50. https://doi.org/10.1111/j.1151-2916.2001.tb01137.x

7. Dong, Yingchao, Xuyong Feng, Xuefei Feng, Yanwei Ding, Xingqin Liu, and Guangyao Meng. 2008. "Preparation of Low-Cost Mullite Ceramics from Natural Bauxite and Industrial Waste Fly Ash." *Journal of Alloys and Compounds* 460 (1–2): 599–606. https://doi.org/10.1016/j.jallcom.2007.06.023

8. Foo, Choo Thye, Mohamad Amran Mohd Salleh, Kok Kuan Ying, and Khamirul Amin Matori. 2019. "Mineralogy and Thermal Expansion Study of Mullite-Based Ceramics Synthesized from Coal Fly Ash and Aluminum Dross Industrial Wastes." *Ceramics International* 45 (6): 7488–94. https://doi.org/10.1016/j.ceramint.2019.01.041

9. Gomes, S., and M. François. 2000. "Characterization of Mullite in Silicoaluminous Fly Ash by XRD, TEM, and 29Si MAS NMR." *Cement and Concrete Research* 30 (2): 175–81. https://doi.org/10.1016/S0008-8846(99)00226-4

10. Schneider, Hartmut, Reinhard X. Fischer, and Jürgen Schreuer. 2015. "Mullite: Crystal Structure and Related Properties." *Journal of the American Ceramic Society* 98 (10): 2948–67. https://doi.org/10.1111/jace.13817

14

Futuristic Approach to Energy in Smart Cities

Shivom Sharma

IMS Engineering College Ghaziabad, Ghaziabad, India

Gaurav Saini

Indian Institute of Engineering, Science and Technology, Shibpur Howrah, India

Krishna Kumar

Indian Institute of Technology, Roorkee, India

Karuna Saini

Gurukula Kangri Vishwavidyalaya, Haridwar, India

CONTENTS

DOI: 10.1201/9781003287186-14

14.1 Introduction

The fast rate of urbanization and population growth have given birth to the concept of the smart city. Different countries around the world are focusing on developing their cities as per the guidelines of the smart city. However, the concept of smart city has not yet been prevalent, but there is a preview of smart cities including the intelligence, efficiency, and comfort for the people. Smart city is a concept for making the lives of residents comfortable, safe, secure, and efficient, at lower costs, by using technology. Simultaneously, the concept of the smart city focuses on the conservation of environment and effective utilization of resources. Smart city includes smart energy, smart transportation, smart security, smart healthcare system, smart drainage system, smart industry, smart education system, smart waste management system, and smart early warning system for natural disasters. The word "smart" stands for making something more useful, user friendly, and environment friendly, at the same or lower cost.

Energy is the driving force of civilization, and without adequate energy, life on earth is unimaginable [1]. In the present chapter, various energy related aspects of smart cities have been considered. The standard of living in any country is usually measured by an index, such as 'Per Capita Energy Consumption' [2]. There is a huge gap between per capita energy consumption in developed and developing countries as shown in Figure 14.1.

Smart city deals with systems which generate, distribute, and utilize energy in an efficient manner. Over recent decades, the energy needs of human beings were catered for by fossil fuels. Due fossil fuel depletion and their detrimental impacts on the environment, we have been forced to use alternative fuels such as renewable energy. People are now more focused on utilizing environmentally friendly and clean sources of energy, which can only be met with renewable sources such as solar, wind, geothermal, and biomass. In recent years, the share of renewable energy in total energy consumption increased.

Although renewable energy is abundantly available at very low operational cost. However, due to higher demand, currently no country is not in the position to completely rely on the renewable energy sources due to some technological and economic factors. Further, the conversion efficiency of systems is also a secondary issue. By keeping in view the aforementioned issues,

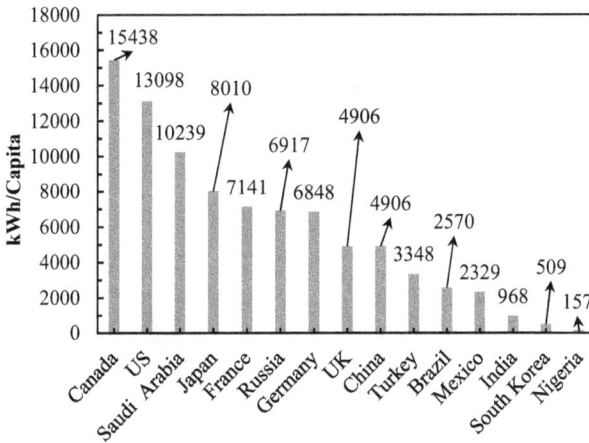

FIGURE 14.1
Per capita energy consumption of various countries [3].

it is observed that extensive research is required to fulfill the need of smart city. This will lead to develop the systems which can completely change the scenario of the world in energy prospects. Furthermore, new methods of capturing the renewable sources of renewable energy are being explored along with the improvement in conversion efficiency. These improvements would be applicable for smart city concept.

14.2 Vision of Smart Energy

The vision of smart energy starts with smart generation, distribution, and smart uses of energy. In the case of generation, state-of-art equipment and technology with better conversion efficiency is involved. Most older energy generation station did not fit in with the smart city concept, due to lack efficiency [4]. Furthermore, the distribution of energy should have very low losses during distribution, requiring dedicated sources of generation nearby. Previously, power distribution took place over long distances, being costly and involving losses, as well as high infrastructure requirement. Further, the fossil fuels-based power plants are more prone to negative environmental impact.

In recent years, the global concern toward ecology and environment leads to the inclination of developed and developing countries toward renewable energy resources for their energy supply. The economic feasibility of power from the RE sector is increasing day by day with the recent research

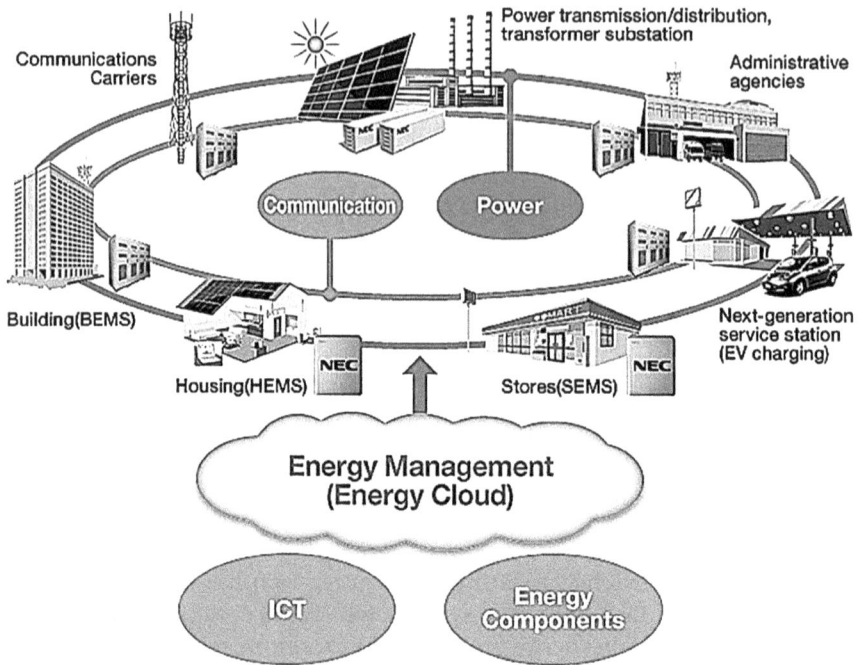

FIGURE 14.2
Next generation smart energy plan [5].

and development in the field of technology and materials advancements. According to energy plan of smart cities (Figure 14.2), huge, centralized plants and big grids can be replaced with large number of small, decentralized units located in households, commercial buildings, societies, farms, and so on. However, complete transformation is a tedious task due to existing huge infrastructure of transmission and distribution, which is totally different and not suitable for a decentralized power generation system. The application of decentralized power generation has less impact on the environment, with high efficiency, and may help to increase the share of renewable energy.

The smart grid and decentralized power system can play a vital role to improve the energy security and clean environment. In connection of smart grid this type of power system can also work as a power bank. It should be connected with internet, where information and power move simultaneously, and in both directions [6]. Such small power units connected with the grid not only fulfill their individual need, but can also supply surplus energy to the grid; or vice versa. This system can generate lage volumes of data and consumption with respect to time and location, which may be

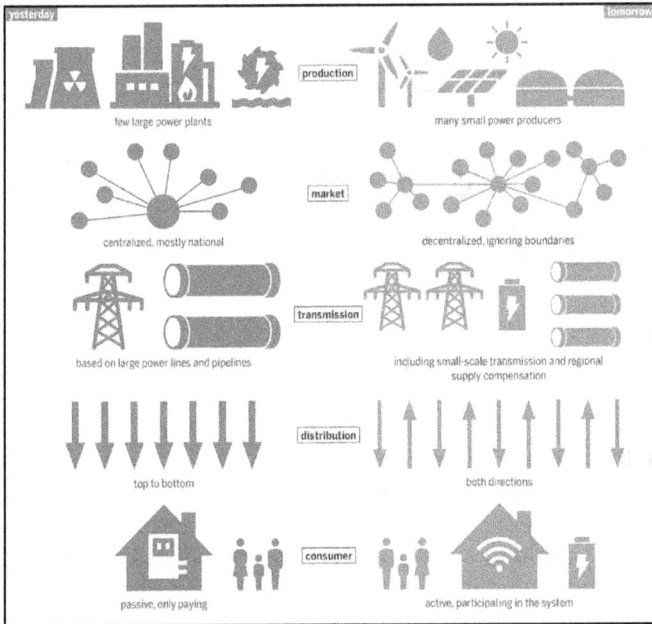

FIGURE 14.3
Comparison of current grid and smart grid system.

useful big tool to satisfy the extra demand of a region, or in the case of adversity. A comparative visualization between smart and existing grid system is shown in Figure 14.3.

14.3 Characteristics of Smart City

In this rapidly changing world, the concept of the smart city is fully-fledged. Smart city is a concept to develop a dwelling place which is intelligent enough in each and every aspect, including energy requirements and usage. A city can be referred to as smart if the systems and users are both efficient and intelligent for its applications [7]. Various characteristics and elements of smart city are shown in Figure 14.4.

14.3.1 Smart Residents

A smart city is not smart without smart residents. People of a city should be aware by knowing the data on a daily basis and by energy usage. In order

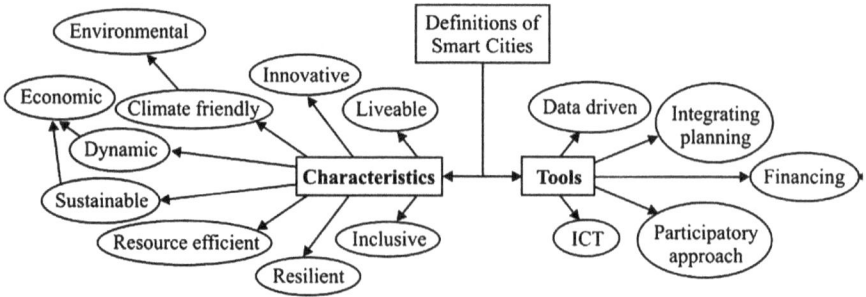

FIGURE 14.4
Characteristics flow chart of smart city [7].

to quantify the energy going to waste per unit time, collected data must be compared with efficient users who live at the same quality standards [8]. In this manner, the system becomes more intangible and transparent, which in turn may help people to spread understanding about energy consumption and possible savings. Smart meters, mobile apps and display meters can also be helpful to motivate the people to consume less energy and save it as far as possible [9]. Smart residents should be well aware of the importance, availability, consumption rate of energy, actual number of users of energy along with various methods of efficient usage.

14.3.2 Smart Rules

Constraint always leads to living with discipline, and several rules need to be formulated for smart residents. Government plays a vital role for the formation of rules which are useful to run the system without affecting the routine and to ensure the facility to residents. Smart rules can also bring big changes in the mindset of people. These rules should be understood as bidirectional; i.e. if a person consume less than the average energy consumption then he or she should be rewarded. On the other hand, those who misuse the energy should be imposed with penalty. In the similar fashion, renewable energy dependent houses should be given additional subsidy for generating the additional energy to add more energy for supply. Provisions should be made to rule out the old technology-based equipment and having poor efficiency.

14.3.3 Smart Transport

Smart transport or mobility means a comfortable movement from one place to another at least cost and with minimal environmental impact. Necessary arrangements/plans should be taken to minimize the movement for daily needs and routine activities like school, office, banks, hospitals, and malls.

Thus, energy consumption in transport can be reduced, secondly public transport can be made comfortable, approachable, and reliable. People using their own vehicles with single sitting should be charged extra tax. Ride sharing apps may help to a great extent for both who offers the ride and the one who avails the ride. Traffic management systems should be enabled with AI to operate according to the quantity of vehicles waiting to cross a signal. It can save time, energy, and simultaneously, reduce air and noise pollution. As per the requirements of electric vehicles, sufficient numbers of charging stations must be developed. These stations could be at workplaces, homes, markets, and city streets, or even on highways. Vehicles not in use must be parked at charging places, and in the role of energy storage. All such vehicles can make use of renewable energy for their charging on the availability.

14.3.4 Smart Lighting

Lighting plays a crucial role in the current scenario, and involves consumption of huge amounts of energy. Lighting needs are various places as on the roadside area and other public places viz. bus stand, railway station, hospitals, schools, and government offices. A careful watch and intelligent usage can save much energy. In case of street lighting, the lamps should be fitted with sensors which can make the lamp lit when a vehicle or person approaches from a defined direction and range. Similarly, fans and lights on railway station, bus stands, waiting halls should be provided with smart sensors to save energy. Secondly, streetlights and other public area lights should be supported with solar panels to reduce the burden for their duration and regular basis applications.

14.3.5 Smart Living

Smart living refers to the practice of making a wise use about energy and technology. Home automation is one example of such technology which can make your home smart by effective utilization of energy. Smart homes provide comfort, security, and most efficient use of energy, using Internet of Things technologies [10]. A huge growth can be anticipated in the market of home automation [11]. Large number of sensor and AI based new appliances are entering in the market on daily basis. Some of these appliances like smart home stereos, smart thermostats, light bulbs, and electric vehicle home chargers have become very popular.

14.3.6 Smart Economy

Economy is considered the engine of growth and development. In context of the Paris agreement, there is a global concord to move our global economy toward a resourceful and low-carbon economy. Nowadays, more importance is given to measure the growth of economy with respect to the impacts on environment [12]. In other words, an economic growth along with healthy environment is usually termed as green GDP. It is a combined indicator of economy and environment both and is more suitable for smart cities. Smart economy focuses on sustainable development therefore, environment will definitely be influenced.

14.3.7 Smart Environment

The idea of smart grid through decentralized, small-sized power stations, can potentially reduce the level of greenhouse gas emissions releasing from existing large capacity power houses. This energy method will be produced and based on the requirement of that region; greenhouse gas emissions should be distributed uniformly. Real-time data of power generation and consumption along with greenhouse gas emission will also be available to take corrective actions whenever needed. The smart grids can be mapped with a safe level of air quality index of that area which predict that power generation should immediately be stopped to save environment. In the case of non-availability, the balance amount of energy should be purchased from the nearby grid. The vehicles should make use of fossil fuel without crossing the line of emission characteristics (safe level).

14.4 Energy Aspects of Smart City

Smart cities are being thought of not only making effective and efficient utilization of available energy resources but also to search and develop feasible techniques to extract energy from waste. The exhaustible form of energy with serious environmental threat is not sustainable. Therefore, fossil fuels-based energy sources cannot be treated as a good choice for smart cities. A clean environment is foremost for human life. Various sources of renewable energy need to be explored. It depends somewhat on the availability of renewable energy, and thus on the location of the smart city. The cities located in coastal area have huge potential to generate wind energy. On the other hand, city lies in desert area are more dependable on solar energy. Similarly, if the smart city is planned to be developed in a hilly region, hydro- and geo-thermal energy can be considered as good options for energy supply.

As a matter of fact, that standard of living is directly dependent on per capita energy consumption. This prospects energy consumption and needs to be revised. In context of the smart city, effective and efficient use of energy decides your living standard. The smart residents should think to develop an energy efficient lifestyle without affecting the level of comfort, safety, security, and happiness. Secondly, research the community need to put high thrust to reduce the dependency of daily routine appliances like mixer, grinder, refrigerator, fan, air conditioner, room heater, cooler, pump, cars, bikes, lights, and oven on electrical energy. These appliances are needed to redesign and redevelop, at the same time they should operate on renewable energy, and they can also make use of low-grade energy like heat.

14.4.1 Energy Needs of Human

Energy is required at every step of life in different forms. It is needed in the form of food and sometimes to maintain the hot or cold environment for better comfort. In this modern era, many areas consume energy. Basically, there are main two categories for energy use: domestic (or direct) demand, and outside (or indirect) demand.

14.4.2 Domestic or Direct Demand of Energy

In case of domestic demand, various daily basis need of energy needs lies such as:

- Cooking.
- Heating water.
- Air conditioning.
- Washing and cleaning.
- Refrigeration the preservation.
- Pumping.
- Entertainment.

14.4.3 Outside or Indirect Demand of Energy

The outside or indirect demand of energy contains the energy need which is useful for secondary purposes:

- Construction of home and roads.
- Manufacturing of home appliances.
- Manufacturing of automobiles.
- Manufacturing of clothes.
- Manufacturing of daily consumables.
- Healthcare services.
- Banking Services.

- Education centers.
- Transportations.
- Shopping centers.

In order to optimize both direct and indirect energy demand of human being, smart buildings/factories can be a promising solution. Smart buildings are key components of the smart city. Smart building refers to a building which helps to generate and plan the efficient use of available energy.

14.5 Smart Building

A smart building is defined as a structure that employs various automated processes to control the building operations including heating, ventilation, air conditioning, lighting, security, and other similar systems. The primary objective of a smart building is to collect data for further research and business functions. Smart building utilizes the sensors, actuators, and microchips. This type of infrastructure may provide the necessary help to the concerned for better management of building as well as residents [13]. The data collected is helpful to operators, facility providers, and owners of the building to improve the asset reliability and performance of the building. Performance of building includes uses of the space availability, efficient energy consumption, and impact of building on environment.

Previously, old, constructed buildings (in the last few decades) not connected with the smart network, provide the essentials like shelter, temperature control, and safety. The performance of these buildings is same or going down day by day. However, newly constructed smart buildings change its requirement frequently. Such buildings need to be connected to an intelligent network and should be adaptive to the software absorb it like a living entity.

The residents of smart buildings should be more vigilant and productive regarding lighting, thermal comfort, air quality, physical security, sanitation, and much more. These facilities need to avail at lower costs and lesser environmental impact than buildings that are not connected with smart technologies.

Nowadays, office buildings, healthcare facilities, hospitals, educational centers, stadiums, and many other similar infrastructures are of smart building type. The conversion of such buildings into smart building will be more beneficial in term of energy saving and safety. Further, due to the limited availability of fossil fuels and lower efficiency of alternative energy sources, the consumption of energy should be controlled, and energy savings should be encouraged.

14.5.1 Construction of Smart Building

During the construction of smart buildings, a proper control system is required to monitor various appliances such as lighting, power and water meters, pumps, heating, fire alarms, and chiller plants. In the case of more advanced levels of automation elevators, access systems and shading can also be a part of the same control system.

To make a smart building there is not just one set of parameters to include; rather the integration of various parameters. Lots of new buildings have 'smart' technologies and are connected with a responsive smart power grid. Honeywell and Johnson Controls provide such competitive automation systems that can convert an ordinary building into smart building. Property owners need not demolish structures and construct from scratch.

14.5.2 Technology Smart Building

Various technologies are simultaneously involved for the development of smart building. In order to make smart buildings, the following pillars are applied:

- Energy and sustainability.
- Building automation and control.
- HVAC installation and optimal working.
- Assurance for fire safety.
- Security issues.
- Efficient services.
- Digital building lifecycle analysis.

In this domain, developers mainly focus on energy efficiency and conversion methods having lower cost. The increment of renewable energy share in the total energy consumption of the building automatically resolves the sustainability issues. Furthermore, the better response of building to the smart grid makes it energy efficient and sustainable.

14.5.3 Benefits of Smart Building

- Construction or conversion of a building to smart building is beneficial for both the owner and the organizations working within, along with the local environment.
- Indirectly smart buildings are good with better energy savings and extends to gain the productivity to improve the sustainability.

- The ultimate goal of smart building is to reduce the cost of energy and increase the productivity with the use of existing facility/staff. In smart buildings, the adoptability of preventive maintenance and decision making of supporting staff becomes effective and responsive.

Energy efficiency is the use of optimal start/stop, allowing building automation systems to respond effectively. The other important aspect is categorization of electrical loads on a priority basis; i.e. critical to high priority to non-essential.

14.6 Energy Conservation and Energy Efficiency of Smart Building

Energy conservation is one of the fundamental laws of science, stating that energy changes form from usable to non-usable over time. Energy conservation exercise should be followed by which the degradation of energy is minimum. Various domestic and industrial processes can be designed in a way to reduce the entropy generation which directly affect the performance of a process. Energy systems should have lower pressure drop in case of heat exchangers and lower temperature differential to minimize the loss to surroundings. By using smart materials, thermal resistance to heat transfer can also be reduced significantly. Most of the energy systems used in daily life and in basic industries like dairy, chemical having heat transfer phenomena are dependent on heat transferring fluid. The thermo-physical properties can be significantly improved using nano-fluid. Nano particles enhance the thermal conductivity of the fluid and hence the heat transfer rate along with overall heat transfer. Proper insulation, intelligent geometrical parameter selections, and the right choice of materials also increases the heat transfer rate [14].

Energy efficiency involves the best use of technology with minimum energy requirement to perform the given task with same level of satisfaction. Lot of examples of energy efficiency are as LED TV sets and computer screens in place of VCR and LED bulbs in place of incandescent light bulbs. The concept of energy efficiency is universal and can be applied to most of the devices as well as processes. The replacement of less efficient equipment with the more efficient one is the approach of energy conservation. Therefore, energy efficiency and energy conservation are said to word hand-in-hand. Energy conservation is not possible without developing energy efficient devices.

Investigations are being carried out on energy conservation. Nowadays, governments are working to create awareness in people about energy efficiency, and to provide the latest technology at relatively low cost. Industries also play vital roles in lowering the production costs of energy efficient technology to reach all sections of society.

Energy consumption through home appliances like fans, lights, TV sets, music, and refrigerator can be controlled to a significant level by using Artificial Intelligence (AI) based technologies. The application of AI, energy system will operate with more working hours and efficiently. For example, the refrigerator, air conditioner will adjust the temperature automatically with the weather and time by automatically analyzing the system and weather data. Lights and Fans will automatically on/off by analyzing the day and night related data. As an individual, our energy choices and actions can result in a significant reduction in the energy requirement in each sector of the economy.

Building layout and location play an important role to construct an energy efficient building. A lot of energy is goes to waste in unproductive activities such as continuous use of exhaust fans, drying of clothes in machines, and the use of air purifiers. These activities can be avoided with the use of natural resources like ventilation, sunlight, and open space.

Leakages of warm air into a room in summer, and cold air in winter, give rise to the need of cooling and heating respectively. Both heating and cooling waste substantial amounts of energy. The requirements of cooling and heating can be reduced by 20% if effective insulation and weatherization products are used. One-time small investments can increase energy efficiency of the house, thus saving in the long run. Insulation materials are rated by R value, and the higher the R value, the higher the insulation capacity of the material. A minimum R-26 with 3 inches thickness material for insulation is recommended under normal climatic conditions.

Even after good insulation air can still leak into the room through small cracks, which raises the energy expenditure. One effective solution to stop this leakage is to use the caulk, seal, and weather strip at all seams, cracks, and openings to the outside which ultimately can reduce the energy bills by 10%.

14.7 Impact of the Smart City on the Environment

Smart cities are the projects in which information and communication technologies are well-collaborated and mapped to the structure to raise the quality of life of residents at lower cost and least environmental impact. Nowadays, developers are trying to incorporate the methods that will be helpful to minimize the wastage of natural and municipal resources. It will lead to a definite positive impact on environment. Activities and aims for the development of smart cities:

- Clean air and clear sky due to reduced emissions are the biggest gain to environment from smart city projects.

- A practice of offering the ride for co-travelers is a long term and consistent benefit to the environment.
- Use of electric vehicles in place of petrol- or diesel-powered vehicles contribute greatly toward lower emissions, and thus cleaner environment and better health.
- A separate pathway to pedestrians and bicycles can be proved as most valuable investment in regard of environment.
- Smart cities can save lot of energy by making smart use of technology. Streetlights of smart cities are calibrated with the intensity of natural light and also enabled with sensors to operate in response to pedestrians.
- By means of the smart grid system, huge amounts of energy can be saved, that would have been lost in transmission and distribution if conventional big grid were used. The additional generated amounts of energy can be supplied to the need-based desired area.
- GPS enabled transport system, use of IOT and coordinated traffic signal can help to enhance the energy efficiency of smart city.
- Smart buildings contribute to the cleaner environment due to utilization of smart and renewable materials along with natural energy to support energy requirements.
- Smart buildings focus on reducing the power requirements for heating or cooling the air for comfort conditions. For such energy needs, solar energy is proved to be a competent alternative which also improves the energy efficiency of the building.
- The roofs of smart buildings should be covered with plants to make natural air purification.
- The extensive use of renewable energy like solar energy, wind energy, and hydro power, has significantly reduced the impact of human actions on the environment.
- The mantra of smart cities depends on the 4 Rs: Reduce, Recycle, Reuse, and Restore.

14.8 Challenges in the Path of Smart Cities

The development of smart cities is a tedious task, and the level of difficulties depends on the social standards. Most of developing countries including India faces various socio-economic challenges such as includes.

- Old and inefficient planning of cities.
- Inadequate waste management systems.
- Improper drainage systems.
- Digital security.
- Transportation.
- Safety and security.
- Pollution.
- Lack of digital awareness.

The concept of the smart city is the amalgamation of innovation and technology for the betterment of people and environment. The smart city cannot be smart if the residents of that city are not aware of and connected to the environmental concerns and energy crisis. The residents should be ready to adopt the challenges involved with the adoption of new technology; i.e. cashless economy. Therefore, without public support and creative involvement, the concept of smart city cannot be completed.

Every technology takes ample of time to get fully developed and adopted by the people in their daily life. In the early phases of development, users find it more complex and cumbersome due their old practice. After adoption, new technology becomes handy and looks very useful. Therefore, it is essential to patiently endure the transition phase from old practices to new. As an example, the issue of traffic jams is common in most of cities. The absence of seamless transportation costs residents in term of money as well as with time and energy. The application of IOT and GPS can significantly reduce this problem by tracking current location of vehicles with respect to real time data of traffic signals. This way alternative route can be suggested to the vehicles to avoid the congestion and to save time, energy, and environment.

The development of new technology and the implementation depends on the curiosity of people and will power of the government. The sense of responsibility toward the public property is also an important trait for the maintenance and developing new facilities for people.

14.9 Conclusions

The present chapter deals with the concern related to the rapid rate of growing population followed by the rate of urbanization across the globe. People are shifting toward cities in the search of jobs, education, and healthcare facilities. In the recent years, the concept of smart city provides a quality life to its residents at lower costs, along with the least impact on the environment.

Smart cities make use of smart systems like smart grids, smart meters, sensors, IOT, and GPS technology for effective utilization of resources. The energy needs of smart cities are met by smart micro grids which are designed for small area unlike conventional big grids for huge power stations. Such small grid cause to reduce transmission and distribution losses.

A city can be characterized as a smart city if it has smart residents, smart governance, smart lighting, smart living, smart transportation network, and smart environment. The people of a smart city must be smart enough to understand the need and adopt the changes with technology. The systems used in a smart city should be energy-efficient and environmentally-friendly. Smart cities cannot be thought of without smart buildings, which are also termed as "green buildings," having negligible impact on the environment. The life of human beings revolves around the consumption of energy in different forms. Smart living is termed as the lifestyle which is energy efficient; i.e. a quality life at the expense of less energy. Thus, participation of residents is of utmost importance to save energy and protect the environment.

Summarizing, this chapter addresses the key benefits to environment by smart city projects along with challenges in the path of developing new smart cities in developing countries like India. The socio-economic challenges like lack of digital awareness, unplanned old cities, and problem of waste management impose a big barrier. The philosophy of smart city is based on the 4-R principles of waste management: Reduce, Reuse, Recycle, and Restore.

References

1. G. Saini, and R.P. Saini, Numerical Investigations on Hybrid Hydrokinetic Turbine for Electrification in Remote Area, in: *All India Seminar Renewable Energy Sustainable Development* (Institution Eng. July 27–28, 2018), 2018.
2. G. Saini, and R.P. Saini, A review on technology, configurations, and performance of cross-flow hydrokinetic turbines, Int. J. Energy Res. 43 (2019) 6639–6679. https://doi.org/10.1002/er.4625.
3. N. Sönnichsen, Consumption of electricity per capita worldwide in 2018, Stat. Energy (2020). https://www.statista.com/statistics/383633/worldwide-consumption-of-electricity-by-country/
4. R. Hodgson, IEA chief economist: 1970s power grids need urgent rethink, Euractiv (2017). https://www.euractiv.com/section/electricity/interview/iea-chief-economist-1970s-power-grids-need-urgent-rethink/ (accessed May 12, 2021)
5. NEC, NEC's smart energy vision, Smart Energy Solut (2020). https://in.nec.com/en_IN/products/smart-energy/index.html (accessed June 10, 2021)

6. H. Servatius, U. Schneidewind, and D. Rohlfing, Smart Energy-Change to a Sustainable Energy System, 1st ed., Springer, Berlin, Heidelberg, 2012. https://doi.org/10.1007/978-3-642-21820-0

7. M. Eremia, L. Toma, and M. Sanduleac, The smart city concept in the 21st century, Procedia Eng. 181 (2017) 12–19. https://doi.org/10.1016/j.proeng.2017.02.357

8. C. Fischer, Feedback on household electricity consumption: A tool for saving energy?, Energy Effic. 1 (2008) 79–104. https://doi.org/10.1007/s12053-008-9009-7

9. S. Darby, Smart metering: What potential for householder engagement?, Build. Res. Inf. 38 (2010) 442–457. https://doi.org/10.1080/09613218.2010.492660

10. K.T. Chui, M.D. Lytras, and A. Visvizi, Energy sustainability in smart cities: Artificial intelligence, smart monitoring, and optimization of energy consumption, Energies 11 (2018) 2869. https://doi.org/10.3390/en11112869s

11. J. John, Global Smart Home Market to Exceed $53.45 Billion by 2022: Zion Market Research, INTRADO-Globenewswire (2018). https://www.globenewswire.com/news-release/2018/01/03/1281338/0/en/Global-Smart-Home-Market-to-Exceed-53-45-Billion-by-2022-Zion-Market-Research.html

12. P. Elkins, Economic Growth and Environmental Sustainability: The Prospects for Green Growth, Green Policy Platf (2002). https://www.greengrowthknowledge.org/research/economic-growth-and-environmental-sustainability-prospects-green-growth#:~:text=EconomicGrowthandEnvironmentalSustainability,Growth%7CGreenGrowthKnowledgePlatform&text=Greengrowthisthepursuit,grow.

13. J. Al Dakheel, C. Del Pero, N. Aste, and F. Leonforte, Smart buildings features and key performance indicators: A review, Sustain. Cities Soc. 61 (2020) 102328. https://doi.org/10.1016/j.scs.2020.102328

14. H. Kim, H. Choi, H. Kang, J. An, S. Yeom, and T. Hong, A systematic review of the smart energy conservation system: From smart homes to sustainable smart cities, Renew. Sustain. Energy Rev. 140 (2021) 110755. https://doi.org/10.1016/j.rser.2021.110755

Index

For Product Safety Concerns and Information please contact our EU
representative GPSR@taylorandfrancis.com
Taylor & Francis Verlag GmbH, Kaufingerstraße 24, 80331 München, Germany

www.ingramcontent.com/pod-product-compliance
Lightning Source LLC
Chambersburg PA
CBHW060442240326
41598CB00087B/2624